Estimating Woody Biomass
in Sub-Saharan Africa

Estimating Woody Biomass in Sub-Saharan Africa

Andrew C. Millington
Richard W. Critchley
Terry D. Douglas
Paul Ryan

With contributions by

Roger Bevan
John Kirkby
Phil O'Keefe
Ian Ryle

The World Bank
Washington, D.C.

The findings, interpretations, and conclusions expressed in this publication are those of the authors and do not necessarily represent the views and policies of the World Bank or its Board of Executive Directors or the countries they represent. Some sources cited in this paper may be informal documents that are not readily available.

The material in this publication is copyrighted. Requests for permission to reproduce portions of it should be sent to the Office of the Publisher at the address shown in the copyright notice above. The World Bank encourages dissemination of its work and will normally give permission promptly and, when the reproduction is for noncommercial purposes, without asking a fee. Permission to copy portions for classroom use is granted through the Copyright Clearance Center, Inc., Suite 910, 222 Rosewood Drive, Danvers, Massachusetts 01923, U.S.A.

The complete backlist of publications from the World Bank is shown in the annual *Index of Publications*, which contains an alphabetical title list and indexes of subjects, authors, and countries and regions. The latest edition is available free of charge from Distribution Unit, Office of the Publisher, The World Bank, 1818 H Street, N.W., Washington, D.C. 20433, U.S.A., or from Publications, The World Bank, 66 avenue d'Iéna, 75116 Paris, France.

The boundaries, colors, denominations, and other information shown on any map in this volume do not imply on the part of the World Bank Group any judgment on the legal status of any territory or the endorsement or acceptance of such boundaries.

The cover shows a representation of the vegetation index for Africa, June 1986, derived from the NOAA satellite's AVHRR sensor.

Library of Congress Cataloging-in-Publication Data

Estimating woody biomass in Sub-Saharan Africa / Andrew C. Millington
 ... [et al.].
    p.   cm.
    Includes bibliographical references and index.
    ISBN 0-8213-2306-7
    1. Fuelwood crops—Africa, Sub-Saharan—Geographical distribution.
 2. Woody plants—Africa, Sub-Saharan—Geographical distribution.
 3. Forest biomass—Africa, Sub-Saharan—Measurement. 4. Vegetation classification—Africa, Sub-Saharan.
 SD536.6.A357E87   1994
 333.95'3911'0967—dc20   93-23481
                                                           CIP

Contents

Part II **Regional Distribution of Land Cover Classes**

Glossary   *175*

Index of Botanical Names   *179*

Index of Place Names   *185*

Figures

Tables

Maps (at end of book)

Preface and Acknowledgments

Woodfuels play a dominant role in the energy balance of countries in Sub-Saharan Africa, and shortages of woodfuel exist—or are perceived to exist—in many areas. This volume describes a first attempt to map the varied vegetation and types of land cover in Sub-Saharan Africa and to assess the growing stock and sustainable yield of woody biomass. It is the result of a project undertaken by the Energy Sector Management Assistance Programme (ESMAP, a joint program of the World Bank and the United Nations Development Programme) to address the lack of data on energy resources for most African countries. This information is essential for calculating energy balances.

The work's usefulness, however, goes far beyond questions of energy sufficiency. Appropriate management of Africa's biological resources, to which it is hoped this work will contribute, will help maintain and perhaps enhance environmental quality on the continent. Moreover, the extent of worldwide forest cover has come to be recognized as a critical variable in predicting global climate change. Together these considerations make the assessment of existing wood resources in Africa a timely intervention.

The impetus for this project came from an ESMAP Household Energy Seminar held in Harare, Zimbabwe, in February 1988. Energy and forestry officials attending the seminar reached the conclusion that data on African woodfuel resources, needed as a basic tool for the planning of woodfuel programs, were in short supply. With generous funding provided by the government of the Netherlands, the mapping of African land cover and assessment of its woody biomass that are the basis of this book were undertaken by ETC (U.K.), under the supervision of Paul Ryan of ESMAP.

The project, which got under way in July 1988, was a collective effort between a team at the University of Reading, led by Andrew C. Millington and consisting of Jon Styles, Richard Saull, Pam Kennedy, and Nick Drake; and a team at the University of Northumbria, led by Terry Douglas and consisting of Roger Bevan,

John Kirkby, Richard Critchley, and Anthony Mellor. Jon Styles' participation was made possible through a leave of absence from research training granted by the Natural Environment Research Council (U.K.). Management and administration of the project were led by Phil O'Keefe of ETC (U.K.).

As a first step, it was decided to conduct a continental overview of woody biomass in the region, from Advanced Very High Resolution Radiometry (AVHRR) data obtained by satellite. Instrumental in this initial stage were the help of Jim Tucker, who permitted access to data at the U.S. National Aeronautics and Space Administration's Goddard Space Flight Center, and John Townshend, then director of the Unit for Thematic Information Services at the Natural Environment Research Council, University of Reading, who provided computational facilities for data interpretation.

Drafts of the maps and the project report were presented at workshops in Nairobi, Kenya, and Abidjan, Côte d'Ivoire, in May 1989. An important part of the collective work was the peer review led and conducted by senior African scholars and professionals at these meetings. These individuals included Jean-Marie Ouadba and Cyrille Kabore (Burkina Faso), Benedict Fultang (Cameroon), Dieudonné Dzimasse (Central African Republic), Me Kouame, N'Zore Kadja, Jean-Claude Anoh, and Aidara Gouesse Largeni (Côte d'Ivoire), Waldikidan Nere and Sultan Tilmo (Ethiopia), Kingsley Ghartey and Gamphi-Aidoo (Ghana), J. Gore, J. Mutie, P. Waciori, H. Mwendwa, and J. Agatsiva (Kenya), Antonio Ferrão (Mozambique), Issa Boubacar (Niger), Reg Cline-Cole (Nigeria), Isatu Deen (Sierra Leone), Ismael Grammadid (Somalia), Mustafa El Haq and Mohamed Elrady (Sudan), Idris Kikula and Edward Mlowe (Tanzania), David Siamuele (Zambia), Peter Gondo (Zimbabwe), B. K. Kaale (Southern African Development Community), and Barry Henrikson (Regional Remote Sensing Centre, Nairobi). Their local knowledge of vegetation and woody biomass statistics was crucial to the project's success. As a result of their critiques, we were able to refine the maps and the vegetation typing.

Thanks also are extended to the librarians, particularly at the World Bank in Washington, D.C., and the Food and Agriculture Organization (FAO) of the United Nations, who diligently pursued our bibliographic inquiries beyond the normal call of duty.

Thanks are due also to Gary Haley of the University of Northumbria, who prepared all of the maps, and to Joëlle Hivonnet, translator for the French edition of this volume.

Revised maps and the final report were presented to ESMAP in January 1990. It is fair to say that the project achieved its objectives. Although the accuracy of the assessment is limited by the resolution of the satellite images and by the paucity of existing data on woody biomass in the various vegetation types, the study resulted in a very useful mapping reference. It not only provides sufficient information on woody biomass resources to enable broad policy and planning decisions regarding woodfuel management but also should help researchers investigating broader issues such as the contribution of carbon sequestering by living plants to the mitigation of global greenhouse warming. We believe this collective effort to be a small step toward understanding, maintaining, and enhancing Africa's wood resources, on which so many people depend for their well-being.

Authors and Contributors

Andrew C. Millington, reader in the Department of Geography, University of Reading (U.K.)

Richard W. Critchley, senior lecturer in Ecology, Department of Environment, University of Northumbria at Newcastle (U.K.)

Terry D. Douglas, principal lecturer in Environmental Management, Department of Environment, University of Northumbria at Newcastle (U.K.)

Paul Ryan, a forestry specialist with the Energy Sector Management Assistance Programme (ESMAP), a joint program of the World Bank and the United Nations

Contributors

Roger Bevan, senior lecturer in the Department of Environment, University of Northumbria at Newcastle (U.K.)

John Kirkby, senior lecturer in the Department of Environment, University of Northumbria at Newcastle (U.K.)

Phil O'Keefe, professor in the Department of Environment, University of Northumbria at Newcastle (U.K.)

Ian Ryle, Education and Training Consultants (U.K.)

Acronyms and Abbreviations

AETFAT   Association pour l'Etude Taxonomique de la Flore de l'Afrique Tropicale

AVHRR   Advanced Very High Resolution Radiometer

CAR   Central African Republic

CCT   Computer compatible tape

CPU   Central processing unit

CTFT   Centre Technique Forestier Tropicale

DN   Digital number

ERS   Earth Resource Satellite

ESMAP   Energy Sector Management Assistance Programme (of the World Bank and the UNDP)

ETC (U.K.)   Education and Training Consultants

FAO   Food and Agriculture Organization of the United Nations

GAC   Global Area Coverage

GIS   Geographic information system

GVI   Global Vegetation Index

HRPT   High Resolution Picture Transmission

IIED   International Institute for Environment and Development

INRAN   Institut de Recherche Agronomique du Niger

IRBET   Institut de Recherche en Biologie et Ecologie Tropicale

LAC   Local Area Coverage

LAI   Leaf Area Index

NASA/GSFC   National Aeronautics and Space Administration/Goddard Space Flight Center

NDVI   Normalized Difference Vegetation Index

NOAA   National Oceanic and Atmospheric Administration

NUTIS   NERC (Natural Environment Research Council) Unit for Thematic Information Services

PSG   Polar Stereographic Projection

SADC   Southern African Development Community (formerly SADCC, Southern African Development Coordination Conference)

SMMR   Scanning Multichannel Microwave Radiometer

SPOT   Satellite pour l'Observation de la Terre

ToE   Tonnes of energy equivalent (metric tons)

UNDP   United Nations Development Programme

UNESCO   United Nations Educational, Scientific, and Cultural Organization

WARDA   West African Rice Development Association

PART I

Application of Remote Sensing for Woody Biomass Assessment and Mapping

1

Introduction and Background

Woody biomass constitutes the main domestic fuel in many parts of Africa. The need for more definitive data on this resource was perceived at a Household Energy Seminar held by the World Bank Energy Sector Management Assistance Programme (ESMAP) in Harare, Zimbabwe, in February 1988. The present project was conceived as a result. It is a first attempt to produce an analysis by type of land cover of the woody biomass present in Sub-Saharan Africa. The project was conducted by ETC (U.K.) under the auspices and supervision of ESMAP, a joint program of the World Bank and UNDP (United Nations Development Programme).

On examining the energy balance for most Sub-Saharan countries, one is struck by the dominance of woodfuel, including fuelwood and charcoal. This is well illustrated in table 1-1, which shows that wood-

Table 1-1. Final Energy Consumption for African Countries, 1980–88
(percent)

Country	Petroleum	Coke or coal	Electricity	Woodfuel and agricultural and biomass residue[a]	Total (1,000 ToE)	Year
Benin	12.5	0.0	1.5	86.0	853	1982
Botswana	28.0	15.0	8.0	49.0	853	1988
Congo	49.0	0.0	4.7	46.3	574	1985
Côte d'Ivoire	31.1	0.0	5.1	63.8	2,764	1982
Ethiopia	9.7	0.0	1.4	88.9	10,905	1982
Gabon	73.1	0.0	10.5	16.4	619	1985
The Gambia	25.4	0.0	1.7	72.9	174	1982
Ghana	23.0	0.0	3.0	74.0	3,100	1985
Liberia	22.7	0.0	8.6	68.7	918	1983
Malawi	13.0	4.0	6.0	77.0	3,389	1980
Mozambique	9.7	0.2	1.6	88.5	3,170	1984
Niger	12.0	0.0	2.2	85.8	964	1981
Nigeria	56.3	0.6	3.0	40.1	18,700	1981
Rwanda	8.7	0.0	2.3	89.0	1,129	1987
Senegal	32.8	0.0	2.8	64.4	1,679	1987
Sierra Leone	13.0	0.0	2.1	84.9	911	1984
Somalia	11.8	0.0	1.2	87.0	1,123	1984
Sudan	17.5	0.0	1.1	81.4	5,606	1984
Swaziland	22.7	19.0	7.7	50.6	595	1985
Tanzania	7.4	0.0	0.6	92.0	9,025	1981
Togo	47.0	0.0	13.0	40.0	214	1982
Zambia	13.0	8.0	13.0	66.0	4,233	1988
Zimbabwe	17.1	24.8	15.1	43.0	4,832	1988
Weighted average	25.6	2.7	4.0	67.8		

a. In many countries data are lacking on the quantity of agricultural residues consumed. In less-arid climates this may amount to little, but in woodfuel-poor Sahelian countries it could attain 10 percent of total energy consumption.

Source: UNDP/World Bank Energy Sector assessments.

fuel and agricultural residue represent 67.8 percent of the final energy consumption on a weighted average basis. The dominance of woodfuel is principally related to households, in which the most important use is cooking. Woodfuel, however, also is important for rural and small industries such as tobacco curing, tea drying, fish smoking, beer brewing, brickmaking, and commercial baking.

Accurate estimates of household energy consumption and overall woodfuel consumption are not available for many African countries. Limited surveys have been conducted, enabling estimation of woodfuel consumption in conjunction with demographic data (table

1-2). Assessment of woodfuel supply, however, has not been undertaken in most countries. Further, spatial extrapolation from the limited reliable data that exist is fraught with ecological and environmental interpretive problems.

Without reliable data on the growing stock and, more important, on the sustainable yield of woody biomass, it is not possible to prepare meaningful woodfuel development and conservation plans within an overall energy strategy. Unfortunately, some programs and projects have been designed and implemented without defining the sustainability of the woodfuel resource in relation to current and pro-

Table 1-2. Woodfuel Consumption in Sub-Saharan Africa, 1990
(million air-dry tonnes)

Region	Country	Consumption	Regional total
West African Sahel	Burkina Faso	4.3	17.8
	Chad	2.5	
	The Gambia	—	
	Mali	3.7	
	Mauritania	0.8	
	Niger	2.3	
	Senegal	4.2	
West African Coast	Benin	2.3	81.3
	Côte d'Ivoire	6.2	
	Ghana	8.6	
	Guinea	2.9	
	Guinea-Bissau	0.3	
	Liberia	3.9	
	Nigeria	51.3	
	Sierra Leone	3.0	
	Togo	2.8	
Horn of Africa	Ethiopia	13.5	31.4
	Somalia	3.3	
	Sudan	14.6	
Central Africa	Cameroon	7.4	40.3
	Central African Republic	2.6	
	Congo	1.1	
	Gabon	1.8	
	Zaire	27.4	
East Africa	Burundi	1.7	82.5
	Kenya	25.3	
	Rwanda	3.6	
	Tanzania	34.9	
	Uganda	17.0	
Southern Africa	Angola	3.1	44.9
	Botswana	1.2	
	Lesotho	0.6	
	Malawi	10.2	
	Mozambique	11.1	
	South Africa	5.2	
	Swaziland	0.4	
	Zambia	6.8	
	Zimbabwe	6.3	

— Not available.

Source: UNDP/World Bank Energy Sector assessments.

jected consumption, including potential changes in consumption.

Estimates of woody biomass derived from field data can be divided into two groups, the classical forest inventory and estimates of above-ground woody biomass. In the classical forest inventory, methodologies used by foresters have generally concentrated on commercial wood in high forests. Consequently, estimates have focused on the stem volume of potentially commercial species and within commercial size limits. Such estimates neglect important sources of woodfuel such as branches, twigs, dead wood, woody undergrowth, shrubs, and noncommercial tree species. These mensuration estimates give little indication of woody biomass stocks and productivity in forest areas, yet these areas, particularly woodlands, are a principal woodfuel source in much of Africa.

Estimates of above-ground woody biomass are more useful and have been made in some parts of Africa. Despite their greater application potential for energy planning, considerably fewer of these studies have been conducted compared with forest inventories, and they are restricted mainly to semiarid areas.

Both estimation methods fail to consider biomass on farms and around villages. This is a significant limitation in the use of such data for energy planning, because recent studies have indicated the importance of wood produced on farmland (Munslow and others 1989).

In this book we describe the initial stages of a multistage mapping and assessment strategy for producing estimates of growing stock and sustainable yield of woody biomass in Sub-Saharan Africa (table 1-3). To achieve this we created a data base of the existing woody biomass estimates just described and extrapolated these data in a framework of spatial land cover. The framework derived from remotely sensed data at a continental scale. The stages of this methodology are shown in figure 1-1.

Clearly, only two types of data can be used to map the woody biomass resource at this scale: existing maps and remotely sensed data. A compilation of previously published maps would have provided an unsatisfactory product for three reasons: the age of the maps in relation to the dynamic vegetation and land cover; the variation in scale of mapping; and the

Table 1-3. Multistage Strategy for Woody Biomass Resource Assessment

Review of Existing Data and Maps for Area

Low-Spatial-Resolution Satellite Imagery
- Advanced Very High Resolution Radiometer (AVHRR)
- Resolution: 1.1 to 15 kilometers
- Environment: large country or multicountry region
- Supported by limited ground verification and existing environmental data
- Provides 1:5,000,000-scale maps with broad land cover zones

High-Spatial-Resolution Satellite Imagery
- Landsat Multispectral Scanning System (MSS), resolution 79 meters
- Landsat Thematic Mapper (TM), resolution 30 meters
- SPOT (Satellite pour l'Observation de la Terre) High-Resolution Visible (HRV) multispectral, resolution 20 meters
- Environment: small country or portion of a country
- Supported by ground verification, aerial photography, airborne video imagery, and existing data on biomass estimates
- Provides 1:250,000-to-1:1,000,000-scale maps with more detailed land cover types

Biomass Reconnaissance Inventory
- SPOT HRV (panchromatic) or aerial photography, or both
- Low-intensity inventory, perhaps with subplots for smaller vegetation types
- Supported by cartographic or image maps and established regressions and correlations between measurable parameters and biomass
- Destructive sampling to establish regressions
- May include nonforest as well as forest areas

Biomass Management Inventory
- High-intensity inventory with subplots supported by maps in forest and nonforest areas

Conversion of Industrial Wood Volumes to Total Above-Ground Biomass
- Supported by maps or aerial photography, or both, and established regressions and correlations between industrial wood volume and total biomass

Note: Elements are not necessarily in sequential order.

Figure 1-1. Project Methodology

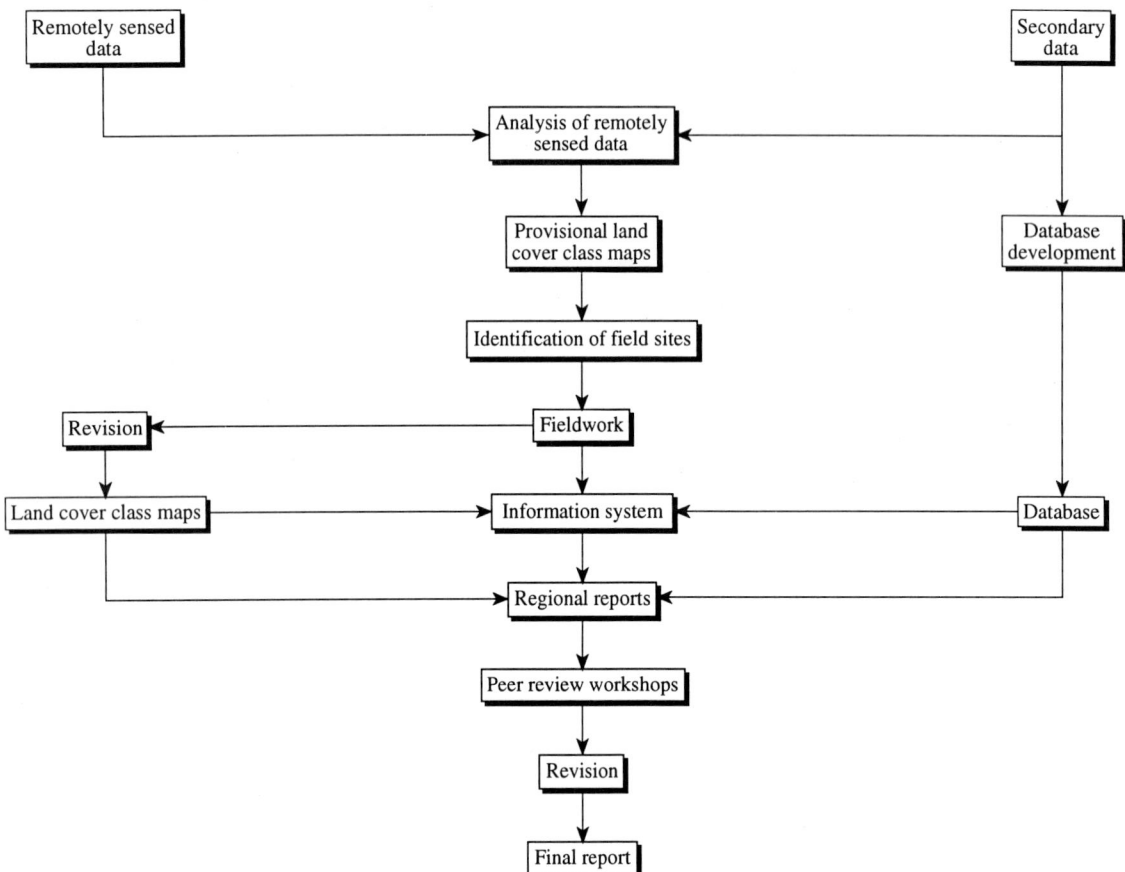

floristic criteria that are the basis for most maps, but which have limited application in biomass studies. These deficiencies were overcome by the use of Advanced Very High Resolution Radiometer (AVHRR) data, which with a spatial resolution of 8 kilometers were appropriate for this project's objective of mapping land cover. For other studies, such data have been used successfully to map land cover at the continental scale in Africa (Townshend and Tucker 1984) and Latin America (Townshend, Justice, and Kalb 1987), and the woody biomass resource in southern Africa (Millington and others 1989).

In consultation with scientists at the National Aeronautics and Space Administration/Goddard Space Flight Center (NASA/GSFC), and after reference to climatic records, it was decided to use the AVHRR data from 1986 for this work. Of the vegetation data for land cover mapping that became available from the AVHRR imagery beginning in 1981, the data for 1986 appear to be the best quality for a relatively normal rainfall year over most of Africa. The use of a single year's data leads to certain problems, which will be addressed later, and the maps produced represent only a situational overview of the woody biomass resource for 1986.

2

Using Meteorological Satellite Data for Vegetation and Land Use Mapping

In this chapter we define remote sensing, stressing its application to the mapping and monitoring of vegetation. We describe the main satellites and sensors, differentiate between Earth resource satellites and meteorological satellites, and discuss the nature of digital imagery. We then examine in detail the Advanced Very High Resolution Radiometer (AVHRR), the use of Normalized Difference Vegetation Index (NDVI) for land cover mapping and vegetation monitoring, and the production of NDVI for this project.

Remote Sensing

Satellite remote sensing is the technique of acquiring data (imagery) that describe the Earth's surface and atmosphere with the use of satellites. Sensors on board satellites provide data at a synoptic scale from a single image acquisition, thereby making the analysis of remotely sensed data a cost-effective tool for resource assessment over large areas.

A further advantage of satellite data is that repeat imagery for any scene usually can be acquired. The repeat period between image acquisitions varies from less than 24 hours for meteorological satellites to between 16 and 26 days for Earth resource satellites. The repeat period can be much longer for Earth resource satellites (for example, 18 days for *Landsat-4* and *Landsat-5*) for various technical and logistical reasons. The capability to provide time-sequential data is of great advantage in both biomass resource assessment and monitoring because of the relatively rapid response of plants to changes in the weather, climatic seasonality, changing climates, and other disturbances.

Sensors and Data Products

Sensors for the acquisition of remotely sensed data are mounted on board various satellites. These satellites can be broadly grouped by function into Earth resource satellites and meteorological satellites. Earth resource satellites include the American Landsat series, the French *SPOT* satellite, the European Space Agency's *ERS-1*, and the Indian Space Agency's *IRS-1*. Meteorological satellites include those of the National Oceanographic and Atmospheric Administration (NOAA) and the European Space Agency's *METEOSAT*. The roster of operational satellite-sensor combinations constantly changes; contemporaneous reviews are available in Cracknell and Hayes (1991) and in Drury (1990).

Data are available either as photographic (optical) products or as computer compatible tape (CCT). The CCT often is more useful for resource assessment because it can be digitally processed to optimize the information that can be extracted. Such digital image processing can be carried out effectively only with a computer-based image-processing system. Interpretation of photographic products is limited because data cannot be manipulated to highlight particular aspects of the image. In some cases, however, photographic prints provide adequate results.

Sensors cover the ultraviolet, visible, infrared, and microwave portions of the electromagnetic spectrum. The Earth's atmospheric gases act as "windows" to allow the transmission of radiation of specific wavelengths, and satellite sensors are constructed to correspond to these windows, maximizing reception of the radiation reflected or emitted from the Earth's surface. Sensors, however, are not built to cover all the available windows between ultraviolet and microwave wavelengths. Although it would be valuable for biomass assessment to have a sensor covering the electromagnetic spectrum from ultraviolet through microwave, thus permitting the simultaneous sensing of all radiation and absorption features, this is not possible for technical reasons. This situation should improve with the anticipated launching of multisensor platforms in the mid-to-late 1990s.

Biomass resource assessment using remotely sensed data mainly concerns the interaction of electromagnetic radiation with the vegetation canopy. Radiation of all wavelengths interacts with this canopy, if one is present, and at any wavelength the radiation is either absorbed, reflected, or emitted. In addition, if the vegetation canopy does not completely cover the ground, some interaction between radiation and the ground surface will occur.

To understand how these interactions affect biomass assessment and mapping, the concept of scale must be considered. At the scale of an individual leaf, radiation interacts with the cells in each leaf of a plant. Although examination of individual leaves does not provide information of direct relevance to a biomass survey, this information is fundamental to understanding the reflectance from vegetation. Of more direct relevance is the interaction of radiation with all elements of a vegetation canopy and with any soil exposed through the canopy. It is this kind of intervention that occurs at the spatial scale of most satellite remotely sensed data and therefore an understanding of it is relevant to the types of measurements made by sensors on board satellites.

It is possible to measure the proportion of radiation reflected at a particular wavelength and to derive a characteristic curve for green vegetation or a particular land cover type. Such curves can be quite distinct (figure 2-1), but the precise nature of a vegetation curve or land cover curve varies according to the type of vegetation, the time of year with respect to plant development, whether the plant is stressed, and the completeness of the vegetation canopy—that is, how much reflection occurs from the soil surface.

The spectral characteristics of vegetation reflectance are such that remotely sensed data are most useful for mapping and monitoring of vegetation and land use when they are acquired in the visible red and near-infrared regions of the spectrum. In the visible red region (0.6 to 0.7 micrometer or μm), radiation is absorbed by chloroplasts for use in photosynthesis. This contrasts with the strong reflection of near-infrared radiation that is used for sensing vegetation (0.8 to 1.1 μm); it is caused by reflection and refraction from leaf cells (figure 2-2, top).

Figure 2-1. Characteristic Land Cover Reflectance Curves and Relation to AVHRR Sensor Bandwidths

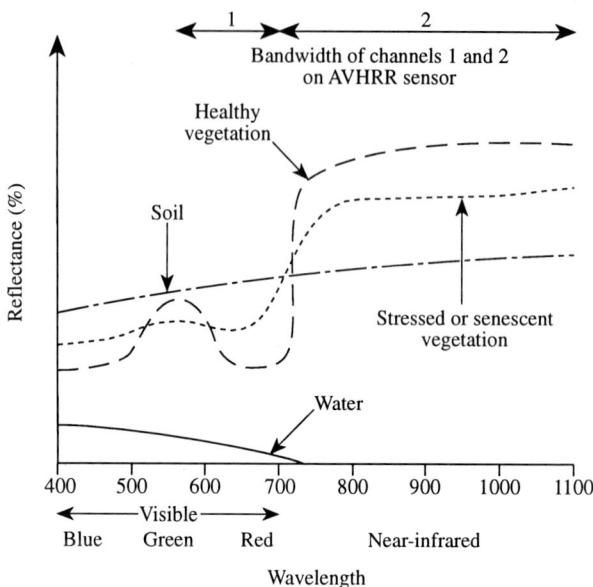

Detailed studies relating the physiology and biochemistry of plants to their spectral characteristics has been undertaken (for example, Sellers 1985, 1986; Tucker and Sellers 1986). These studies conclude that spectral reflectance of vegetation is related to the physiological processes that drive plant growth, namely, photosynthesis and respiration.

Shifting our perspective from the scale of a single leaf to the scale of a vegetation canopy, it is clear that radiation is reflected from both the vegetation and the underlying soil (figure 2-2, bottom). Sensors for both parts of the spectrum, 0.6 to 0.7 μm and 0.8 to 1.1 μm, are used in most Earth resource and meteorological satellites.

Advanced Very High Resolution Radiometer (AVHRR) Data

The data used in this work were acquired by the Advanced Very High Resolution Radiometer (AVHRR), a sensor carried on board the American NOAA series of meteorological satellites. Data derived from the AVHRR imagery were supplied in CCT format by the National Aeronautics and Space Administration/Goddard Space Flight Center (NASA/GSFC).

Of six NOAA satellites launched between 1979 and 1988, two remain operational: *NOAA-10*, launched October 1986, and *NOAA-11*, launched September 1988. The AVHRR sensors on board each satellite have the following spectral channels:

Channel	Wavelength range, upper to lower (μm)
1	0.58–0.68
2	0.725–1.10
3	3.55–3.93
4	10.30–12.50
5	11.50–12.50

Each satellite has an average orbital elevation of 833 kilometers and a period of 102 minutes, affording complete coverage of the Earth's surface every 9 days. Ground coverage is an area of 2,700 kilometers, with a ground resolution of 1.1 kilometers at nadir.

Data are acquired for any part of the Earth's surface by the AVHRR sensor on board each NOAA satellite every 12 hours. In every 24-hour period, an image for a specific area is acquired once during daylight and once during the night. For example, at the equator, the local time of image acquisition for *NOAA-9* was 1430 and for *NOAA-10* was 0730.

For work on vegetation mapping only the daytime imagery is used because it reflects the interaction between incoming solar radiation and the vegetation that is the basis for differentiating vegetation types, cover, and status. The daily times quoted here are those when the polar-orbiting NOAA satellites cross the equator. These times, however, become later during the life of each satellite because of orbital drift, affecting the overall reflectance. Gutman (1991) offers further discussion of this.

The short time interval between daily acquisitions of imagery creates a high probability that a pixel or ground resolution element (the area of Earth's surface viewed by the sensor at any time during its scan) will be free of clouds.

Normalized Difference Vegetation Index (NDVI)

In the absence of clouds, the AVHRR sensors, particularly those for Channel 1 (0.58 to 6.8 μm) and Channel 2 (7.25 to 1.10 μm), will sense vegetation, bare ground, or open water. Consequently, useful ecological information can be obtained about land surfaces. This information enables calculation of an index of vegetation status for each ground resolution element. A vegetation index is derived to reduce measurements of the vegetation canopy, ground cover, and biomass to a single number (Perry and Lautenschlager 1984).

Many different vegetation indices are available to scientists who use remotely sensed data, but the Normalized Difference Vegetation Index (NDVI) has been used successfully and extensively with AVHRR data principally because the normalizing nature of the

equation diminishes atmospheric effects. The NDVI is produced routinely by NASA/GSFC and is calculated by:

$$NDVI_i = \frac{DN\,(\text{Channel2})_i \; - \; DN\,(\text{Channel1})_i}{DN\,(\text{Channel2})_i \; + \; DN\,(\text{Channel1})_i}$$

where DN is the digital number representing the reflectance from channel n for the ith pixel.

Vegetation and Land Cover Mapping from NDVI Data

Since 1981 a number of researchers have taken advantage of such ecological data from meteorological satellites for mapping of vegetation and land cover over large areas. These include Danaher and others (1992); Eidenshink and Haas (1992); Gatlin, Sullivan, and Tucker (1983); Goward, Tucker, and Dye (1985); Gray and McCrary (1981); Koomanoff (1989); Norwine and Greegor (1983); Schneider, McGinnis, and Gatlin (1981); Tarpley, Schneider, and Money (1984); Townshend and Tucker (1984); Townshend and others (1991); Tucker, Holben, and Goff (1984); Tucker, Gatlin, and Schneider (1984). This body of work has led to a number of natural resource applications in the tropics, including:

- Monitoring Lake Chad (Schneider, McGinnis, and Stephens 1985)
- Mapping biomass production for grazing in Senegal (Tucker and others 1985)
- Monitoring semiarid rangeland in Africa (Prince 1986, 1991; Diallo and others 1991; Franklin and Hiernaux 1991; Justice and Hiernaux 1986; Prince and Tucker 1986)
- Monitoring semiarid rangeland in Australia (Barber 1992; Flemons 1992)
- Monitoring cropping patterns in Southeast Asia (Malingreau 1986)
- Assessing the woody biomass resource for wood energy planning in southern Africa (Millington and others 1989)
- Drought early warning systems (Hendrickson and Durkin 1986)
- Drought monitoring in Australia (Brook and others 1992)
- Vegetation management in Australia (Roderick and Smith 1992)
- Monitoring tropical deforestation in Amazonia (Malingreau and Tucker 1987)
- Monitoring land cover changes associated with desertification in Pakistan (Saull, Millington, and Crosetti 1991).

Figure 2-2. Small-Scale and Large-Scale Perspectives of Radiation Absorption and Reflection

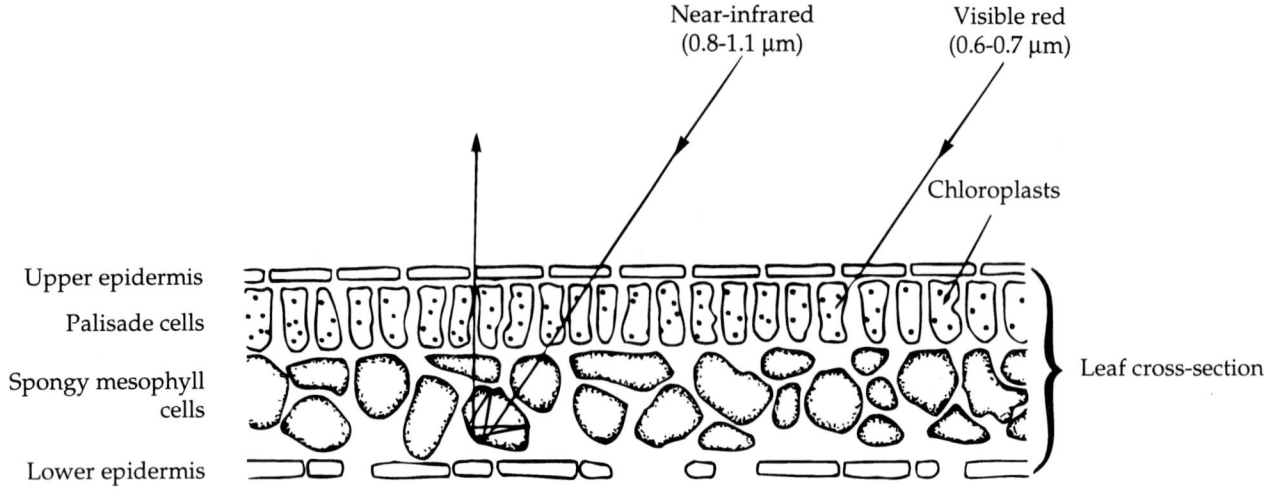

Reflection and absorption of visible red and near-infrared radiation in part of a single leaf

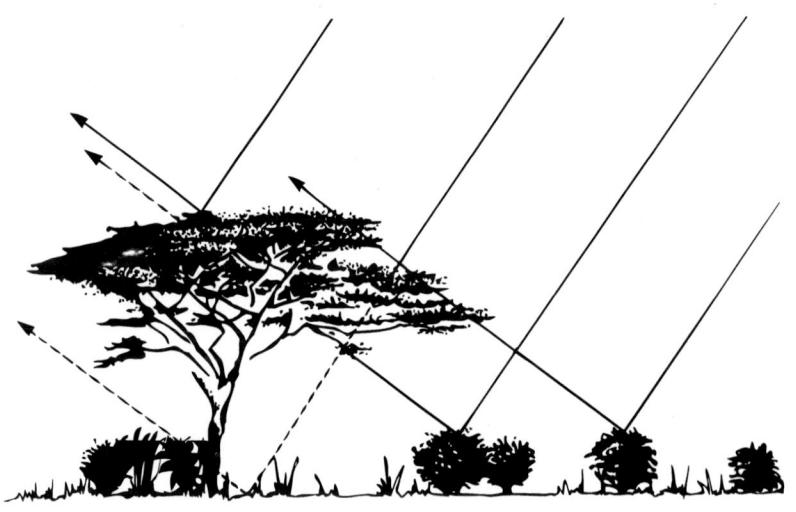

Radiation pathways in an open canopy typical of savanna woodlands

Some studies have correlated the NDVI with biomass parameters such as above-ground biomass (figure 2-3, left) and vegetation cover (figure 2-3, lower right). Although none of the studies relates specifically to tree or shrub biomass, they do suggest that similar correlations should be discovered for woody biomass.

Production of NDVI Data by NASA/GSFC

Early mapping of vegetation carried out using AVHRR-derived NDVI data so vividly illustrated the potential of these data that the NDVI is now routinely calculated by NASA/GSFC and the Food and Agriculture Organization of the United Nations (FAO), using the above equation.

The NDVI is available at different spatial resolutions, including, for example:

- NDVI at a spatial resolution of 8 kilometers is known as the Global Vegetation Index (GVI) (Tarpley 1991). These data also are available at 15 kilometers resolution after resampling to a polar stereographic projection.
- NDVI at a spatial resolution of 4 kilometers is known as Global Area Coverage (GAC) data.
- NDVI at a spatial resolution of 1.1 kilometers is known as Local Area Coverage (LAC) data or High Resolution Picture Transmission (HRPT) data, depending on how it is obtained. If it is obtained by a

Figure 2-3. Relation Between NDVI and Above-Ground Biomass and Vegetation Cover

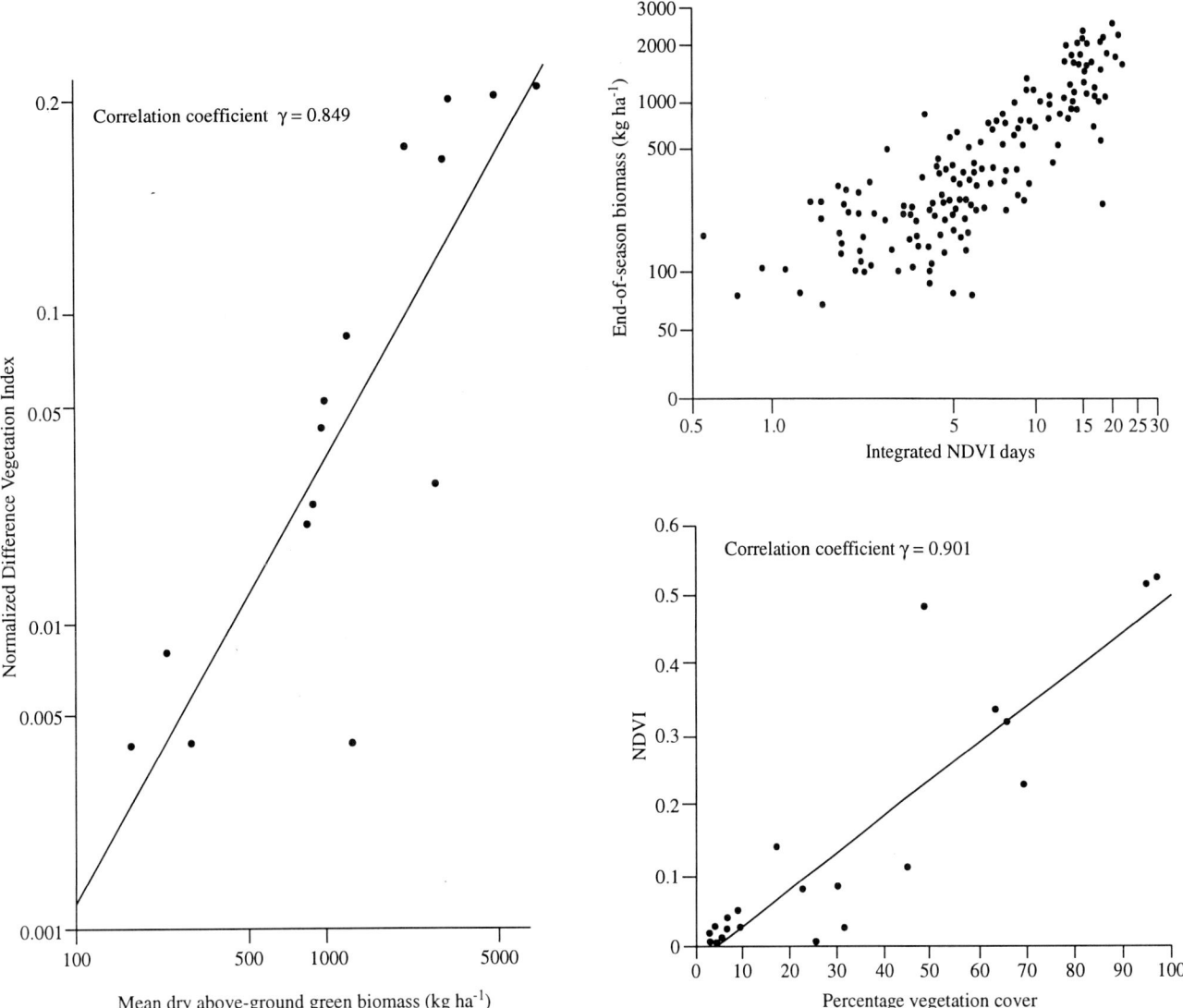

ground receiving station in real time, it is known as HRPT data. If it is obtained beyond the radio horizon of a ground station it is recorded by on-board tape recorders and then transmitted to the ground when the satellite is again in contact with a ground station; it is then known as LAC data.

The NDVI data available from NASA/GSFC have been "cloud-screened." In this procedure, pixels with clouds in the AVHRR images are identified using thermal infrared data and excluded from the later stages of analysis (that is, the production of temporal composites of the NDVI). This procedure has been acknowledged to work well except in areas where thin cirrus clouds are prevalent (Henderson-Sellers and others 1987). Problems were encountered with clouds in the data used for this study (table 3-1 and figure 3-1). After cloud-screening, daily NDVI data are composited over 10-day intervals. For any 10-day period, each pixel is examined and the day with the greatest NDVI for that pixel is used to represent the entire 10-day period (Holben 1986). This procedure reduces the effects from the atmosphere, clouds, and solar angle on calculation of the vegetation index.

3

Mapping of Land Cover Class

To identify and map classes of land cover, we interpreted Normalized Difference Vegetation Index (NDVI) digital data derived from the Advanced Very High Resolution Radiometer (AVHRR). The work was performed in three phases:

Phase I   Data inspection and preprocessing
Phase II   Initial image interpretation, derivation of the NDVI temporal profiles (NDVI phenologies), provisional mapping of land cover classes, and limited field verification of provisional maps of land cover class
Phase III Automatic classification and mapping of land cover classes.

The following sections describe the equipment, data specifications, errors, preprocessing, identification of land cover classes, mapping, and interpretation of the NDVI data.

Phase I—Data Inspection and Preprocessing

All image processing was undertaken on mainframe and microcomputer-based image-processing systems at the University of Reading. The image data initially were evaluated on-screen, and hard copy imagery was produced for intermediate stages of the interpretation process and for field verification.

Global Vegetation Index (GVI) data at 8 kilometers spatial resolution for 1986 were provided by NASA/GSFC for 10-day and monthly intervals. These data were registered to one another and were provided on the Hammer-Aitoff Conic Equal Area Projection. Initial inspection of these data disclosed two types of errors: pixels lacking NDVI values, and misregistered data.

The digital data recorded on tape are configured into pixels by the image-processing system. An absence of data in a pixel can occur because of sensor, transmission, or processing errors. Unless attended to before processing, these errors remain in the processed data. In the data set used for this work, errors involved only individual pixels. We constructed a digital filter to solve the problem of empty single pixels. The filter was a modified 3×3 pixel average filter that replaced any pixel which fell outside the range of possible DN (digital number) values (120 to 210) with the average of the DN values of its eight neighboring pixels. A pixel also was replaced if it fell within this range but the sum of the differences between the pixel and its neighbors exceeded a certain threshold. This was done so that all images that subsequently were processed would not have any pixels that lacked data.

Larger areas that lacked data entirely also were observed in the imagery. These occurred in areas where cloud cover persisted during the imaging periods and meant that, during some months, no pixels free of clouds could be found from which to calculate the NDVI. These areas did not occur in every month of the year at a particular location, but because a continuous sequence of months is needed to define land cover classes, the presence of cloud in pixels clearly could affect the accuracy of the final land cover map. These problem areas occurred mainly in the coastal humid tropics, in the high mountains of eastern and central Africa, around Lake Chad, and in the vicinity of Table Mountain in South Africa.

The smaller areas of cloud were eliminated after classification (compare with Phase III) using an iterative median-filtering method. The original NDVI data and the filtered data then were combined using a logical *or* operation, so that the original data were altered only when a zero value (cloud) occurred in the original data. Some very large areas of cloud, however, could not be eliminated entirely by filtering; these occurred in Cameroon, Ethiopia, Nigeria, South Africa, and Zaire. These areas were excluded from the final classified images because they would produce classes that were impossible to interpret.

Consequently, we produced a cloud mask which was overlaid on all of the final images. We constructed the mask by extracting information on cloud occurrence pixel-by-pixel for each month and then summing the monthly data. The mask shows the pixels where cloud restricted the calculation of the NDVI for at least one month during 1986. Areas under this cloud mask thus lack information on land cover, and these areas are not included in the calculations of area, growing stock, and yield. The extent of cloud-affected pixels is summarized country-by-country in table 3-1, and the distribution of cloud cover is mapped in figure 3-1.

Data for one of the 10-day periods in March were misregistered with the other images, offset to the east by six pixels. To solve this problem, we simply reregistered these data to the other 10-day-period data. The March NDVI image then was recalculated and these data were used in the Phase III analysis.

Phase II—Initial Image Interpretation

In this phase, we derived images from the data and identified and mapped provisional land cover classes.

Imagery Derived from AVHRR Data

Various image products were derived from the monthly NDVI imagery during this initial interpretation phase, in which provisional biomass classes were identified and mapped. The images used in this phase were:

- Individual monthly NDVI images
- Mean annual NDVI images
- Difference images between two monthly NDVI values
- Unsupervised classification images.

All of the images were produced at various scales, ranging from the entire continent to individual countries, to facilitate interpretation and mapping. Individual GVI images were prepared for each month. Because the data were supplied directly from NASA/GSFC, image production required only tape unloading, inspection for errors, and color coding of the image using the standard NASA color scale (see the section in Chapter 2 titled "Vegetation and Land Cover Mapping from NDVI Data").

A mean annual GVI image was created through pixel-by-pixel summing of all monthly GVI values for corresponding pixels. The sum of the GVI values for each pixel then was divided by 12 (months). This mean-annual image provided a good indication of annual vegetation productivity and the spatial variation within it. It is analogous to the integrated NDVI images produced by other workers (see Chapter 2), but differs because this image is calculated from the area under the annual NDVI curve.

Difference images were created by subtracting one set of monthly NDVI data from another and adding a constant. The constant was added pixel-by-pixel to corresponding pixels to ensure that all values were positive and could be displayed. The months used to create the difference images were chosen on the basis of known phenological characteristics of African vegetation and land use.

Table 3-1. Cloud Cover in 1986 Imagery

Region or country	Area (km²)	Area (percent)
West African Sahel		
Burkina Faso	0	0
Chad	0	0
The Gambia	0	0
Mali	0	0
Mauritania	0	0
Niger	0	0
Senegal	0	0
West African Coast		
Benin	0	0
Côte d'Ivoire	0	0
Ghana	0	0
Guinea	0	0
Guinea-Bissau	0	0
Liberia	11,362	1
Sierra Leone	263	0
Togo	0	0
Central Africa		
Cameroon	14,649	3
Central African Republic	0	0
Congo	0	0
Equatorial Guinea	1,634	7
Gabon	0	0
Zaire	66,818	3
Southern Africa		
Angola	1,634	0
Botswana	0	0
Lesotho	211	1
Malawi	896	1
Mozambique	5,744	1
Namibia	1,475	0
South Africa	2,371	0
Swaziland	0	0
Zambia	2,002	0
Zimbabwe	0	0
Horn of Africa		
Djibouti	0	0
Ethiopia	15,071	1
Somalia	0	0
Sudan	0	0
East Africa		
Burundi	1,686	6
Kenya	4,216	1
Rwanda	4,058	2
Tanzania	7,325	1
Uganda	3,636	1

Figure 3-1.   Distribution of Cloud Cover in 1986 Imagery

Two types of difference images are particularly useful in the interpretation of land cover classes:

1. Images in which data for the low-NDVI period (vegetation senescence) are subtracted from data for the high-NDVI period (maximum greenness) on a pixel-by-pixel basis. Such images provide a good indication of vegetation dieback in response to declining seasonal rainfall, varying levels of residual soil moisture,

or the onset of winter conditions (such as cooler temperatures or the increased frequency of frosts).

2. Images in which data for the high-GVI period are subtracted from data for the low-NDVI period on a pixel-by-pixel basis. These images provide information on the greening of vegetation.

In this phase of the analysis, classification of images was restricted to unsupervised classification. In this context, unsupervised classification is the clustering of pixels into statistical groups, defined by the temporal distribution of their NDVI values. Four months were chosen for the unsupervised classification. These represented periods of high NDVI and low NDVI and the two intermediate periods during which the NDVI values were increasing or decreasing at varying rates, depending on the type of vegetation and land use. These data then were clustered such that pixels having a statistically similar range of values over the 4 months were grouped into the same class.

In addition to the images just described, temporal profiles of the NDVI values were plotted; these are called NDVI phenologies. These profiles indicate the seasonal fluctuations of the NDVI, which can be used to analyze and interpret seasonal variations in photosynthesis, leaf area index, and biomass production. Each point represents the average NDVI over the same 3×3 pixel square area on each monthly image. Representative temporal profiles are provided in figures 3-2, 3-3, and 3-4. Although they relate to some of the land cover classes mapped in Phase III, they also are typical of the profiles derived in Phase II and were used in the provisional identification of land cover class and in field verification.

Identification and Mapping of Provisional Land Cover Classes

All of the images were evaluated on-screen, and hard copy imagery was generated. The hard copy images were used to identify and map land cover classes on 1:5,000,000-scale base maps for the six areas where field checking was undertaken: (a) South Africa and Swaziland, (b) western Zaire and Congo, (c) West Africa—Senegal and the Gambia, (d) West Africa—Mali, (e) West Africa—Côte d'Ivoire and Ghana, and (f) Ethiopia.

Identification of vegetation types in these areas employed the criteria of:

- Vegetation phenology, obtained from temporal profiles of the NDVI values and the seasonality (difference) images
- Spatial patterns of vegetation on monthly and mean annual GVI images
- Secondary data sources, particularly vegetation, land use, and forestry maps, but also including environmental information such as geological, soil, and topographic maps and statistics on climate.

We decided to undertake the initial mapping of provisional land cover classes at the scale of 1:5,000,000 for two reasons. First, this scale had been used successfully for woody biomass mapping in the

Southern African Development Community (SADC) region (Millington and others 1989). Second, the United Nations Educational, Scientific, and Cultural Organi-

Figure 3-2. NDVI **Profiles, Summary Land Cover Classes 0, 1, 2**

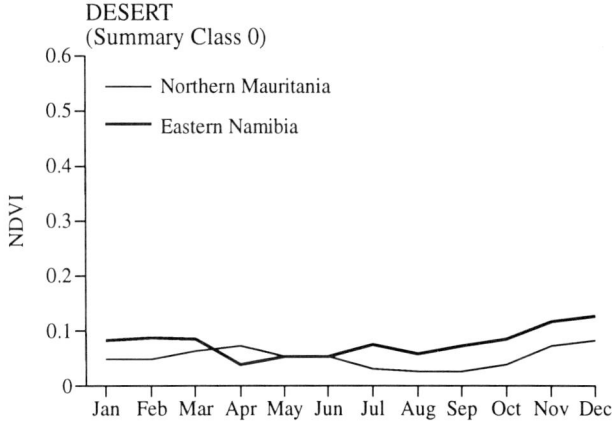

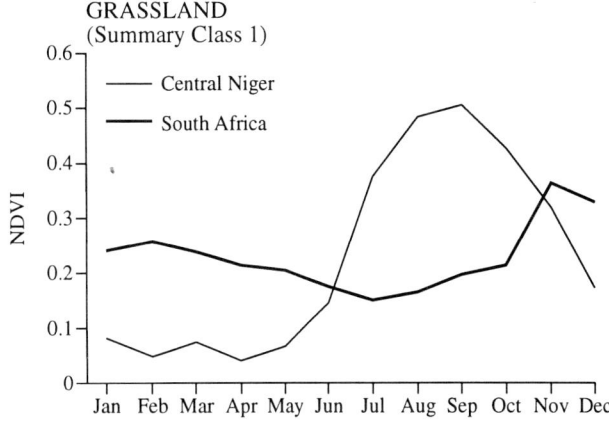

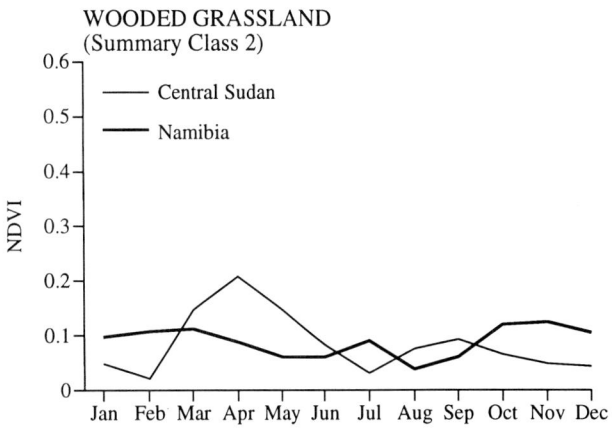

Figure 3-3. NDVI **Profiles, Summary Land Cover Classes 3, 4, 5**

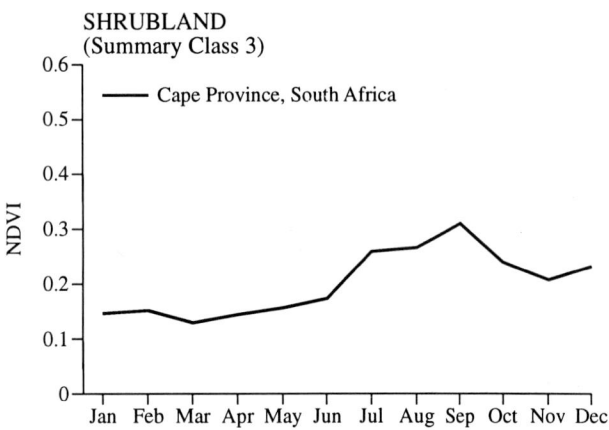

SHRUBLAND
(Summary Class 3)

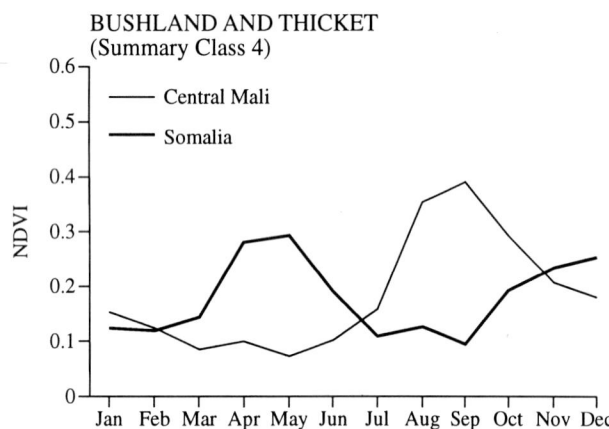

BUSHLAND AND THICKET
(Summary Class 4)

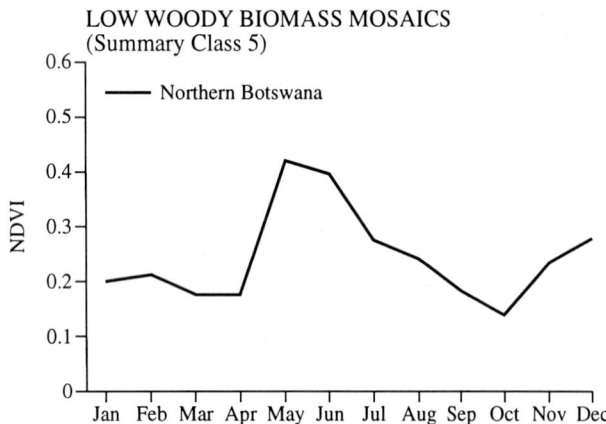

LOW WOODY BIOMASS MOSAICS
(Summary Class 5)

Figure 3-4. NDVI **Profiles, Summary Land Cover Classes, 6, 7, 8**

WOODLAND
(Summary Class 6)

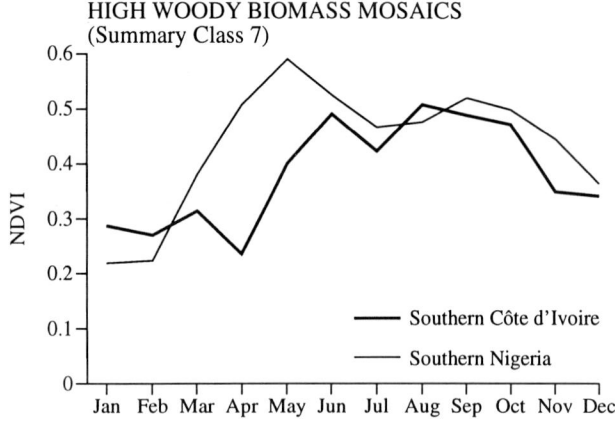

HIGH WOODY BIOMASS MOSAICS
(Summary Class 7)

FOREST
(Summary Class 8)

zation (UNESCO) Vegetation Map of Africa (White 1983) was produced at this scale.

The UNESCO map is one of the most recent vegetation maps covering all of Africa and provides the most consistent reference source for checking image inter-

pretations. This map was compiled from previous floristic maps, as well as expert review, but is not a land cover map. Nevertheless, we adopted the broad mapping units devised by White (figure 3-2). Adoption was possible because the classes of land cover are related to different vegetation communities (for example, forest, woodland, shrubland, bushland, grassland, and desert) or types of land use. Consequently, their distribution does not exactly correspond to any one previous botanical or ecological study of the region, although they are broadly comparable to most regional vegetation maps. The structure adopted is discussed in Phase III.

Phase III—Automatic (Supervised) Classification and Mapping of Land Cover Classes

The final biomass class maps were produced using supervised classification. Supervised classification is a particularly powerful image-processing procedure because the value of each pixel in each spectral channel is compared with various sets of pixel values (called training statistics) established from preselected, known areas (called training sites). The individual pixels are assigned to the most suitable class, based on the training site statistics. In this study, the data used for supervised classification were not from different spectral changes but were GVI from different months. As a consequence, the classification identified and mapped pixels having similar NDVI phenologies. Different algorithms can be used in supervised classification, but in this project a maximum likelihood classifier was used.

Training sites were identified from Phase II interpretations. These were located in areas where provisionally interpreted land cover classes attained maximum areal extent, or where they were known to be distinctive. Training sites were selected from field team reports and previous research. Training statistics were generated for all the types of land cover identified, and the data were classified on this basis. This produced an automated classification of sixty land cover classes for all of Sub-Saharan Africa.

We decided to use data spanning 6 months for the supervised classification. This decision was based on two criteria. First, in the woody biomass assessment carried out for the SADC region using the NDVI data, the main technique for map production was supervised classification (Millington and others 1989). Evaluation of the imagery showed that a supervised classification of land cover classes based on the monthly GVI data was more meaningful than other techniques.

Second, significant temporal autocorrelation exists in the NDVI data for adjacent months, which could lead to data redundancy if all 12 months were used in the classification. We tried classifying separately two sequences of data for alternate months—January to November and February to December—and for the entire 12 months of data. The visual differences among the three classified-image maps were insignificant.

The final classification was based on the 6 months of January, April, June, July, October, and December. These months were chosen by a process of eliminating the other 6 months of the year. August was eliminated because it had high-altitude cloud cover north of the equator. The months of February, March, May, September, and November were eliminated because correlation coefficients >0.9 were discovered for the adjacent-month pairs of January-February, April-May, May-June, September-October, and October-November, creating a virtual data redundancy. A by-product of this decision was that central processing unit (CPU) time for the classification using only 6 months of data was reduced dramatically compared with that using 12 months of data.

Previous work on land cover mapping has shown that using more months in the analysis improves the classification accuracy when compared with training site statistics (Townshend, Justice, and Kalb 1987). Although we do not question this finding, the decision to use 6 months of data was based on operational criteria rather than scientific. It should not be assumed that accuracy necessarily would increase if more months in 1986 were used in the classification, but accuracy would increase significantly if data from other years were included. The reason is that the NDVI is very sensitive to rainfall and soil moisture conditions, which vary considerably from year to year across much of Africa.

The most accurate map of land cover that could be produced would use a data set back to 1981, when the NDVI data first became available. We strongly recommend that future use of the NDVI data for application of natural resources employ long-term data sets, rather than data sets for single years.

Land Cover Mapping

The land cover classes were reinterpreted using the same criteria employed to identify them in Phase II. At this stage, classes having strong similarities were merged, producing forty-three land cover classes. These are listed in table 3-2 and shown on the four regional maps at the end of this volume. The classes were reviewed by experts in the workshops held in Abidjan and Nairobi, and recommendations were made for further merging of classes.

Most of these mergers are of classes having statistically similar NDVI curves which, on the basis of floristic and ecological criteria, were incorrectly split in the Phase III analysis. The special case of cultivation mosaics in West Africa and Ethiopia, however, presented a more difficult problem. In the West African coastal zone, areas originally classified as different types of humid tropical forest and forest regrowth were merged

Table 3-2.   Land Cover Classes for Sub-Saharan Africa

Class number	Land cover class
0	*Desert*
1	*Grassland*
11	*Veld* Grassland
12	Hydromorphic Grassland
13	Ethiopian Montane Steppe
14	Montane Grassland and Heathland
2	*Wooded Grassland*
21	Semidesert Wooded Grassland
22	*Acacia* Wooded Grassland
23	Plateau Wooded Grassland
24	Transitional Wooded Grassland
25	Edaphic Wooded Grassland
3	*Shrubland*
31	*Veld* Shrubland and Cultivation
32	Hill Shrubland
33	Bushy Shrubland
34	Kalahari Shrubland
35	Wooded Shrubland
4	*Bushland and Thicket*
41	Dry *Acacia-Commiphora* Bushland and Thicket
42	*Fynbos* Thicket
43	Moist *Acacia-Commiphora* Bushland and Thicket
44	Sahel-Sudanian *Acacia* Wooded Bushland
45	Escarpment Wooded Thicket
5	*Low Woody Biomass Mosaic*
51	*Acacia* Woodland Mosaic
52	East African Low Woody Biomass Mosaic
6	*Woodland*
61	Open Woodland
62	Dry Sudanian Woodland
63	Sudan-Ethiopian Woodland and Thicket
64	Sudanian Woodland
65	Moist Sudanian Woodland
66	Seasonal *Miombo* Woodland
67	Wet *Miombo* Woodland
7	*High Woody Biomass Mosaic*
71	Evergreen Woodland Mosaic
72	Cultivation and Forest/Woodland Mosaic
73	Cultivation and Forest Regrowth Mosaic
74	Guinean Woodland
75	High-Productivity West African Cultivation and Forest Mosaic
76	Medium-Productivity West African Cultivation and Forest Mosaic
77	Highland Cultivation Mosaic
8	*Forest*
81	Mangrove
82	Evergreen Forest
83	Coastal and Gallery Forest
84	Montane Forest
85	Mesophilous Humid Tropical Forest
86	Humid Tropical Swamp Forest
87	Ombrophilous Humid Tropical Forest

into two classes: High-Productivity West African Cultivation and Forest Mosaic and Medium-Productivity West African Cultivation and Forest Mosaic.

This merger was based on the realization that very little humid tropical forest remains in this region. The remaining forest is in small blocks and reserves interspersed with rubber, cocoa, and coffee plantations, farms, and a seral succession forest regrowth. This pattern of land use occurs at scales that cannot be mapped with any accuracy from the GVI data.

In the highlands of Ethiopia, a similar problem exists with small land parcels having different biomass levels and NDVI profiles. Here altitude is an additional factor, so we introduced altitude into the mapping procedure. The 2,000-meter and 3,500-meter elevation contours were digitized and overlaid on the classified image. Except for Montane Forest, all classes occurring between 2,000 and 3,499 meters in the area were merged to form a class called Highland Cultivation Mosaic. Those above 3,500 meters were merged to form a class called Ethiopian Montane Steppe.

Classes of land cover are defined as areas having types of vegetation or land use that are comparable in overall biomass, productivity, and seasonality. They may, however, combine more than one type of ecological structure (for example, forests and woodlands) and many floristic units. In this respect they differ significantly in places from the UNESCO Vegetation Map of Africa (White 1983). For the purposes of wood energy planning, however, a classification based on parameters that have been correlated with growing stock and productivity elsewhere is of more use than a floristic-ecological map.

The maps of land cover classes are one of the final products of this project. The maps include:

- Four regional maps depicting land cover classes, appearing at the end of this volume.
- A regional summary map of land cover classes, constructed by categorizing the land cover classes into eight groups on the basis of the dominant ecological structures (figure 3-5). These groups are defined by White (1983).
- Maps of growing stock and sustainable yield, appearing in Chapter 7, figures 7-1 and 7-2.

Interpreting AVHRR NDVI Data for Woody Biomass, Stock, and Sustainable Yield

As has been shown, we identified classes of land cover by using imagery derived from the AVHRR data together with secondary ecological, environmental, and forestry data. The principal characteristics used to distinguish the classes were (a) differences in vegetation cover, as indicated by their monthly NDVI values;

Figure 3-5.　Regional Summary Land Cover Classes

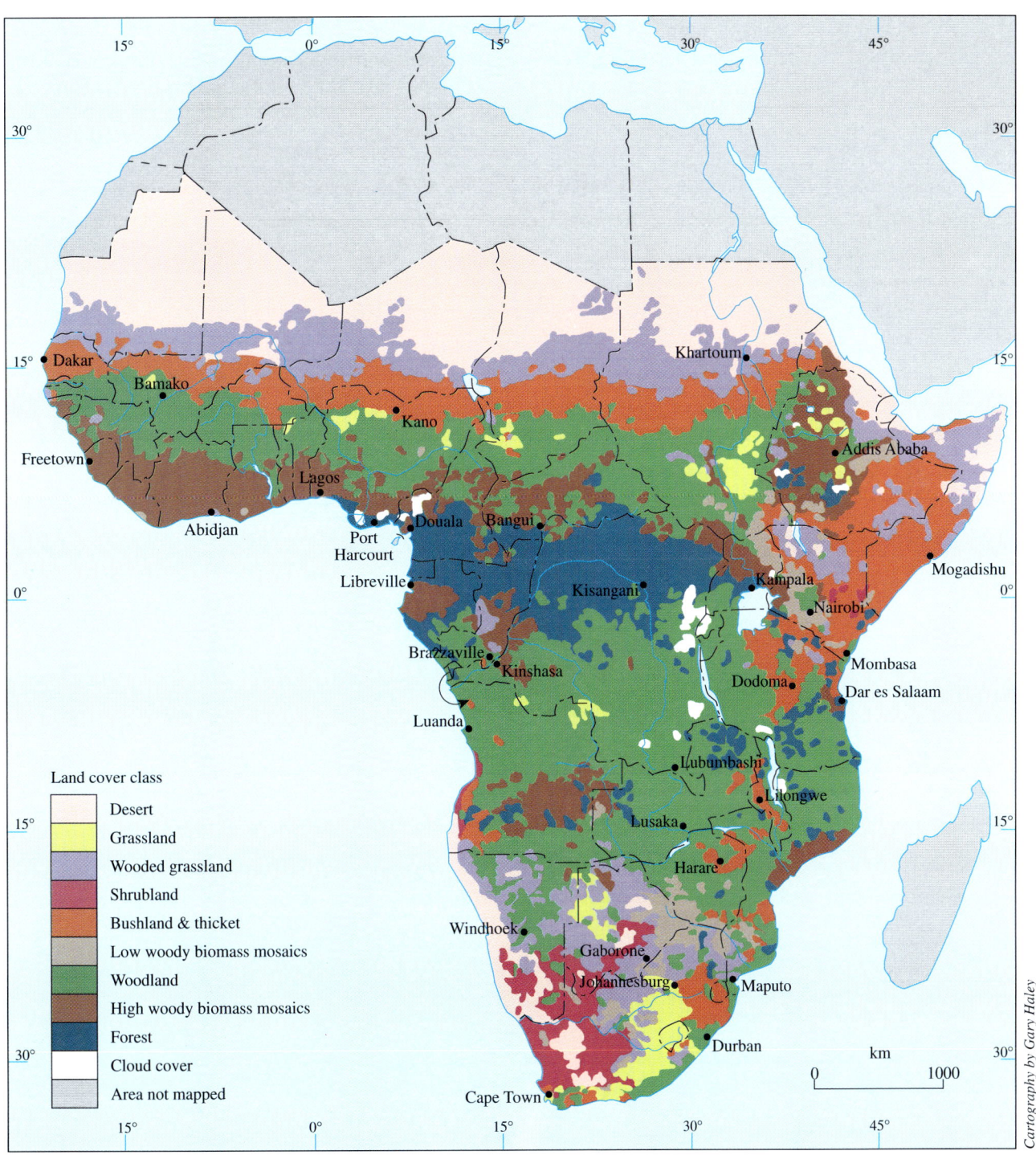

Land cover class

- Desert
- Grassland
- Wooded grassland
- Shrubland
- Bushland & thicket
- Low woody biomass mosaics
- Woodland
- High woody biomass mosaics
- Forest
- Cloud cover
- Area not mapped

Cartography by Gary Haley

(b) differences in annual primary production, as indicated by the annual sums of their monthly NDVI values; and (c) differences in phenology, as indicated by their temporal profiles of NDVI.

The basic arguments for interpretation of the AVHRR-derived NDVI data in relation to primary production are reviewed by Tucker and others (1985) and Townshend, Justice, and Kalb (1987). In these works, the authors show that classifications from analysis of data from satellite remote sensing correspond closely to vegetation maps and field observations of principal vegetation types of Africa. In summary, the reasons for this are (a) the strong relation between the NDVI and Leaf Area Index (LAI) and (b) the regular acquisition of cloud-free imagery using the AVHRR, which enables the phenology of the vegetation to be observed across the entire continent.

The most important factor that determines the type and productivity of vegetation across *most* of Sub-Saharan Africa is rainfall. However, in the far south and at high altitude, temperature fluctuations are also important (for example, in East Africa and the Ethiopian Highlands). Strong correlations between rainfall and vegetation often have been observed in Africa in relation to vegetation community, composition, and production—in southern Africa by Lamotte and Bourlière (1978) and Rutherford (1978), in Zaire by Malaisse (1978), and in West Africa by Menaut and César (1979, 1982). Such patterns depicted by NDVI imagery have been related to rainfall patterns in the Sahel (Malo and Nicholson 1990).

A very strong seasonality in the vegetation also exists, caused by the distinct dry and rainy seasons (Sarmiento and Monasterio 1983; Rutherford 1978). The seasonality of rainfall affects the vegetation directly by affecting primary production (Woodward 1987) and indirectly by the prevalence of fires in the dry season. The latter has selected species that have various forms of dormancy (Sarmiento and Monasterio 1983). The phenology of African vegetation is easily detected by the NDVI calculated from the AVHRR data, and this contributes significantly to the information content of the multitemporal classifications used in this project (Tucker and others 1985).

4

Woody Biomass Assessment

In this chapter we review problems in the data base, the procedure for area calculation, and our method of interfacing data to produce the maps of growing stock and sustainable yield.

Data Base for Biomass Estimation

We define growing stock as air-dried, above-ground woody biomass, in tonnes per hectare. Sustainable yield we define as the mean annual increment of air-dried, above-ground woody growth, also in tonnes per hectare per year.

Problems in the Data Base

Numerous problems exist in the data base and we review these difficulties in this section. In summary, the problems include:

- Extremely variable data from secondary sources
- Insufficient methodological detail
- Merchantable timber studies exclude fuelwood and understory
- Studies of selected species and classes may be unrepresentative
- Studies assess volume, not mass
- Incompatible terms and no classification criteria
- Inconsistent or missing data for classes and regions
- Unrepresentative data from limited sampling and no seral information
- Studies do not relate biomass and population distributions
- Studies do not identify nonfuel uses of selected species
- Extensive nonforest vegetation near culture is excluded.

The data base used for estimating growing stock and sustainable yield for each land cover class was derived from a number of secondary sources. Conse-

quently, the quality of the data base is extremely variable. Ideally, data for each land cover class in each country should be used, but such data either do not exist or are not accessible. Even at the regional level, these data frequently are unavailable. Before we could interpret imagery, we were forced to derive estimates for some classes and countries through generalization, extrapolation, or interpolation.

Also, many of the data collected are not accompanied by sufficient methodological detail to permit satisfactory evaluation. In the absence of other data, however, these estimates had to be used. Where no data were found, estimates were based on values in the existing literature, bearing in mind empirical data from similar land cover classes and geographical areas.

Our main sources of literature were the FAO Forestry Department Library in Rome; Department of Forestry Library at Edinburgh University; The World Bank Library and reference data file in Washington; the SADC data file held at the Department of Geography, University of Reading; and the Biomass User Network at King's College, University of London. Much of the material at the first two sources concerns forestry projects, both local and countrywide, and consists of forest inventories or other estimates of commercial wood, often to the exclusion of fuelwood and omitting estimates of natural and seminatural vegetation. Further, these estimates usually are of woody biomass of merchantable size, such as timber with a diameter greater than 10 centimeters, or in many cases 30 centimeters. This clearly omits much wood of great significance with regard to fuelwood.

In particular, most studies exclude the understory, a resource of great significance in fuelwood supply. Brown and Lugo (1984) suggest that the total biomass for the understory in closed forests is less than 2 percent, whereas in open forest the percentage may be much higher. They further point to large underestimates that are likely to occur if only stem wood is considered. They suggest that the ratio of total woody biomass to stem wood biomass in closed forest may be 1.6 and as great as 2.9 in open forest, with the increased proportion in the lower stories. Clearly, where merchantable timber and stem wood biomass are the focus, and the understory and standing dead wood are ignored, a significant underestimate of the woody biomass resource is likely.

These problems are generally recognized, as is the fact that the studies are concerned with a limited number of land cover classes and tree species—those capable of producing commercial yields, and which are not necessarily representative of an area.

Also, these estimates generally are concerned with the volume of wood rather than its mass, and this is particularly true for estimates in francophone Africa. In these cases, conversion to air-dried mass within the data base was necessary. This contrasts with the situation noted for the southern African states (Millington and others 1989), in which the commonest data available are estimates of the mass of dry material. Although oven-dried biomass is preferred for energy studies, so few make reference to it that air-dried biomass is used here.

Adjustment from volume to mass in the data base used the following conversion factors:

Wood type	Average air-dried density
Acacia-Commiphora woodland	0.6 tonnes/m^{-3}
Miombo woodland	
(*Brachystegia-Julbernardia*)	0.9
Rain forest	0.65
South African woodland	0.6

These coefficients were derived from the literature and discussion of the problem within ESMAP. In a few studies, stacked volume of wood (*stères*) is indicated, for which we used the conversion of 0.25 tonnes per *stère*. Volume-to-biomass conversions for commercial timber species in high forests do not necessarily represent the mensuration statistics for woodfuel because variations occur due to species differences and their use. For many species having less-dense wood, no data were available at all.

A further difficulty was the incompatibility of descriptive terms used for vegetation types and the lack of clear definitions and criteria for their classification. In some instances this made problematical the assignment of data to a particular land cover class.

Inconsistent data recording is another problem in creating a standardized data base. Such inconsistencies reflect the particular interests and objectives of the forester or researcher with the result that many references provide estimates of, for example, growing stock but not of growth rates, or vice versa. Despite an extensive literature search, data of good quality are lacking for particular land cover classes, such as high forests and the woody biomass mosaics, and for some critical regions, such as the Sahel and East Africa.

Yet another problem is whether the data are representative. Factors such as the successional state of a system can substantially affect growth rates of vegetation, but most studies give no indication of seral stage. Many studies also rely on very small sampling areas, creating the possibility of a large sampling error because the data do not represent land cover classes mapped on a continental scale.

No attempt was made in the present study to relate the distribution of woody biomass supplies to population distribution. (We do note that many of the studies have been in areas far from population centers, possibly making them inaccessible as woodfuel resources.) This shortcoming can be addressed by addition of current population estimates at the provincial or district level to existing biomass data in a geographic information system (GIS).

There also has been no attempt to distinguish tree species that are used locally for specific purposes, which would exclude them from use as fuelwood. This problem needs local research and more data on the quantity of individual species present. At this time little detailed information is available for the data base.

It was not possible to take account of the extensive nonforest trees and shrubs that occur in close association with villages and farmland in many areas. The spatial resolution of the images was inadequate for their assessment, and consequently almost no meaningful data exist on the growing stock and growth rates of such vegetation. Nevertheless, these trees and shrubs may form an important source of woodfuel for many rural people who sustainably collect woody and nonwoody biomass near their homes (Leach and Mearns 1988; Munslow and others 1989).

The combined effect of these factors produces a high degree of variability and unreliability in the data for each country and each land cover class. The final tabulated data must be regarded with these limitations.

Limitations of Data from Literature Sources

Table 4-1 indicates the growing stock and sustainable yields derived from the data base. The median value reported for each class represents the central value of a range of estimates from the literature pertinent to that class. Where no estimates were available, values of growing stock and sustainable yield were derived from ecologically similar classes. Few estimates are

Table 4-1. Growing Stock and Sustainable Yield Data, 1953–90

Land cover class	Growing stock air-dry t ha⁻¹ range	Median	Sustainable yield air-dry t ha⁻¹ yr⁻¹ range	Median	Source
1 Grassland					
11	0.63–3.90	(2.27)	—	[0.10][b]	Huntley 1978
12	—	[2.27][a]	—	[0.10][b]	n.a.
13	—	[2.27][a]	—	[0.10][b]	n.a.
14	—	[2.27][a]	—	[0.10][b]	n.a.
2 Wooded Grassland					
21	1.80–4.80	(3.3)	0.04–0.15	(0.10)	Openshaw 1982; Andeke Lengui 1987
22	—	[3.3][c]	—	[0.10][b]	n.a.
23	—	[3.3][c]	—	[0.10][b]	n.a.
24	0.23–1.00	(0.62)	—	[0.10][b]	Rutherford 1978
25	—	[3.3][c]	—	[0.10][b]	n.a.
3 Shrubland					
31	6.10–11.00	(8.55)	—	[0.50][e]	Huntley 1978; Rutherford 1978
32	4.50	4.50	—	[0.50][e]	Rutherford 1978
33	5.00–15.00	(10.00)	—	0.50[e]	Rutherford 1978
34	—	[10.00][d]	—	[0.50][e]	n.a.
35	3.70–7.50	(5.60)	—	[0.50][e]	Rutherford 1978
4 Bushland and Thicket					
41	7.80–20.00	(13.9)	—	[0.21][g]	Huntley 1978
42	2.00–35.01	(18.51)	—	[0.21][g]	Rutherford 1978; van Wilgen, Higgins, and Bellstedt 1990
43	3.03–30.80	(16.92)	0.04–0.38	(0.21)	Rutherford 1978
44	0.46–2.35	(1.41)	0.35	0.35	Daus, Guero, and Ada 1986; Andeke Lengui 1987
45	—	[18.51][f]	—	[0.21][g]	n.a.
5 Low Woody Biomass Mosaic					
51	22.30	22.30	—	[0.63][h]	Rutherford 1978
52	9.44–33.55	21.50	0.24–1.01	0.63	Stomgaard 1985; ETC (U.K.) 1987; Persson 1975
6 Woodland					
61	0.92–56.87	(29.40)	0.02–0.85	(0.44)	FAO/PNUD 1978; Bianchi 1986; Andeke Lengui 1987
62	9.00–44.40	(26.70)	0.16–1.38	(0.77)	FAO 1984; FAO/PNUD 1978; Clément 1982
63	—	[26.70][i]	—	[0.44][j]	n.a.
64	—	[26.70][i]	0.15–0.77	(0.46)	Bianchi 1986; Clément 1982
65	—	[26.70][i]	0.45–0.51	(0.48)	Clément 1982
66	2.25–109.00	(55.63)	0.45–1.32	(0.89)	Persson 1975; Jackson 1971; ETC (U.K.) 1987; Guerreiro 1966; Chidumayo 1987
67	16.00–222.00	(119.0)	0.41–2.25	(1.33)	Guy 1970, 1981; Stomgaard, 1985; Lundgren 1975; ETC (U.K.) 1989; Persson 1975; Kennard and Walker 1973; Guerreiro 1966
7 High Woody Biomass Mosaic					
71	2.88–7.38	(5.13)	—	[0.19][l]	Malleaux 1980
72	—	[16.84][k]	—	[0.19][l]	n.a.
73	16.84	16.84	—	0.19	Bianchi 1986
74	—	[16.84][k]	—	[0.19][l]	n.a.
75	—	[16.84][k]	—	[0.19][l]	n.a.
76	—	[16.84][k]	—	[0.19][l]	n.a.
77	—	[16.84][k]	—	[0.19][l]	n.a.

(Table continues on the following page)

Table 4-1 (continued)

Land cover class	Growing stock air-dry t ha⁻¹ range	Median	Sustainable yield air-dry t ha⁻¹ yr⁻¹ range	Median	Source
8 *Forest*					
81	246.80	246.80	29.49	29.49	Christiansen 1978; Golley and others 1971
82	60.00	60.00	—	[4.98]ⁿ	Baines 1980
83	88.78	88.78	4.98	4.98	Christiansen 1978
84	99.00	99.00	1.73–6.21	3.97	Chapman and White 1970; ETC (U.K.) 1987
85	22.75–232.80	(127.78)	0.263–0.65	[14.8]º	Pecha 1986; FAO/PNUD 1978; Marsch 1978; Bianchi 1986; Greenland and Kowal 1960; Nye 1961; John 1973
86	—	[127.78]ᵐ	—	[14.80]º	
87	69.78–140.00	(104.89)	14.50–15.10	(14.80)	République du Togo 1987; Bartholomew, Meyer, and Landelot 1953

— Not available.

n.a. Not applicable.

( ) Median value.

[ ] Value to use in calculations where no data exist.

a. Derived from values for Class 11; b. Derived from values for Class 21; c. Derived from values for Class 21; d. Derived from values for Class 33; e. Derived from data given for differences and stands of 33/34 (*fynbos* heath) in Rutherford (1978) and translated to Classes 31, 32, 34, and 35; f. Derived from values for Class 42; g. Derived from values for Class 43; h. Derived from values for Class 52; i. Derived from values for Class 62; j. Derived from values for Class 61; k. Derived from values for Class 73; l. Derived from values for Class 73; m. Derived from values for Class 85; n. Derived from values for Class 83; o. Derived from values for Class 87.

available in the summary classes of Grassland (1), Wooded Grassland (2), Low Woody Biomass Mosaic (5), and High Woody Biomass Mosaic (7).

The values for all Grassland classes (11–14) are based on a sample of estimates for growing stock for *Veld* Grassland (11). The several studies quoted by Rutherford (1978) on grasslands in southern Africa are mainly concerned with herbaceous biomass. No values for sustainable yield were found for these classes and therefore values have been inferred from data gathered for wooded grassland systems. In practice, given the lesser biomass present in the grasslands, the sustainable yield would probably be lower than that for wooded grasslands. A similar lack of data exists for these wooded grassland systems, however, and because these classes have low woody biomass components, they are likely to become areas of critical wood supply if demand should increase.

For Shrubland classes (31–35), Bushland and Thicket (41–45), and Woodland (61–67), the levels of data available are generally better. Nevertheless, data on growing stock are notably absent for some important woodland classes (63–65) in East Africa and the Sahel. In addition, far fewer data are available on sustainable yield for Shrubland and for Bushland and Thicket classes than are available for the equivalent growing stocks. Only one value of sustainable yield for Shrubland was obtained from the literature, and that had to be estimated from studies of stands of varying age of burned *fynbos* heath (Rutherford 1978).

Furthermore, only two of the Bushland and Thicket classes had any data on sustainable yield.

The estimates for the Forest classes are usable in the data base, although four of the seven classes had only one value for growing stock or sustainable yield. Of particular importance are the estimates for Mangrove forests (81), which are very small because they came from ecologically constrained areas such as Somalia and Guinea-Bissau. These low values clearly do not represent the bulk of mangrove forests in Africa, so we decided to use the value of 246.8 tonnes per hectare with an equivalent sustainable yield value of 29.49 tonnes per hectare per year. These values come from estimates of mangrove stocks and productivity in Puerto Rico (Golley and others 1969) and Thailand (Christiansen 1978).

Few quantitative data were available on the woody biomass stocks and yields for the Low and High Woody Biomass Mosaics (5, 7). Of the two, more data exist for Low Woody Biomass Mosaics, mainly because they are most widespread in southern Africa where significantly more ecological studies have been conducted and estimates have been made of woody biomass for planning fuelwood energy. Unfortunately, the same cannot be said of the High Woody Biomass Mosaics anywhere in Africa.

We recognize that the ratio between growing stock and sustainable yield shows a level of inconsistency that can be attributed to the use of median (central) figures from different studies.

Area Calculations of Land Cover Class

A map of national boundaries for Sub-Saharan Africa was digitized and warped to the land cover class map derived from the supervised classification of AVHRR imagery. In addition, the outlines of the main lakes and reservoirs in Sub-Saharan Africa were digitized and warped to the GVI imagery. This enabled the maps of the land cover classes and summary classes (see Chapter 3) to be overlaid with national boundaries, lakes, and reservoirs.

A program was written to count the pixels in each land cover class within each country across the entire image. This produced a table of pixel counts by land cover class for all Sub-Saharan African countries. The table was imported to a spreadsheet and used to calculate growing stocks and sustainable yields.

The area of each country was calculated by summing its pixels and multiplying them by the area of each pixel. This was possible because the map projection used to display the GVI data, the Hammer-Aitoff projection, is an equal area projection, and therefore each pixel occupies the same area across the entire image. The areas thus calculated were compared to areas of each country taken from standard sources (for example, Stonehouse 1985). In no instance were the differences between pixel-derived areas and those given in the literature greater than 1 percent of the country area.

These small differences are easily explained by the raster nature of the international boundaries and coastlines from the overlays on the remotely sensed imagery, which depict "smooth" curves and lines on maps as steplike boundaries. Note, however, that the summed areas of land cover classes for most countries are less than the recorded area of the countries. These discrepancies arise for three reasons:

1. The area of lakes and reservoirs is included in the total country areas, but these have no data for land cover class or woody biomass and consequently are omitted from the tables.

2. Some countries had significant cloud that affected the classification of land cover (table 3-1). These areas were masked from the final imagery (Chapter 3) and are not recorded in the tables.

3. Supervised classification establishes a class of pixels which were rejected because they were outside the statistical limits of the land cover classes. These pixels were not assigned to any land cover class and constitute a further proportion of the area of each country that is not mapped or used in the woody biomass data base.

Interfacing Area Growing Stock and Data on Sustainable Yield

Data on biomass growing stock and sustainable yield, assigned to land cover classes, were transferred into the spreadsheet directly from the data base. Macros were written within the spreadsheet to calculate the following information from the interfaced data base files:

- Growing stock by biomass class and summary biomass class
- Sustainable yield by biomass class and summary biomass class.

The growing stock and sustainable yield for each land cover class were used to produce maps for all of Sub-Saharan Africa. Mapping was achieved by examining the data on stock and yield (table 4-1) and identifying natural breaks in both, which defined five classes of stock and yield. Each land cover class in the final supervised classification image was assigned to a color-coded stock class and a yield class to produce the maps of growing stock (figure 7-1) and sustainable yield (figure 7-2).

5
Description of Biomass Classes

On a continental scale, the vegetation of Sub-Saharan Africa forms fairly distinct bands that span the continent latitudinally, although this zonation disintegrates in the extensive highlands associated with the Rift Valley of East Africa, especially in Ethiopia. The map of growing stock (figure 7-1) shows a similar pattern, with the greatest density of woody biomass centered on the Guineo-Congolian rain forest belt, which extends through much of coastal West Africa and across northern Zaire as far as Uganda. To the south, biomass stocks decrease steadily from the extensive *miombo* woodlands between about 5° S and 15° S to the open woodlands, shrublands, and eventually the *veld* grasslands and deserts of southern Africa.

North of the rain forest belt, a similar decrease in growing stock occurs through rather narrower zones of Guinean and Sudanian woodland, dry *Acacia-Commiphora* bushland and thicket, and the semidesert wooded grassland, which eventually merges into the Sahara Desert at about 20° N. East of the Rift highlands, much of Kenya and the Horn are covered by dry bushland and semidesert vegetation, giving way to desert in coastal Ethiopia, Djibouti, and parts of northern Somalia, and to a lesser extent in northern Kenya. In total, Sub-Saharan Africa contains an estimated 65,689 million tonnes of growing stock, producing a sustainable yield estimated at 3,549 million tonnes a year.

The following paragraphs summarize the growing stock and sustainable yield for the most significant land cover classes listed in table 3-2.

0 Desert dominates the northern portion of the study area, with the Sahara Desert extending into large areas of Sudan, Chad, Niger, Mali, and covering most of Mauritania. Desert also covers coastal regions of the Horn as far south as northern Somalia. Apart from a few woody species at the southern fringe of the Sahara, most biomass is in the root systems of drought-resistant plants.

In southern Africa, the Namib Desert stretches along the west coast from southern Angola through Namibia and into South Africa. The other significant desert areas are parts of the Kalahari Desert in Botswana. In total, the Desert class covers nearly 4.1 million square kilometers in Sub-Saharan Africa. Despite the dwarf shrubland on the fringes of the Namib Desert and a few widely scattered trees in the Kalahari Desert of southwestern Botswana, the overall contribution of Africa's desert areas to growing stock is negligible.

11 *Veld* Grassland covers only small areas over much of Africa, but it is extensive in southern Africa, covering nearly 6 percent. The largest expanse covers most of Lesotho and much of the Orange Free State and southern Cape Province. Outliers are found throughout northwestern Botswana and around the northeastern border of Angola. Woody biomass is largely restricted to riparian woodlands and to small, widely spaced trees in the Kalahari thornveld of northwestern Botswana. The total contribution of *Veld* Grassland to growing stock in Africa is negligible.

21 Semidesert Wooded Grassland is most extensive along the southern edge of the Sahara Desert, stretching eastward from southern Mauritania through the central areas of Mali, Niger, Chad, and Sudan. The southernmost areas are in northern Kenya, and the class is also important in Somalia and the eastern half of Ethiopia. The vegetation is grassland with bushes and small bushy trees (White 1983). The grassland is dominated by annual species, and the cover is ephemeral except where shaded by larger trees.

The woody component is largely of *Acacia* spp., with a crown cover of less than 10 percent. Woody biomass stocks are low—for example, Semidesert Wooded Grassland covers more than 18 percent of the West African Sahel region but provides less than 6 percent of the region's growing stock. Population density is low, increasing toward the south in the belt. In Somalia, however, the semidesert area supports more than 40 percent of the population, and their demand for fuelwood has caused severe stress on the trees (Kamweti 1984). Semidesert Wooded Grassland has a growing stock of nearly 364 million tonnes, with a sustainable yield of less than 14 million tonnes. The severe social consequences of depleting this resource make sustainable management vital.

24 Transitional Wooded Grassland covers 11 percent of southern Africa, including 40 percent of Botswana, 20 percent of Namibia, and areas of northern and central South Africa which add up to 14 percent of that country. Over much of this area, the vegetation is similar to that of the *Veld* Grassland (11). In parts of Botswana and the Orange Free State, this class consists of grassland dominated by *Aristida*, with a number of *Karroid* shrub communities up to 5 meters in height. In the Transvaal and southern Mozambique, this class includes Zambezian woodland dominated by *Acacia* in a mosaic with *veld* grassland.

Despite its large area, the Transitional Wooded Grassland class includes only 1.23 percent of the growing stock for southern Africa. Nonetheless, these areas are heavily cultivated and grazed. Because the majority of biomass is nonwoody, and access is restricted because of land tenure, Transitional Wooded Grassland is a poor source of fuelwood. This class has an estimated growing stock of nearly 146 million tonnes, producing a sustainable yield of about 5.6 million tonnes.

33 Bushy Shrubland occurs mainly in East Africa and southern Africa, covering 1.8 percent and 1.96 percent respectively. In Kenya, this class consists mainly of small bushes and stunted trees, dominated by *Acacia reficiens* ss. *misera*, although it also accounts for areas of montane vegetation in the eastern Aberdare Mountains. In southern Africa, the largest areas are found along the west coast of Cape Province and on the northern fringe of the Namib Desert along the coast of Angola. In Cape Province, the vegetation is sclerophyllous shrubland, with a few taller bushes but very few trees. In coastal Angola, this class contains a rather open vegetation of small shrubs, with occasional taller *Acacia-Commiphora* shrub communities.

Once again, grass forms a large proportion of the biomass, and wood production is low. From a total growing stock estimated at 96.7 million tonnes, Bushy Shrubland produces a sustainable yield of less than 1.6 million tonnes. Thus, the potential for large-scale, sustainable exploitation of fuelwood is extremely limited.

34 Kalahari Shrubland covers 4 percent of southern Africa, including 10 percent of South Africa, 9 percent of Botswana, and 8 percent of Namibia. In Botswana, Namibia, and northern Cape Province, this class consists largely of rather sparse sand dune vegetation, with scattered trees on the dune crests and shrubs restricted to the troughs. In the hills bordering the Namib Desert, tall succulents and bushy trees up to 5 meters are scattered in an open shrub layer of about 2 meters height. (See Class 35, Wooded Shrubland, for growing stock and sustainable yield statistics.)

35 Wooded Shrubland is similar to Kalahari Shrubland (34). Although its distribution is similar to that of Kalahari Shrubland, it occurs in slightly moister areas, often at higher altitudes. Each of these classes has a growing stock of less than 150 million tonnes, although the sustainable yield of more than 30 million tonnes for Kalahari Shrubland compares to only 2.5 million tonnes for Wooded Shrubland. In such barren surroundings, these classes represent a relatively poor source of fuelwood, and the present rate of fuelwood extraction easily exceeds annual production.

41 Dry *Acacia-Commiphora* Bushland and Thicket occurs in the countries of East Africa and the Horn, extending over much of southeastern Ethiopia and inland areas of southern Somalia and eastern Kenya. Significant outliers are found in central and northwestern Kenya, northern Tanzania, northeastern Uganda, and the extreme southeastern area of Sudan. The bushland is dense, often forming impenetrable thickets attaining 5 meters in height, with scattered emergent trees up to 10 meters. With a growing stock of nearly 944 million tonnes, this class produces a sustainable yield of only 10 million tonnes. Overpopulation and overgrazing have severely degraded large areas of the bushland, often leading to desertification around boreholes. The considerable demand for fuelwood exacerbates the problem.

43 Moist *Acacia-Commiphora* Bushland and Thicket is spread in patches across southern and East Africa in a roughly arc-shaped distribution. The most northerly, large area extends down the coastal strip of East Africa from about 4° N in Somalia to the south coast of Kenya. A large triangular area of moist bushland covers much of central Tanzania as far as Lake Malawi, and the arc continues through Malawi and parts of Mozambique and Zambia, eventually reaching the coast of southern Angola. Large outlying areas are found in the south of Mozambique and especially in Swaziland and the Transvaal.

In East Africa, this class is principally dense bushland up to 7 meters height, with grass cover growing to 1.5 meters. In southern Africa, however, this class spans a wide range of floristic regions. The major similarity between these areas is in their seasonal phenology. Productivity peaks in the summer, sometime between December and May, depending on the area of Africa examined. A decline sets in from May, with the lowest productivity occurring toward September, the end of the dry season. After this, production increases steadily.

The moist bushland is rather more productive than the Dry *Acacia-Commiphora* Bushland and Thicket (41), having a sustainable yield of nearly 77 million tonnes on 1,685 million tonnes of growing stock. Nonetheless, the pronounced dry season ensures that productivity remains relatively low, limiting the potential for fuelwood exploitation.

44 Sahel-Sudanian *Acacia* Wooded Bushland forms a belt across North Africa centered on about 12° N, from Senegal in the west to central Sudan and northern Ethiopia in the east. This belt lies between Semidesert Wooded Grassland (21) to the north and Dry Sudanian Woodland (62) to the south. Sahel-Sudanian *Acacia* Wooded Bushland is generally found in areas receiving between 250 and 500 millimeters of rainfall a year. There is a herbaceous layer of mainly annual grasses, although perennial species are more common toward the south of the zone. The ground layer is more persistent than that of semidesert, surviving most of the dry season. The main woody species are *Acacia*, increasing in height from 4 meters in the north to 8 meters in the south.

The growing stock of nearly 4,203 million tonnes is the third largest of any class and produces a sustainable yield exceeding 178 million tonnes. Processes such as agricultural and pastoral intensification (Graham 1969) and overgrazing around tube wells have destroyed a great deal of woody biomass. Careful management is needed if this class is to supply a sustainable yield of fuelwood.

51 *Acacia* Woodland Mosaic occurs mostly in southern Africa, where it covers large areas of eastern Botswana and the Namibian interior, as well as northern Transvaal and the fringes of South Africa's *veld* grasslands. Extensive areas are also found in the southern regions of Mozambique and Zimbabwe. Generally this class consists of rather open woodland with trees up to 10 meters height, most commonly *Acacia*, *Commiphora*, *Combretum*, and *Terminalia* species. These woodlands have frequently undergone scrub invasion.

In southern Africa this class is dominated by agriculture. Pastoralism has destroyed much of the woody biomass and ranching has restricted access. Productivity is fairly low, with 35 million tonnes of sustainable yield from nearly 1,100 million tonnes of growing stock, so once again, exploitation requires management.

52 East African Low Woody Biomass Mosaic occurs mostly in a belt from northeastern Uganda to the highland region of southwestern Kenya, although small outliers exist in the countries of the Horn. Much of this vegetation is *Combretum* small tree savanna, becoming evergreen and semigreen bushland in the Kenyan highlands. With a growing stock estimated at 399 million tonnes, producing nearly 32 million tonnes of sustainable yield, this class is locally important.

61 Open Woodland is scattered among the *miombo* woodland of Tanzania, where its composition is similar to that of the surrounding vegetation, and in Burundi, where the class is represented by moist *Acacia-Commiphora* bushland and *Acacia* wooded savanna. The vast majority of this class, however, lies in the southern African region. Here it covers much of southeastern Angola and northeastern Namibia, northern Botswana, southwestern Zambia, much of Zimbabwe, and parts of southern Mozambique. This forms an irregular belt running latitudinally between the *miombo* woodland to the north and shrubland to the south. Open Woodland encompasses a variety of vegetation, including woodlands dominated by *Brachystegia* spp., *Baikiaea* spp., and *Burkea africana*.

The growing stock is estimated at 743 million tonnes, producing a sustainable yield of nearly 63 million tonnes. Despite being an important source of fuelwood in South Africa and Lesotho, the low productivity of this class makes high rates of sustainable exploitation impossible in most areas.

62 Dry Sudanian Woodland lies in a latitudinal belt immediately south of the Sahel-Sudanian *Acacia* Wooded Bushland (44) previously described. From west to east, it occurs in southern Senegal and the Gambia; southern areas of Mali and Burkina Faso; northern Ghana, Togo, and Benin; northern Nigeria, Cameroon, and Central African Republic; southern parts of Chad and Sudan; and in northern Ethiopia. The vegetation is mainly a rather open tree savanna of 15 to 25 meters in height, or shrub savanna. Both *Acacia* and *Combretum* species are common. A ground layer of annual and perennial grasses exists, which burns easily during the dry season.

The growing stock for this class is less than 1,050 million tonnes, with a sustainable yield exceeding 213 million tonnes. Agriculture and grazing have degraded large areas of woodland. Fuelwood collection for cities such as Khartoum has left considerable areas almost treeless (Lewis and Berry 1988). The current fuelwood shortages and associated environmental degradation will almost certainly become worse at present rates of extraction.

64 Sudanian Woodland forms another latitudinal belt across Africa, lying immediately to the south of the Dry Sudanian Woodland (62). The belt is rather scattered, but broadest in the region of the West African coast. From the west, this class covers areas of Guinea-Bissau, Guinea, Mali, Côte d'Ivoire, Burkina Faso, Ghana, Togo, Benin, Nigeria, Cameroon, southern Chad, northern Central African Republic, southern Sudan, and small areas in the highlands of Ethiopia. This class corresponds to areas that White (1983) classified as "Sudanian woodland with abundant - *Isoberlinia*" and a mosaic of lowland rain forest and secondary grassland. Much of the area may be described as derived savanna, ranging from closed canopy savanna woodlands to thicket savannas.

Sudanian woodland contains a growing stock of nearly 1,184 million tonnes, producing a sustainable yield exceeding 391 million tonnes, the second highest of all the classes. Disturbance has been low and the prospects for sustainable fuelwood extraction at present rates are good.

65 Moist Sudanian Woodland is scattered along the south of the Sudanian woodland belt in coastal West Africa, spreading from central Guinea into central Africa. The largest area of this class straddles the border between Sudan and Central African Republic. The zone ends in northern Uganda. Vegetation is generally an open woodland savanna with trees attaining 15 meters in height. The most usual dominant tree species is *Isoberlinia doka*, and an understory of shrubs sometimes is present. Growing stock is 1,887 million tonnes, with a sustainable yield exceeding 282 million tonnes.

66 Seasonal *Miombo* Woodland covers huge areas of central and East Africa between about 15° S and the equator. It is most extensive in northern Angola, southern Zaire, Zambia, Zimbabwe, Tanzania, Malawi, and Mozambique, and extends along the Indian Ocean coast from Beira in southern Mozambique to Port Elizabeth on the southern Cape. This class covers a wide range of woodland types, the most common including *Brachystegia*, *Julbernardia*, and *Isoberlinia* woodland, often to 20 meters height, and frequently with a woody understory.

Seasonal *miombo* contains the largest growing stock of any class, exceeding 6,836 million tonnes. Sustainable yield is nearly 138 million tonnes. This represents a considerable potential for fuelwood exploitation.

67 Wet *Miombo* Woodland is most extensive in southern Zaire, northern and central Angola, northern Zambia, southeastern Tanzania, and central Mozambique. Generally found in wetter areas with a lower seasonality than Seasonal *Miombo* Woodland (66), Wet *Miombo* Woodland has a similar species composition, but with rather dense thickets that often form the understory. Wet *Miombo* Woodland has a growing stock of nearly 2,540 million tonnes, producing a sustainable yield of more than 43 million tonnes. With few exceptions, this class represents an important fuelwood resource in the countries where it occurs.

74 Guinean Woodland has a similar distribution to the Moist Sudanian Woodland (65) already described. It is most extensive in three main areas: (1) in coastal West Africa, where it occurs in every country except Liberia; (2) in Central African Republic and adjoining areas of Cameroon, northern Zaire, and southern Chad; and (3) in the border region of southern Sudan, northwestern Uganda, and northeastern Zaire. Rainfall is fairly high in these areas, so trees and

shrubs tend to form relatively dense stands separated by grasslands or herbaceous wooded savanna (Laclavère 1980).

Guinean woodland has a growing stock estimated at nearly 3,123 million tonnes, greater than all but four of the other classes, producing a sustainable yield of 256 million tonnes.

76 Medium-Productivity West African Cultivation and Forest Mosaic is the most extensive land cover class in coastal West Africa. It occurs in every country of the region and forms a broad band from Guinea-Bissau to southeastern Nigeria. This class contains remnants of semideciduous humid tropical forest in a mosaic with areas cleared for agriculture and timber. The mosaic typically contains both rain forest and savanna species at its northern limits, where it may represent an ecological transition from humid tropical forest and Guinean Woodland (74).

Generally, forest and woodland dominate in areas of low population density and clearance is more extensive in heavily populated regions, especially around towns and cities. The mosaic has the fourth-largest growing stock of any class, exceeding 3,641 million tonnes, coupled with the largest sustainable yield, nearly 561 million tonnes. Ongoing forest clearance makes this class an important supplier of fuelwood both locally and to deficit areas.

77 Highland Cultivation Mosaic is confined to Ethiopia, of which it covers about 17 percent. Found on plateau areas between 2,000 and 3,000 meters elevation, it is the result of thousands of years of deforestation due to fuel collecting, grazing, and cultivation. Large areas are now treeless, although land uses range widely from dense cultivation to tree plantations, as well as remnants of forest and woodland. Total growing stock is less than 1,350 million tonnes, but with a sustainable yield less than 15 million tonnes. The region has long suffered from critical fuelwood shortages, and it appears that these will continue and worsen.

81 Mangrove grows extensively on the coasts of East, central, and West Africa, especially around river estuaries. Local pressure on mangrove swamps is intense in many areas. Mangrove wood is excellent fuel and is used for smoking fish, so many mangrove stands are heavily exploited and depleted, despite general protection by law. The inability of the satellite images to resolve very narrow fringes of mangrove swamp, notably on the East African coast, means that the figures for growing stock (193 million tonnes) and sustainable yield (68 million tonnes) are almost certainly underestimates.

82 Evergreen Forest mainly represents montane forest vegetation. It occurs in a number of highland areas, including the mountains of southern Tanzania and eastern Transvaal in South Africa, as well as the rift escarpment in eastern Zaire. These areas typically experience high rainfall and humidity and encompass a wide range of both montane and lowland rain forest species. The growing stock for the class is 1,310 million tonnes, with a sustainable yield of more than 90 million tonnes. Clearance has occurred in some areas, with only the steepest slopes remaining forested.

85 Mesophilous Humid Tropical Forest is most extensive on the northern fringe of the equatorial and Ombrophilous Humid Tropical Forest (87), where it covers large areas of Cameroon, Río Muni, Gabon, Congo, Central African Republic, and especially Zaire. South of the equatorial forest, this class occurs in a broad swath along the west coast from Cabinda to southern Gabon. The mesophilous forests are rather more seasonal than the equatorial forests, showing a more pronounced dry season, and correspond partly to the drier Guineo-Congolian rain forest of White (1983). The range of species is wide, as is the degree of species dominance. This class contains about 2,850 million tonnes of growing stock, producing a sustainable yield of more than 72 million tonnes.

86 Humid Tropical Swamp Forest forms a narrow strip near the equator, with areas around the western shores of Lake Victoria in Tanzania and Uganda, and in Zaire, Congo, southern Cameroon, and Gabon. Swamp forest covers up to 6 percent of central Africa and constitutes a growing stock of about 1,540 million tonnes. Sustainable yield is a little more than 39 million tonnes.

87 Ombrophilous Humid Tropical Forest extends along the equator, covering vast areas in Zaire, Congo, Gabon, Cameroon, Central African Republic, and Río Muni. There is little seasonality, with the mean annual rainfall of 1,500 to 2,100 millimeters distributed fairly evenly through the year, and productivity is consistently high. The forest has a tall, closed canopy which may reach 45 meters in height. Most species are evergreen, although a few deciduous species shed their leaves during the brief dry season.

The ombrophilous forests have a growing stock of nearly 5,932 million tonnes, second only to the much more extensive Seasonal *Miombo* Woodland (66) to the south, and a sustainable yield exceeding 150 million tonnes. Much of the remaining forest area throughout Africa is protected within reserves. Much of the present equatorial forest area has been cultivated in the past, and the vegetation is secondary. Within some reserves, older secondary forest may be indistinguishable from primary forest. This enormous biomass resource has been largely removed from potential fuelwood reserves in the interest of science and tourism.

6

Regional Summaries by Class of Biomass

This chapter summarizes physical characteristics, population, resources, and the condition of woody biomass for each of the six regions examined in this study. For each region, a table presents (a) the estimated surface area and percentage of the nine summary classes of land cover, (b) the growing stock by summary land cover class, and (c) the sustainable yield by summary land cover class. For a more detailed review of the distribution and character of the woody biomass in each region, please refer to Part II where the 43 land cover classes are discussed.

The West African Sahel

For the purpose of this study, the West African Sahel is bounded by the latitudes of 10° N to 23° N and the longitudes of 18° W to 24° E. The region includes the seven countries of Mauritania, Senegal, the Gambia, Burkina Faso, Mali, Niger, and Chad. The area is delimited to the north by the Sahara Desert, and all countries have significant desert territory except the Gambia, Senegal, and Burkina Faso.

Physical Characteristics, Population, and Resources

Rainfall in the region is markedly seasonal, with the humid season peaking in July and August and the principal biological production occurring soon after. In the semidesert regions, annual rainfall is only 100 to 500 millimeters a year. Evaporation rate is rapid and water-holding capacity of the soils is small. In the south of the region, annual rainfall may exceed 1,000 millimeters. This, combined with high summer temperatures, produces favorable conditions for photosynthetic activity and results in the marked seasonality of the biomass.

The north of the region is subject to periodic fluctuations of climate, of which the great drought of 1968–73 was an extreme. This significantly affects water availability. As a consequence, a mosaic of vegetation patterns exists, reflecting areas in which land is stressed.

In the northern countries, distribution of the sedentary human population is restricted to more productive areas having greater and more predictable rain-

Table 6-1. Distribution of Summary Classes and Estimated Woody Biomass, West African Sahel

Summary class	Area km²	Area Percent	Growing stock Million tonnes	Growing stock Percent	Sustainable yield Thousand tonnes per year	Sustainable yield Percent
0 Desert	2,590,342	49.31	0.00	0.00	0.00	0.00
1 Grassland	40,997	0.78	9.31	0.37	409.97	0.41
2 Wooded Grassland	981,080	18.68	323.76	12.96	9,810.80	9.84
3 Shrubland	0	0.00	0.00	0.00	0.00	0.00
4 Bushland and Thicket	879,957	16.75	130.85	5.24	30,722.52	30.80
5 Low Woody Biomass Mosaic	211	0.00	0.45	0.02	13.29	0.01
6 Woodland	730,777	13.91	1,951.17	78.09	50,396.70	50.52
7 High Woody Biomass Mosaic	8,326	0.16	14.02	0.56	158.19	0.16
8 Forest	2,793	0.05	68.93	2.76	8,236.56	8.26
Lakes	18,233	0.35	0.00	0.00	0.00	0.00
Total	5,252,716	100.00	2,498.49	100.00	99,748.03	100.00

Note: Details may not add to totals because of rounding.
Source: Tables 8-1 to 8-7.

fall. Land use in the semidesert regions is largely pastoral, although significant permanent populations exist along the Senegal and Niger rivers and their associated irrigation schemes. According to the FAO, population density varies throughout the region, with 1985 statistics showing 1.8 persons per square kilometer in Mauritania, contrasting with 33.2 in Senegal and 56 to 59 in the Gambia (World Resources Institute 1986). These mean figures, however, cannot truly reflect the possible demographic stress caused by the population distribution.

Relief and soil clearly have implications for biomass production. The plateaus of the southern Sahara and the Manding region of Guinea and Mali are reflected in the classification produced from the imagery. Red-brown soils of the south are endowed with greater capability of retaining water, contrasting with desert soils and sands farther north.

A review of the classes shown in table 6-1 discloses that Desert (Class 0) occupies 49 percent of the region, including large areas of Mauritania, Mali, Niger, and Chad. Grassland (1) covers less than 1 percent but Wooded Grassland (2) covers 19 percent as a broad latitudinal belt. No Shrubland (3) is recognized in this region but Bushland and Thicket (4) covers 17 percent. Woodland (6) covers 14 percent of the region and locally is important at the southern edge adjacent to the West African Coast region. Low Woody Biomass Mosaic (5), High Woody Biomass Mosaic (7), and Forest (8) are almost completely absent from the West African Sahel region.

Condition of Woody Biomass

Assessment of woody biomass in the West African Sahel is problematic because of several difficulties in interpreting Normalized Difference Vegetation Index (NDVI) values in relation to biological activity. One problem is that we cannot distinguish between herbaceous and woody biomass. In a region like the Sahel, which includes areas of limited tree cover and vast areas of seasonal grassland, this is obviously significant. For example, in a study of the regeneration of degraded Sudanian savanna during a 10-year period, Bonkoungou, Bortoli, and Oudba (1988) show that the grass *Andropogon gayanus* forms 50 percent of the biomass. Another problem is our inability to distinguish between seasonal crops and natural wooded grassland, which produce similar phenologies.

Human population density is much greater in the southern portion of all the countries, exacerbated both by desertification and population growth. United Nations figures show that all Sahelian countries except the Gambia have a population growth rate exceeding 2 percent, and all rates are projected to increase (World Resources Institute 1986). (The region does not exhibit the rapid growth rates of other parts of Africa, such as Kenya.) This increase, coupled with greater than 40 percent of the population being 14 years of age or younger, means that demographic pressure can only increase. The demand for greater space for food production results in degradation of land already exploited to its full potential. This can be seen in areas of Senegal or in extensive migration like that occurring in Burkina Faso.

Discussion with the director of the Institut de Recherche en Biologie et Ecologie Tropicale (IRBET), based in Burkina Faso, confirms that the migration is particularly significant. Areas of Sudanian woodland that previously provided a more-than-adequate fuelwood resource now are being cleared for agriculture.

Research within the Sahelian zone by the University of Kano in northern Nigeria suggests three management strategies for sustainable production of fuelwood (Cline-Cole and others 1987): (a) farmed parkland in which trees are retained, (b) shrubland or fallow, and (c) forest reserves. This work also notes the diversity of use and calorific values of different species.

Accessibility to fuelwood in the most densely populated areas varies through the region. Where trees are retained to a significant extent, management is possible that recognizes diverse use for food and fodder, construction, pharmaceuticals, heating, and cooking. An assessment of woody biomass is not necessarily an assessment of fuelwood, although all trees become fuelwood if no alternative exists.

A significant contribution to growing stock comes from the Bushland and Thicket class, of which the dominant type is Sahel-Sudanian *Acacia* Wooded Bushland (44). It accounts for more than 130 million tonnes of growing stock and a sustainable yield of 30 million tonnes, about 30 percent of the regional total (table 6-1). The Sudanian Woodlands (62, 64, and 65) are highly significant, at more than 1,900 million tonnes of growing stock and 50 million tonnes sustainable yield. This group is most vulnerable to the present population migration in the south of the region.

It should be remembered that estimates for this region are particularly crude because of the paucity of data for herbaceous biomass and noncommercial wood. The Sudanian Woodlands in particular merit closer, coordinated study, with a more intensive local examination of woody biomass estimates linked to a finer resolution of remotely sensed imagery. Much work is under way within organizations such as IRBET, the Centre Technique Forestier Tropicale (CTFT), and the Institut de Recherche Agronomique du Niger (INRAN).

The West African Coast

The West African coastal region includes all of the countries bordering the Atlantic Ocean from Guinea-Bissau to Nigeria. Strong ecological and agricultural arguments exist for this division. First, this region contains the humid tropical forest zone of West Africa and wet (Guinea) savanna woodland. Second, cultivation in the southern two-thirds of the region is dominated by root and tuber crops, rice of both upland and swamp varieties, and plantation tree crops such as cocoa, coffee, rubber, and oil palm. Grazing systems and dryland crops are restricted mainly to the northernmost parts of the region.

Physical Characteristics, Population, and Resources

Estimated woodfuel consumption in 1990 was approximately 81 million air-dried tonnes (table 1-2). Sustainable yield significantly exceeds this; therefore the region has a positive overall balance of woody biomass supply. Two problems exist, however, with both the actual supply and the potential supply. First, the northern part of the zone is dominated by dry savanna woodland, from which it is difficult to transport wood to large cities such as Kano and Sokoto. Second, much wood from the Sudanian woodland is converted into charcoal and transported to the large coastal cities.

In the wetter areas of higher productivity to the south, significant extraction of timber occurs. Such forest degradation probably has little negative impact on the wood energy market and in fact may lead to a local surplus of "waste" wood that is sold or freely collected. Of course, local supply problems exist, such as those in central Sierra Leone, related to particularly dense population concentrations, edaphic factors that restrict tree growth, and reservation of forest and other types of land.

A review of the classes in table 6-2 shows that Desert (0) and Shrubland (3) are virtually absent from this region. Grassland (1) represents less than 2 percent and Wooded Grassland (2) less than 1 percent of the region's land cover. Bushland and Thicket (4) covers 8 percent of the region, principally in northern Nigeria. Low Woody Biomass Mosaic (5) is relatively unimportant and covers just 1 percent. Woodland (6) is an important category, covering 40 percent of the region in a broad belt from Guinea in the west to central Nigeria in the east. Bordering this zone to the south is the most extensive summary class, High Woody Biomass Mosaic (7), which represents 44 percent of the regional land cover and is present in all countries. Finally, Forest (8) covers only 3 percent of the region, mainly in small coastal blocks.

Condition of Woody Biomass

The West African coastal states generally have adequate woody biomass, particularly for fuelwood and charcoal. Two countries, Liberia and Sierra Leone, have only local problems of supply, due either to dense population concentration, edaphic restriction on tree growth, or land reservation.

The other countries all have a general decline northward in growing stock and sustainable yield. This pattern results from the latitudinal disposition of land cover classes across the region. The high-productivity Forest and High Woody Biomass Mosaic classes are adjacent to the coast, the wet (Guinea) savanna woodland is farther inland, and the drier (Sudanian) savanna woodland thrives in the northern part of the region. This pattern is a function of the mean annual rainfall and the length of the wet season.

The majority of growing stock and sustainable yield is in the High Woody Biomass Mosaic adjacent to the coast, the Woodland, and the highly significant Mangrove class. Repeated cutting of the wood in the High Woody Biomass Mosaic means that many trees and shrubs continue to grow rapidly. To counter this rapid annual growth, however, forest reservation and plantations probably restrict access to land more for two classes than for any others in the region: High-Productivity West African Cultivation and Forest Mosaic (75) and Medium-Productivity West African Cultivation and Forest Mosaic (76).

Guinean Woodland (74) includes 7.6 percent of the growing stock and 1.8 percent of the sustainable yield. Apart from Guinea, where extensive clearance of these woodlands has been conducted in the Fouta Djallon region, this zone is the least exploited of the main biomass classes in the region. This situation could change because of increasing population density or increasing exploitation of the trees for fuelwood and charcoal in the Guinean woodland.

The Sudanian Woodlands (62 to 65) encompass only 40 percent of the growing stock and 17 percent of the sustainable yield (table 6-2). This suggests a currently high level of exploitation, probably attributable to charcoal production for cities along the coast and to local demand. Comparison of growing stock with the area covered by these woodlands reveals that they are understocked. Consequently, woody biomass supply problems undoubtedly exist in these woodlands, particularly where the population is concentrated and the cultivation is intensive.

The most northerly large biomass class in this region is Sahel-Sudanian *Acacia* Wooded Bushland (44), which is typically Sahelian. Occurring only in northern Nigeria, it represents just 0.5 percent of the growing stock and about 2 percent of the sustainable yield.

Table 6-2. Distribution of Summary Classes and Estimated Woody Biomass, West African Coast

Summary class	Area		Growing stock		Sustainable yield	
	km²	*Percent*	*Million tonnes*	*Percent*	*Thousand tonnes per year*	*Percent*
0 Desert	475	0.02	0.00	0.00	0.00	0.00
1 Grassland	38,415	1.88	8.72	0.16	384.15	0.15
2 Wooded Grassland	17,706	0.87	5.84	0.11	177.06	0.07
3 Shrubland	0	0.00	0.00	0.00	0.00	0.00
4 Bushland and Thicket	167,832	8.22	30.79	0.57	5,808.49	2.26
5 Low Woody Biomass Mosaic	18,447	0.90	42.82	0.80	1,254.83	0.49
6 Woodland	809,981	39.69	2,162.52	40.34	43,683.43	16.99
7 High Woody Biomass Mosaic	906,252	44.40	1,526.13	28.47	17,218.79	6.70
8 Forest	64,615	3.17	1,583.73	29.54	188,656.05	73.35
Lakes	17,284	0.85	0.00	0.00	0.00	0.00
Total	5,252,716	100.00	5,360.55	100.00	257,182.80	100.00

Note: Details may not add to totals because of rounding.
Source: Tables 9-9 to 9-17.

This area has dense population concentrations (for example, the Kano and Sokoto "close-settled" zones), and in these areas fuelwood supply is a significant problem, exacerbated by an overall shortage of woody biomass. This class offers the poorest woodfuel supply in the region.

The Horn of Africa

Four countries constitute this region: Djibouti, Ethiopia, Somalia, and Sudan. Although the population is overwhelmingly dependent on subsistence agriculture, commercial agriculture, or pastoralism, woody biomass is vital to everyone in these countries, even those in urban areas. Woody biomass is essential as a fuel (it is locally available and relatively inexpensive), as building and construction material, as material for the construction of equipment, and as browse for ani-

mals. This is quite apart from its value in environmental maintenance of soil quality, as shade trees, and as a habitat for economically valuable wildlife.

Physical Characteristics, Population, and Resources

Woody biomass in this region reflects variations of climate. These are caused by seasonal variations in the position of subtropical high-pressure cells; by the latitudinal extent of the region; by variations in altitude, particularly in Ethiopia; and to a limited extent by higher precipitation, which may be caused by proximity to the sea. Although temperatures significantly influence the highland areas of Ethiopia and the higher parts of the Equatoria region in Sudan, the main determinants of vegetation patterns are precipitation and seasonal variability.

Table 6-3. Distribution of Summary Classes and Estimated Woody Biomass, Horn of Africa

Summary class	Area		Growing stock		Sustainable yield	
	km²	*Percent*	*Million tonnes*	*Percent*	*Thousand tonnes per year*	*Percent*
0 Desert	1,115,190	25.65	0.00	0.00	0.00	0.00
1 Grassland	133,952	3.08	30.41	0.65	1,339.52	0.98
2 Wooded Grassland	761,815	17.52	251.40	5.38	7,618.15	5.56
3 Shrubland	685	0.02	0.69	0.01	34.25	0.02
4 Bushland and Thicket	1,053,007	24.22	976.66	20.90	28,002.47	20.43
5 Low Woody Biomass Mosaic	106,656	2.45	229.51	4.91	6,719.33	4.90
6 Woodland	755,016	17.37	2,015.89	43.15	47,645.08	34.76
7 High Woody Biomass Mosaic	363,598	8.36	612.30	13.11	6,908.36	5.04
8 Forest	52,695	1.21	555.24	11.88	38,797.79	28.31
Lakes	4,690	0.11	0.00	0.00	0.00	0.00
Total	4,347,304	100.00	4,672.10	100.00	137,064.95	100.00

Note: Details may not add to totals because of rounding.
Source: Tables 10-1 to 10-4.

Except for deserts and small areas of tropical forest, almost all vegetation classes reflect the pronounced seasonality of precipitation, which includes one or two dry seasons of differing length. In many of the classes, precipitation variations from year to year may profoundly influence plant cover, particularly when prolonged drought is exacerbated by heavy human interference.

Variation in soil and relief is important in some classes of land cover because such variations determine the overall nature of the class. For example, vegetation is influenced by the lateritic soils of southern Sudan, the dark, cracking clays and sands of central Sudan, and the escarpments of Ethiopia. Within each class, variations of soil, relief, and drainage also may locally affect biomass.

A review of the classes (table 6-3) discloses that Desert (Class 0) occupies about 26 percent of the region, including a significant portion of each country. Grassland (1) covers 3 percent of the region and is important mainly in Sudan. Wooded Grassland (2) of various types occupies nearly 18 percent of the region and is present in each country, but is most extensive in Sudan. Shrubland (3) is nearly absent. Bushland and Thicket (4) occupies 24 percent of the area and is important in Sudan, Ethiopia, and Somalia.

Low Woody Biomass Mosaic (5) is confined to the extreme south of the region and covers just 2 percent. Woodland (6) covers 17 percent of the area and is most important in Sudan and Ethiopia. High Woody Biomass Mosaic (7) covers 8 percent of the region and is most important in Ethiopia. Forest (8) covers only 1 percent of the region, largely in Ethiopia; it is significant because it may be associated with high biomass potential.

Condition of Woody Biomass

For some decades throughout the region, concern has grown among both experts and laypersons about the condition of natural biomass, both woody and non-woody. In the drier regions of each country, biomass condition relates to desertification, details of which are considered in the following discussion of specific land cover classes. In general, desertification is the degradation and possible elimination of biomass caused by sequences of unusually dry years, which have occurred approximately every 20 years in the present century.

This climatic stress on vegetation is compounded by intensified consumption of biomass through grazing and browsing by the animals of herders who have been displaced from other degraded areas. Increasing population, both human and animal, no doubt contributes to the process. Tube wells focus grazing near watering points, leading to severe deterioration of

vegetation in a zone radiating from the wellhead, possibly extending several kilometers. This phenomenon has been noted in central Sudan and in Somalia. One unfortunate effect of heavy grazing has been to encourage development of unpalatable species.

Clearance of trees for agriculture, either bush fallowing or permanent agriculture, has reduced woody biomass in damper areas of Sudan, Ethiopia, and Somalia. This may lead to virtual elimination of tree species unless trees are deliberately incorporated into the farming system. In bush-fallowing systems, certain desired tree species may be protected, but a reduced period of fallow may suppress the successful regeneration of trees.

Commercial timber extraction is selectively destructive of larger straight-boled trees which are most common in forested areas. Complete removal of all large trees may follow in accessible areas. Areas of higher population also may suffer loss of trees and bushes through the extension of agriculture; this is notable in areas of coffee plantation in southwestern Ethiopia.

Burning of grassland as a method of encouraging grass growth is widespread in wooded savanna grasslands. Consequently, trees and bushes that are not fire resistant are killed off, whereas resistant species are encouraged.

Fuelwood collection is the other main cause of tree cover loss. It occurs in rural areas, but may be greatly increased by urban demand for wood or charcoal, which may be transported hundreds of kilometers (for details, see the discussion of Class 22 in Chapter 10). Unless a deliberate policy of management of fuelwood plantations is introduced near towns, tree removal spreads outward, particularly along motorable roads. This has been done in Ethiopia, where *Eucalyptus* has been planted for almost a century, and in parts of Somalia.

Kamweti (1984) has investigated the probable effect of fuelwood collection up to the year 2000 in the East African region, which includes Ethiopia and Somalia. Most of Ethiopia and Somalia already suffer fuelwood deficit, with the most serious effect in the desert, dry savanna, and especially the densely populated highlands of Ethiopia. In Somalia, the problem is serious throughout the country.

In Ethiopia, at the present rate of deforestation, 1.76 million hectares of woodland will be lost by the year 2000, and 2.27 million hectares of shrubland will be degraded. In Somalia, the main problem is degradation of shrub; however, 13,000 hectares per year of productive forest is being lost. Comparable information is not available for Djibouti and Sudan, but clearly, what little forest remains in the former is being devastated, and in the latter the effect of wood removal in the central area is catastrophic (for details see the discussion of Classes 21, 22, 44, and 62 in Chapter 10).

Whitney, Dufournaud, and Murck (1987) calculate that 31,000 square kilometers of woodland are consumed as fuelwood each year in Sudan. Additional pressures have occurred in parts of Sudan, Ethiopia, and Somalia as the result of famine and political unrest. Refugees from both threats have concentrated in small areas of the three countries, causing additional stress on fuelwood resources. In Sudan, areas particularly affected are the eastern plains, the Red Sea Hills, east of El Obeid, western Darfur, and the Juba area. The Horn of Africa has experienced a great influx of refugees: 1.5 million people have moved from Ethiopia to Somalia, Sudan, Kenya, and Djibouti since the mid-1970s. Civil wars in recent years in Sudan and Somalia have displaced more than 800,000 refugees to neighboring countries (Black 1989).

In the famine areas of eastern Ethiopia (Tigre, Welo, and Ogaden) and Sudan (Darfur, Kordofan, and Red Sea Hills), additional stress on vegetation is caused by increased animal browsing and by human consumption of food derived from trees during famine (for example, see FAO/UN 1984). It is unclear if the present resettlement camps will permanently stress local biomass, as might occur when host governments attempt to make the camps self-supporting by giving only land as a contribution to refugee self-sufficiency. This is, of course, an additional pressure on biomass resources. For example, near Qala 'en Nahal, the largest Sudanese refugee settlement, the rate of deforestation has increased dramatically during the past decade.

The previous Ethiopian government's policy of relocating people to new settlements, particularly in the southern part of the country, undoubtedly did contribute to deforestation of land cover classes which at present have reasonable reserves, such as Ombrophilous Humid Tropical Forest (87).

Central Africa

Six countries constitute the central African region: Cameroon, Central African Republic, Congo, Equatorial Guinea (Río Muni), Gabon, and Zaire. Woody biomass is generally plentiful in central Africa, although important local exceptions occur, and a strong geographical pattern reveals significant variation between the relatively plentiful supplies of the Zaire basin and the less-endowed areas of northern Cameroon, Central African Republic, and southern Zaire. Locally, large urban populations can stress the supply of woody biomass, a situation recognized in cities such as Kinshasa, Brazzaville, and Pointe Noire along the Zaire River and Kolwezi and Lubumbashi in the southern part of Zaire (Malaisse and Binzangi 1985). Only in a few areas like these, however, is woody biomass depletion likely to exceed the locally available sustainable supply.

Physical Characteristics, Population, and Resources

Estimating woody biomass in this large region is difficult, so it is reassuring that the classes of land cover that we describe often are similar to previously identified ecological zones. In the humid tropical forest zone, areas exhibiting both sparse population and plentiful woody biomass often are a conjunction of cause and effect. Because domestic fuel is the main end use of woody biomass in this region, our discussion focuses on areas of scarce or potentially scarce woodfuel.

Woody biomass of the region reflects variations of climate associated with distance from the equator. Thus, the largest forest areas of Africa occur in this region, within a narrow belt extending 2 degrees north and 2 degrees south of the equator. This area receives less rainfall than many other tropical rain forests (for example, Amazonia and Southeast Asia), but still supports extensive tracts of closed forest. This narrow equatorial zone has a brief dry season of about 2 months.

North and south of this zone, where the dry season extends to 3 or 4 months, thinner and usually deciduous forests and woodlands occur. These are *miombo* woodlands in the south (part of the Zambezian phytogeographical region) and Guinean and Sudanian woodlands to the north. At the northern extremity of the central African region (10° N in Central African Republic; 12° N in Cameroon), the dry season lasts at least 6 months, and the vegetation is that of the Sahel belt. At the southern extremity of the region in the high plateau of Shaba (12° S in Zaire), where the dry season also extends to 6 months, the vegetation is open woodland, often described as wooded savanna.

On this general pattern of woody biomass imposed by variation of the equatorial climate, further variability results from relief, soils, and human interference. The effect of relief is most notable on the high ground of the western Rift Valley in eastern Zaire and on the mountains of western Cameroon. (Here, persistent cloud cover precluded some small areas from being classified from the satellite imagery.)

Because of its varied relief and latitudinal extent, Cameroon provides a complete spectrum of the vegetation of intertropical Africa, with humid tropical forest in the south, open woodland in the center, grassland and bushland in the north, and mountainous forest (Laclavère 1980). Relief also plays a role in the extensive plateau areas to the south of the Zaire Basin, where temperature reflects the extensive terrain above 1,000 meters. Away from the narrow belt of humid tropical forest, the interplay of the relief with the pre-

Table 6-4. Distribution of Summary Classes and Estimated Woody Biomass, Central Africa

	Area		*Growing stock*		*Sustainable yield*	
Summary class	km^2	*Percent*	*Million tonnes*	*Percent*	*Thousand tonnes per year*	*Percent*
0 Desert	105	0.00	0.00	0.00	0.00	0.00
1 Grassland	17,336	0.44	3.94	0.01	173.36	0.01
2 Wooded Grassland	41,575	1.05	12.05	0.04	415.75	0.02
3 Shrubland	7,430	0.19	7.43	0.03	371.50	0.02
4 Bushland and Thicket	49,797	1.26	71.08	20.90	1,200.66	0.05
5 Low Woody Biomass Mosaic	5,111	0.13	11.03	0.04	321.99	0.01
6 Woodland	1,620,272	40.93	9,311.56	33.25	137,550.85	5.78
7 High Woody Biomass Mosaic	645,098	16.30	1,058.39	3.78	12,256.86	0.52
8 Forest	1,550,884	39.18	17,525.69	62.59	2,226,336.80	93.60
Lakes	20,868	0.53	0.00	0.00	0.00	0.00
Total	3,958,476	100.00	28,001.17	100.00	2,378,627.77	100.00

Note: Details may not add to totals because of rounding.
Source: Tables 11-1 to 11-6.

cipitation variation from year to year can produce much greater spatial and temporal variation in the plant cover.

Soil types and conditions can influence extensive tracts where the woody biomass is less than would be expected for central Africa. Examples include *Veld Grassland* (11) on the Kalahari sands of Quaternary age in southern Zaire and *Edaphic Wooded Grassland* (25) in Congo on "droughty" plateau soils, despite receiving more rainfall on average than the adjacent humid tropical forest.

A review of the classes in table 6-4 discloses that the lower woody biomass summary classes are either not represented or occur only as very restricted land covers. Desert (0) is not present in the region at all; Grassland (1) covers less than half of 1 percent; Wooded Grassland (2) covers 1 percent; Shrubland (3) and Low Woody Biomass Mosaic (5) are nearly absent; and Bushland and Thicket (4) covers just over 1 percent. By contrast, Woodland (6) covers 41 percent of the region and is an important fuelwood reserve in many of the more populated areas. High Woody Biomass Mosaic (7) covers 16 percent and Forest (8) covers 39 percent in a broad swath along the equator, presenting Africa's best-endowed woody biomass reserve.

Condition of Woody Biomass

Even here, in the African region having the most plentiful woody biomass, reference often is made to the profound influence of human interference on the land cover. This is illustrated by data collected by Malaisse and Binzangi (1985) for an area 1 degree of latitude by 1 degree of longitude, centered on Lubumbashi in southern Zaire. These authors estimate that, since the beginning of the century, the retreat of woodland and forest involved 21.6 percent of the area (212,162 hectares), with an associated diminution of wood reserves

of 18.8 percent. They observe that some vegetation types had almost disappeared locally, including dense riparian forest.

Detailed studies like this are rare, but the extent of human impact is not. Even in Gabon, 80 percent of which formerly was covered with forest, large areas now are recognized as High Woody Biomass Mosaic. Here, small-scale clearance by traditional means of shifting cultivation has significantly changed the nature of the forest. However, no disquiet need yet be felt about Gabon regarding woody biomass supply (Catinot 1978). Cultivation and forest mosaic is an important component of land cover, accounting for 16 percent of the area of central Africa (table 6-4), commonly occurring at the margin of the forest zone where it represents human encroachment and modification.

East Africa

Five countries constitute this region: Rwanda, Burundi, Uganda, Kenya, and Tanzania.

Physical Characteristics, Population, and Resources

Household fuel accounts for 90 percent of all wood use in this region. With an annual average per capita consumption of 1 cubic meter a person, coupled with an annual average population growth rate exceeding 3 percent, it is not surprising that concern exists regarding dwindling woody biomass. At first, attention focused on semiarid regions of obvious scarcity, particularly along roadsides. More recently, attention has focused more logically on high-potential arid areas where the majority of the population lives and where the greatest demand exists. This refocusing has led to the design of projects to reduce the problem of fuelwood scarcity within areas of high potential.

Table 6-5. Distribution of Summary Classes and Estimated Woody Biomass, East Africa

| | Area | | Growing stock | | Sustainable yield | |
Summary class	km²	Percent	Million tonnes	Percent	Thousand tonnes per year	Percent
0 Desert	14,913	0.83	0.00	0.00	0.00	0.00
1 Grassland	6,059	0.34	1.38	0.02	60.59	0.03
2 Wooded Grassland	142,908	7.95	38.87	0.60	1,429.08	0.62
3 Shrubland	34,462	1.92	34.24	0.53	1,723.10	0.75
4 Bushland and Thicket	490,118	27.25	752.02	11.61	10,296.90	4.50
5 Low Woody Biomass Mosaic	146,914	8.17	320.85	4.95	9,255.58	4.04
6 Woodland	526,003	29.24	3,581.07	55.27	50,304.67	21.97
7 High Woody Biomass Mosaic	140,220	7.80	194.17	3.00	2,664.18	1.16
8 Forest	202,455	11.26	1,556.89	24.03	153,265.44	66.93
Lakes	94,588	5.26	0.00	0.00	0.00	0.00
Total	1,798,640	100.00	6,479.49	100.00	228,999.54	100.00

Note: Details may not add to totals because of rounding.
Source: Tables 12-1 to 12-5.

The situation, however, is not simple. First, critical fuelwood shortages exist primarily around towns in the more arid areas, including the rain shadow of the coast, rather than in any particular rural production system. Second, although wood scarcities are obvious in landscapes as one drives through them, what one sees is wholesale cutting for charcoal production in response to the demand for urban energy—an arterial process closely related to principal transport routes. Third, although fuelwood constitutes most of the wood consumed, this wood is waste wood—it is a by-product of woody biomass being used for many other purposes, including fodder. Finally, significant evidence exists that land privatization or greater control by village communities leads to significant increases in woody biomass stocking because it ceases to be a free good accessible to all. This occurs in poorly controlled and managed government "forests," which often are treated as commons.

Estimating woody biomass production in East Africa is particularly difficult, especially from large-scale imagery, for two reasons. First, the East African biomass landscape is made up of ecological niches in which altitude, not spatial variation, determines growing stock and annual productivity. Second, on this altitudinal mosaic, which ranges from equatorial glaciers to desert, a complex series of production systems exists, from pastoralism to modern monocropping agriculture. These systems involve sophisticated traditional technologies of dry land farming as well as intensive irrigation for high value-added export crops. The phenology of woody biomass generally shows a marked seasonality, which reflects the bimodal distribution of annual rainfall.

The use of satellite imagery created problems in mapping of the forested areas in this region. The most extensive type of forest in East Africa is Evergreen Forest (82), mostly situated in the highlands of Kenya.

Because the satellite images are intended to give a detailed view of vegetation on a regional scale, the resolution is insufficient to record the often dramatic changes in forest type caused by changing altitude. Hence, the images fail to record Montane Forest (84) in such areas as Mount Kilimanjaro and Mount Kenya.

Although these ambiguities deserve recognition, Montane Forest is of little value as available woody biomass. It not only is physically inaccessible, but remaining Montane Forest generally is strictly protected by government. Fortunately, in light of their immense scientific value, these sites have been well documented elsewhere.

A similar problem arose in the mapping of extensive Mangrove (81) forest, which lines much of the East African coast. For most of its length, Mangrove forest is too narrow to be resolved at the mapping scale used here, apart from the area around large river inlets. Although more accessible than Montane Forest, as well as being an excellent source of wood for charcoal manufacture, the value of mangrove as a source of fuelwood is also limited because of frequent protection.

A review of the classes in table 6-5 discloses that many different land covers are important in this region. Desert (0) and Grassland (1) together cover only about 1 percent of the region, but Wooded Grassland (2) covers 8 percent and is important in northern Kenya. Shrubland (3) covers 2 percent of the region, mainly in eastern Kenya. Bushland and Thicket (4) occupies a broad area in eastern Kenya and central Tanzania, covering 27 percent of the region. Low Woody Biomass Mosaic (5) covers 8 percent, mainly in western Kenya and Uganda. Woodland (6) forms the most extensive class at 29 percent and is distributed in the west and south of the region. High Woody Biomass Mosaic (7) covers 8 percent and Forest (8) covers 11 percent of the region; they are confined mainly to areas immediately north and west of Lake Victoria.

Condition of Woody Biomass

At a general level, the figures for growing stock reveal the characteristic vegetation of the three larger countries of East Africa—woodland in Tanzania, bushland in Kenya, and the mosaic of cultivation, forest remnant, and derived wooded grassland in Uganda. Of the main classes of land cover, the first significant growing stock is Semidesert Wooded Grassland (21), which dominates the arid north of Kenya. Despite its large extent, the growing stock of this class is only 34 million tonnes, less than 1 percent of the regional total. An important class in southern Kenya and northern Tanzania is Transitional Wooded Grassland (24). It has a growing stock of nearly 1.8 million tonnes, more than two-thirds of which is in Tanzania. This represents 0.03 percent of the regional growing stock, and the class produces 0.13 percent of the regional sustainable yield.

Bushy Shrubland (33) is locally quite important, although the greatest area is largely within the protection of Aberdare National Park in the highlands of southern Kenya. These relatively small areas have a combined growing stock of more than 32 million tonnes, with two-thirds in Kenya and most of the remainder in Tanzania. Sustainable yield is 0.7 percent of the regional total.

Dry *Acacia-Commiphora* Bushland and Thicket (41) covers huge dry areas of the region, especially eastern Kenya and the steppes of northern Tanzania. Of a growing stock of 359 million tonnes, 298 million are in Kenya, with 50 million of the remainder in Tanzania. This class constitutes 5.5 percent of the regional growing stock. A rather smaller area is covered by Moist *Acacia-Commiphora* Bushland and Thicket (43), which forms a large triangular block in the center of Tanzania, with significant outliers in all of the other countries. Despite its size, the growing stock of this land cover class is almost twice that of the drier bushland, with 294 million tonnes in Tanzania and 66 million tonnes in Kenya. Similarly, the sustainable yield of this moister bushland is much greater, accounting for a higher percentage of the regional sustainable yield.

The East African Low Woody Biomass Mosaic (52) covers much of northeastern Uganda and much of southwestern Kenya. This class has most of its growing stock in Kenya—about 112 million tonnes, with the remaining 70 million tonnes in Uganda. The productivity is similar in the two countries, with the combined sustainable yield nearly 5 percent of the regional total.

Perhaps the most important biomass resource is the extensive *miombo* woodland that dominates much of

Tanzania and significant areas of Rwanda and Burundi. One woodland type, however, Moist Sudanian Woodland (65), occurs predominantly in the northern part of Uganda. These open woodland areas have a growing stock of more than 40 million tonnes, nearly 1 percent of the regional total. The remaining woodland classes occur principally in Tanzania.

Eastern Tanzania has the main block of Wet *Miombo* Woodland (67), with outliers scattered throughout Tanzania as well as Rwanda, Burundi, and southwestern Uganda. The growing stock of Wet *Miombo* Woodland occurs mainly in Tanzania (1,272 million tonnes), followed by Uganda (126 million tonnes), Rwanda (28 million tonnes), and Burundi (22 million tonnes).

Seasonal *Miombo* Woodland (66) dominates the western third of Tanzania, again with outliers throughout the country and, to a lesser extent, in the other countries of East Africa. With more than one-fifth of the entire regional growing stock, Seasonal *Miombo* Woodland accounts for more biomass than any other class in East Africa. The vast majority of the growing stock is in Tanzania (1,872 million tonnes), although the class is locally important in Burundi (60 million tonnes), southwestern Uganda (31 million tonnes), Kenya (25 million tonnes), and Rwanda (12 million tonnes).

The High Woody Biomass Mosaic is most important in Uganda and southwestern Kenya, although significant areas exist in the highlands of Tanzania. The Guinean Woodland (74) is especially common in northwestern Uganda, whereas the center of the country is predominantly a Cultivation and Forest Regrowth Mosaic (73). Evergreen Woodland Mosaic (71) is restricted to the moister parts of the region, notably the Kenyan highlands, the mountains north of Lake Malawi, the Lake Victoria Basin, and the coastal and offshore islands of Tanzania.

Guinean Woodland has a growing stock of 35 million tonnes, producing only 0.45 percent of the sustainable yield in Uganda, with 3.25 percent of the growing stock. The Cultivation and Forest Regrowth Mosaic of central Uganda is a little more productive, with 1 percent of the Ugandan sustainable yield on 7.4 percent of the growing stock. Uganda holds 72 million tonnes of growing stock, with the remainder in Tanzania.

The forests of East Africa occur largely in the Lake Victoria Basin of southern Uganda and in the highlands of the other countries. The Montane Forest class (84) is most extensive in the highlands of Kenya, where the growing stock is 29.7 million tonnes. The remainder of this class is in Uganda, where the Montane Forest of the Lake Victoria escarpment contains a growing stock of 13.5 million tonnes. Productivity is high, as in most of the forest regions, which account for 67 percent of the regional sustainable yield.

Coastal and Gallery Forest (83) occurs mostly in Kenya and Tanzania, with much smaller areas in Uganda and Burundi. The growing stock of nearly 150 million tonnes includes 93 million in Kenya and 49 million in Tanzania. Overall, they represent 3.5 percent of the regional sustainable yield and 2.3 percent of the growing stock. Most Evergreen Forest (82) growing stock is in Tanzania, with 660 million tonnes, although the other countries each have stocks between about 17 million and 41 million tonnes. Much of this is physically inaccessible, and large areas of East Africa's dwindling forests are protected.

The majority of Mesophilous Humid Tropical Forest (85) is in Uganda, with a growing stock of about 172 million tonnes. The remaining 10 million tonnes is in Kenya, largely within the Kakamega forest northeast of Lake Victoria, which is unique because it is the only remaining moist tropical forest in Kenya. The Lake Victoria Basin has been heavily cultivated, and the indigenous forest, representing the eastern periphery of the Guineo-Congolian rain forest belt, exists mainly in a mosaic of secondary savanna and farmland. The growing stock is 2.8 percent of the regional total and sustainable yield is 9.3 percent.

Humid Tropical Swamp Forest (86) is restricted to Uganda, where it occurs extensively on the western shore of Lake Victoria and elsewhere. The growing stock exceeds 18 million tonnes, representing 0.2 percent of the East African total and 0.3 percent of the sustainable yield for the region. The Ombrophilous Humid Tropical Forest (87) is much more extensive, with a growing stock of 347 million tonnes in Uganda and lesser amounts of 37 million in Kenya and 9 million in Tanzania. In Uganda, this class is a mosaic of cultivated land and remnants of the drier peripheral semi-evergreen Guineo-Congolian rain forest of White (1983).

Southern Africa

For the purpose of this study, southern Africa is that portion of the continent south of Zaire and Tanzania, excluding the small Angolan enclave of Cabinda. Its land area is nearly 6 million square kilometers.

Physical Characteristics, Population, and Resources

All eight of the summary classes of land cover used in this study occur in southern Africa, where we have divided them further into nineteen subdivisions. These nineteen classes are based on vegetation phenology, productivity, and land cover, the floristic composition being less important. Consequently, some classes incorporate a variety of vegetation types that have been identified as distinct floristic units by authors such as Werger (1978) and White (1983) but which we have related by phenology, productivity, and biomass. Our land cover classes, however, are more appropriate for appraisal of fuelwood resources than previously defined floristic mapping units.

A review of the classes in table 6-6 shows that Desert (0) occupies nearly 6 percent of the region, being confined largely to the western coastal strip from southern Angola to South Africa. Grassland (1) covers 6 percent of the region and Wooded Grassland (2) covers 11 percent; both are important classes in South Africa, Botswana, and Namibia. Shrubland (3) covers nearly 11 percent and is important in the same countries. Bushland and Thicket (4) covers 8 percent and Low Woody Biomass Mosaic (5) covers 9 percent. Woodland (6) is by far the most extensive class, blanketing 38 percent of the region with a broadly northern and eastern distribution from Angola to Mozambique. High Woody Biomass Mosaic (7) and Forest (8) each cover 5 percent of the region.

Table 6-6. Distribution of Summary Classes and Estimated Woody Biomass, Southern Africa

	Area		Growing stock		Sustainable yield	
Summary class	km^2	Percent	Million tonnes	Percent	Thousand tonnes per year	Percent
0 Desert	350,529	5.91	0.00	0.00	0.00	0.00
1 Grassland	356,008	6.00	80.81	0.43	3,560.08	0.79
2 Wooded Grassland	667,018	11.24	43.88	0.23	6,670.18	1.49
3 Shrubland	636,665	10.73	514.69	2.76	31,833.25	7.10
4 Bushland and Thicket	471,255	7.94	800.45	4.29	9,896.36	2.21
5 Low Woody Biomass Mosaic	519,891	8.76	1,160.41	6.21	32,753.13	7.31
6 Woodland	2,243,923	37.82	13,789.88	73.83	195,602.60	43.63
7 High Woody Biomass Mosaic	329,029	5.54	359.96	1.93	6,251.55	1.39
8 Forest	317,803	5.36	1,928.39	10.32	161,784.40	36.08
Lakes	41,681	0.70	0.00	0.00	0.00	0.00
Total	5,933,802	100.00	18,678.47	100.00	448,351.55	100.00

Note: Details may not add to totals because of rounding..
Source: Tables 13-1 to 13-10.

Condition of Woody Biomass

This study extends the work on biomass in the SADC area described in ETC Foundation (1987) and Millington and others (1989). As explained in the methodology, the NDVI were used as the basis of classification for both of these studies and in the present work. The NDVI indicate more than just wood production, considering all plant material, including leaves, grasses, herbs, and crops, as well as wood. Large areas of southern Africa are cultivated, often by large commercial agricultural enterprises, and this poses problems in estimating available woody biomass growing stock and productivity. The risk also exists of overestimating fuelwood yields if the accessibility factor is not considered, for example in the case of areas reserved by government and national parks.

Commercial forestry, such as that in South Africa and Swaziland, also may give a false impression of the availability of fuelwood. Ideally, availability should be evaluated using detailed local or regional information and fieldwork. Although some of this type of data is incorporated herein, the regional information is limited, and the scale of this study meant that field checking could be undertaken only in selected areas of fuelwood having high and low potential.

On a broad, regional scale, the pattern of land cover groupings in this study resembles that of biogeographical studies (for example, Werger 1978) because many of the influencing factors are similar. For example, the regional pattern of land cover groupings closely resembles the regional pattern of mean annual precipitation (figure 7-1). These corresponding patterns range from the Namib Desert (Class 0), which has less than 100 millimeters of precipitation a year, to the extensive biomass and high-productivity areas in the north and east, where precipitation exceeds 1,000 millimeters annually.

Numerous environmental factors also influence the patterns displayed, however. Werger (1978) points out that climate and topography are the basic determinants of vegetation structure, biomass, and productivity in southern Africa. This is reflected in this study, as attested by a comparison of the maps depicting southern African biomass, rainfall, and topography. The subdivision of land cover classes, however, is further affected on a smaller scale by local factors, of which soils, drainage, and land use are the most important.

The impact of human population growth is evident from the different land use pressures and from the clearance of woody vegetation. In rural areas, people rely mostly on wood from trees and bushes for household fuel, and the demand for urban energy promotes cutting for charcoal production on a much more devastating scale. Woody biomass is also part of the integrated production system of the peasant farmer, pro-

viding dry-season fodder, construction poles, furniture, tools, habitat for game, mulch to enrich the soil, and protection against wind and water erosion. This exerts considerable pressure on trees and smaller woody plants in all densely populated parts of southern Africa. It is especially noticeable in the semiarid arc of low woody biomass that stretches from southwestern Angola through Namibia, across southern Zambia and Botswana, and into southern Zimbabwe and South Africa.

Another zone of lesser woody biomass resource, although less obvious, is discernible from western Mozambique through Zimbabwe, Zambia, and Malawi into Tanzania. This wide "dry-zone corridor" has a basis in topography and climate; for example, southwestern Mozambique is drier and the Luangwa Valley in eastern Zambia is lower and drier. Much of this zone's raison d'être, however, seems to be in local soils and drainage and, more important, in the population pressure on woody resources. Therefore, this zone needs to be examined more closely at a national and regional level.

The following list synopsizes the distribution of summary classes in the region:

0 Desert covers nearly 6 percent of southern Africa and occurs in the southwest as the Namib Desert, as part of the Great Karroo of South Africa, and in the driest parts of the Kalahari Desert in Botswana.

1 Grassland also covers nearly 6 percent of the region and includes the *Veld* Grassland (11) of the interior South African plateaus and the Montane Grassland and Heathland (14) of the southern Drakensberg Mountains.

2 Wooded Grassland is represented by Transitional Wooded Grassland (24), covering 11 percent of the land area but containing only 0.23 percent of the woody growing stock (table 6-6). It occurs mostly on fine-textured soils in areas experiencing approximately 400 to 800 millimeters of rainfall a year, in widely varied locations but mainly in Namibia, eastern Botswana, southern Zimbabwe, southern Mozambique, northern Transvaal, and in patches throughout the remainder of South Africa, often on the fringes of the *veld*.

3 Shrubland is relatively important in southern Africa, covering nearly 11 percent of the region but containing only about 2.8 percent of the growing stock. More significantly, Shrubland accounts for 7.1 percent of the region's sustainable yield. It includes the small class *Veld* Shrubland and Cultivation (31) of Cape Province; the large classes of Hill Shrubland (32) and Bushy Shrubland (33), both of which largely fringe the Namib Desert area; the Kalahari Shrubland (34), which occurs mostly in southeastern Namibia, northern Cape Province, and southwestern Botswana; and Wooded Shrubland (35), which occurs largely in central Cape Province, on higher ground east of the Namib Desert, and on the slightly wetter fringes of the Kalahari Desert.

4 Bushland and Thicket covers nearly 8 percent of southern Africa, and represents about 4 percent of the growing stock. It is mostly represented by Moist *Acacia-Commiphora* Bushland and Thicket (43). This class occurs as a broken arc from southern Angola and northern Namibia through parts of Botswana, Zambia, Malawi, Zimbabwe, and Mozambique to South Africa and Swaziland, with its largest continuous occurrence in Transvaal and Natal. Bushland and Thicket is often related to particular soils or land forms, or to regenerative phases following exploitation. The much smaller area of *Fynbos* Thicket (42) is confined to small areas of the coastal ranges of Cape Province in South Africa.

5 Low Woody Biomass Mosaic covers nearly 9 percent of southern Africa and includes about 6 percent of the growing stock. It occurs in Namibia and Botswana and is scattered throughout southern Angola, eastern and southern Zambia, Zimbabwe, and southeastern Mozambique, as well as in South Africa.

6 Woodland dominates nearly 38 percent of the continent south of Zaire and Tanzania and has large woody biomass reserves representing 74 percent of southern Africa's growing stock. It extends across the continent from Angola to Mozambique, generally modulating density according to conditions, from Open Woodland (61), through Seasonal *Miombo* Woodland (66), to Wet *Miombo* Woodland (67). *Miombo* woodland is related to moist, frost-free or nearly frost-free tropical conditions, and it dominates the northern part of this region, with density varying according to seasonality of precipitation, edaphic condition, and degree of human interference. The group, however, also includes warm, temperate woodland (part of Class 67) and tropical coastal woodlands (part of Class 66). These occur respectively in the areas of

greater rainfall along the southeastern coast of South Africa and Mozambique.

7 High Woody Biomass Mosaic is represented by two classes, covering about 5 percent of southern Africa but including only about 2 percent of the subcontinent's growing stock. Cultivation and Forest/Woodland Mosaic (72) includes many high-productivity (although not always very woody) biomass areas, associated with areas of greater moisture and cultivation zones, often occurring in areas of unusual drainage characteristics such as the Okavango Delta in Botswana. Evergreen Woodland Mosaic (71) occurs mainly in south-central Angola and along the coast of central Mozambique.

8 Forest has only one representative in southern Africa—Evergreen Forest (82), which covers more than 5 percent of the land area and includes dense montane forest as well as commercial forested areas. These areas represent about 10 percent of the growing stock of southern Africa, and 36 percent of the sustainable yield.

The proportion of woody biomass in these summary classes mostly increases from Class 0 through Class 8, although exceptions exist. Classes 0 to 3 clearly have very limited fuelwood potential. In Class 4, the rate of exploitation can easily exceed productivity. The Low Woody Biomass Mosaic (Class 5) can be important locally, often coincident with dense rural populations and in need of careful management.

The large Woodland group (6) clearly includes a large proportion of the subcontinent's woody biomass resources, but great variation occurs in actual productivity and availability of fuelwood. The High Woody Biomass Mosaic (7) includes some cultivated areas of little woody biomass stock, but also includes large areas of dense woodland having high productivity and fuelwood potential. Forests are among the most productive woody biomass classes of southern Africa, but often include areas that have accessibility problems because of remoteness or commercial ownership.

7

Conclusions and Future Directions

The products of this study are (a) maps showing classes of land cover (defined by their woody biomass), (b) a data base of woody biomass growing stock and sustainable yield, and (c) maps of growing stock and sustainable yield for Sub-Saharan Africa.

Throughout this volume we have stressed that this study has been a first attempt to cost-effectively produce a continent-wide view of the woody biomass situation in Sub-Saharan Africa. The maps and data base are available for strategic planning, including the priority designation of areas requiring woodfuel supply enhancement. This should enable use of such data for making better-informed decisions about stressed areas. These areas will require more intensive assessment, coupled with surveys of woodfuel demand, to yield a better understanding of how to manage the fuelwood resource.

A benefit of conducting a survey at this scale is the ability to compare areas at reconnaissance level, so that extrapolations based on ecology can be made from areas of plentiful data to areas where data are scarce or absent. From this study it is clear that a pressing need exists to develop among nations a more harmonized strategy for woody biomass estimation.

Another benefit of this study will be increased awareness of the potential of AVHRR imagery to allow biomass assessment at the regional level, when combined with secondary data or field surveys. This method affords an impression of the woody biomass situation in relation to consumption. It also indicates where more intensive investigations are needed to support meaningful intervention to manage the woody biomass resource.

Clearly, use of AVHRR NDVI data at a spatial resolution of 8 kilometers restricted our identification and mapping to broadly defined classes of land cover. Small classes having a distribution restricted to narrow belts or small parcels of land use were unlikely to be identified. Examples include the very narrow belt of mangrove along the Indian Ocean coast,

riparian vegetation on narrow flood plains, narrow belts of land cover controlled by altitude in East Africa and the Ethiopian Highlands, and the individual elements of cultivation mosaic (Summary Classes 5 and 7). The level of detail in our study is adequate for an initial situational overview intended for determining policy objectives, but for planning purposes it is not adequate.

An extensive search of the secondary literature confirmed the lack of data on woody biomass supply. Much of this literature has its origin in forestry and therefore concerns commercial woody species. But other end uses in addition to energy and commercial timber are important in supply-side estimates and should be considered.

Also, this work does not consider agricultural residues and herbaceous biomass. Not only are these an energy resource, particularly in arid and semiarid areas such as the Sahel, but crops and herbaceous plants contribute to the NDVI. Future work to upgrade the quality of data bases such as these should involve such considerations.

A positive outcome of our research is that the integration of the data base and land cover maps to produce maps of growing stock and sustainable yield (figures 7-1 and 7-2) does begin to define areas in which problems of supply may exist or be nascent. The presence of low estimates within or adjacent to areas of greater biomass suggests areas where early action should be taken to confirm the level of supply. Examples of such areas are the West African Sahel and the dry zone corridor adjoining parts of eastern and southern Africa.

Early action could take the form of further mapping from imagery that has both a finer spatial resolution and a continuous time sequence of data. Other early response could mean improving the quality of the field data for the data base. Studies based in regions like those just identified, where improvement in the estimate of wood supply is needed, should include at

41

Figure 7-1.   Growing Stock for Sub-Saharan Africa, 1986

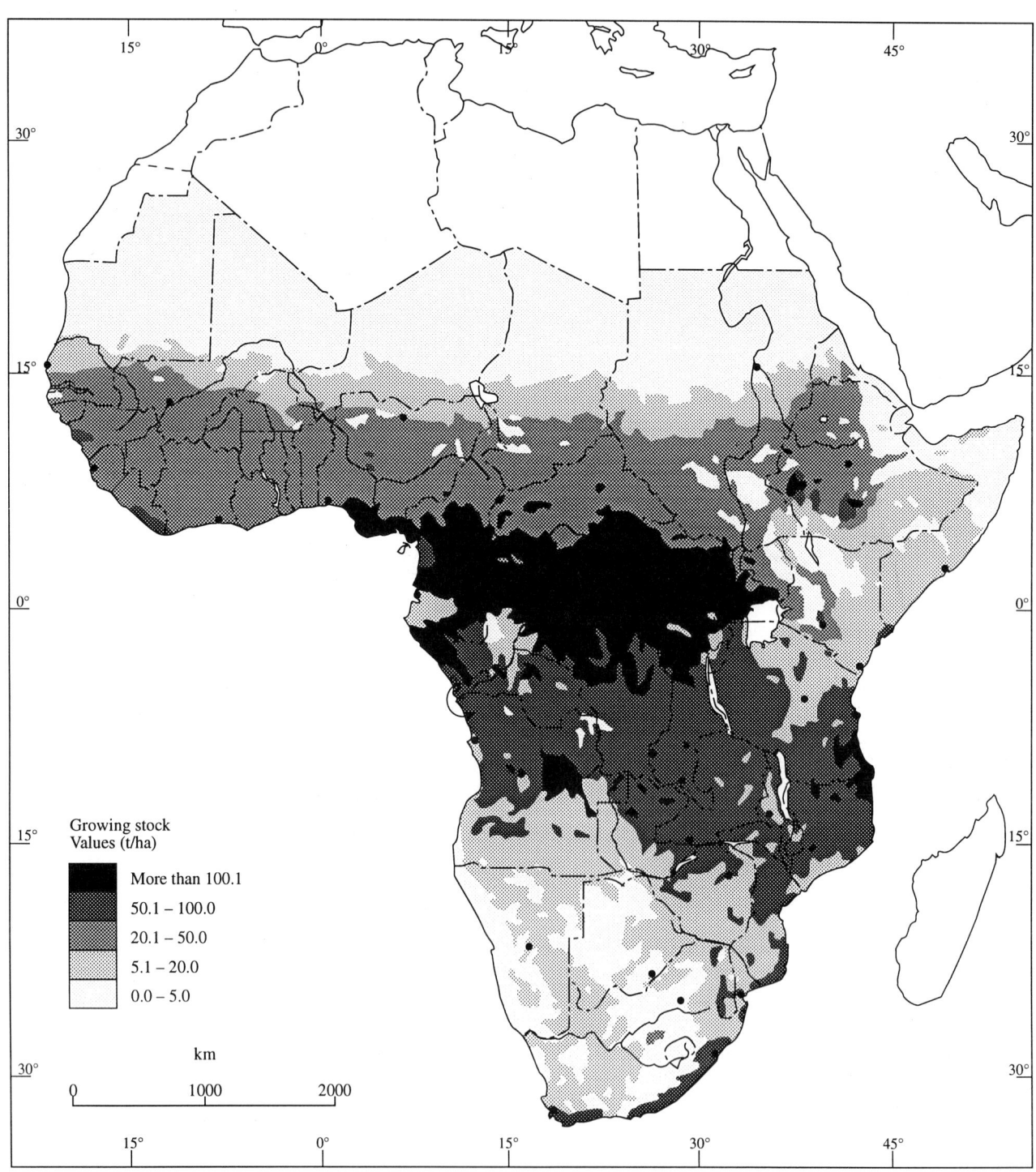

Growing stock
Values (t/ha)

More than 100.1
50.1 – 100.0
20.1 – 50.0
5.1 – 20.0
0.0 – 5.0

km

0   1000   2000

the design stage a common methodology derived from the findings of this project.

We have identified four specific issues for future mapping of land cover for woody biomass assessment. They are:

* *Spatial resolution of the maps needs to be improved for planning purposes.* We are not advocating the repli-

cation of the present study using AVHRR data having a finer resolution. Analysis at the *regional* level, however, requires multisensor land resource satellite imagery of finer resolution (for example, Landsat Thematic Mapper) to provide maps of better detail for energy planners. Such work is already being carried out in Pakistan by ESMAP in conjunction with the Ministry of Planning (Energy Wing).

Figure 7-2. Sustainable Yield for Sub-Saharan Africa, 1986

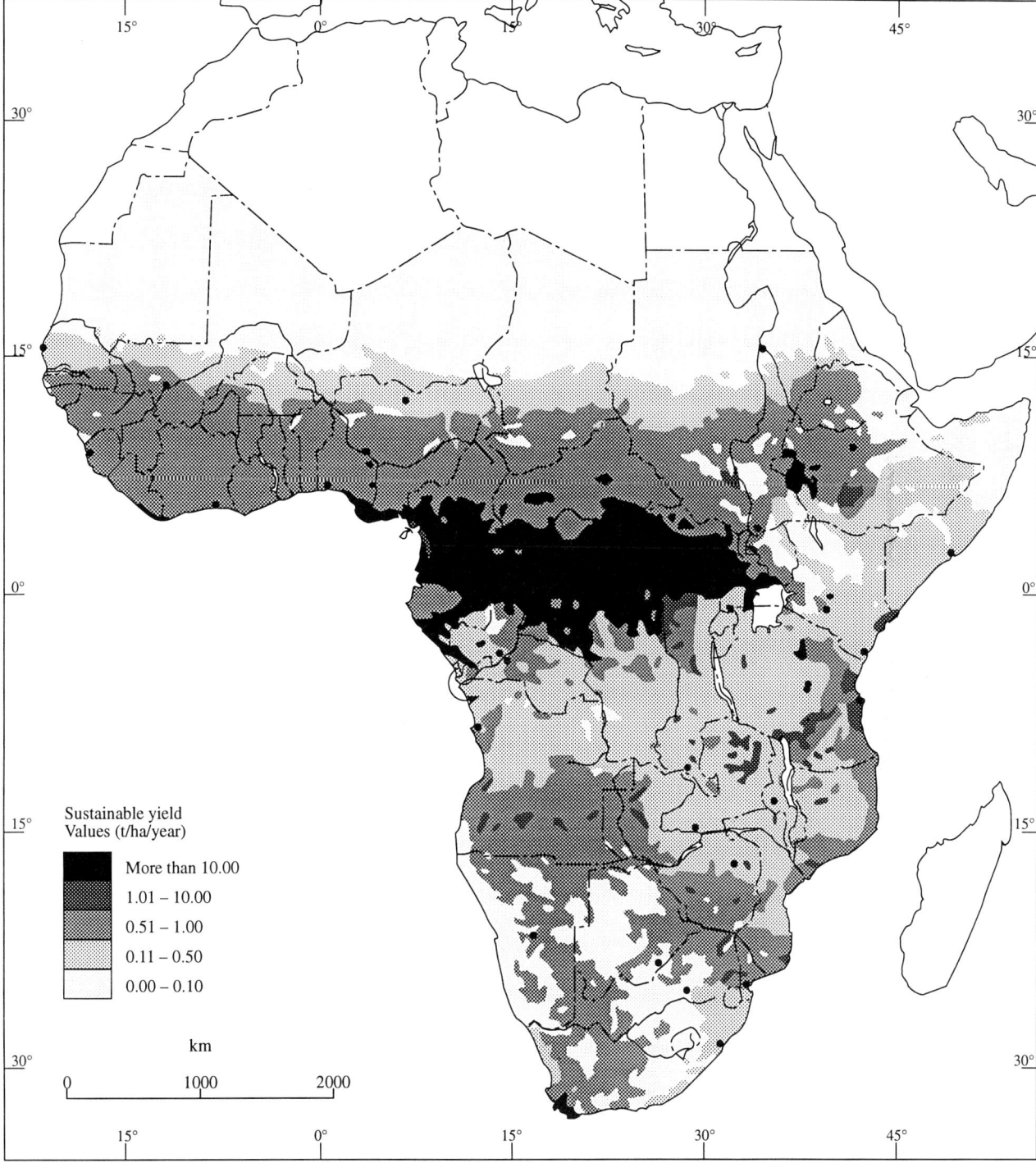

Sustainable yield
Values (t/ha/year)

More than 10.00
1.01 – 10.00
0.51 – 1.00
0.11 – 0.50
0.00 – 0.10

km

0   1000   2000

- *Reliance solely on NDVI products from the AVHRR imagery for woody biomass mapping is questionable.* It has been shown that thermal infrared data, also available from the AVHRR, provides additional information that can be used to refine land cover maps produced using NDVI products. It is particularly useful for identifying and mapping land cover disturbance. In addition, the applicability of Scanning Multichan-

nel Microwave Radiometer (SMMR) data from the microwave portion of the spectrum should be considered in future studies.
- *Year-to-year variations in vegetation cover, photosynthetic activity, and the resultant NDVI response should be considered.* These were not considered in the present study, and the present maps are only a situational overview of the woody biomass resource for

They are subject to change caused by human intervention and climatic variation. Significant improvement to the maps of land cover, growing stock, and sustainable yield would occur with use of the complete AVHRR data set spanning 1981 to 1988.

- *Medium-term monitoring of the woody biomass resource, for example, spanning a 5-to-10-year period, requires (a) a high-quality baseline data set, (b) data of the appropriate spatial resolution to detect changes, and (c) a robust methodology.* We doubt that AVHRR data of 8-kilometer spatial resolution would allow recognition of changes in land cover class caused by human activity that spans periods briefer than a decade. The NDVI is very sensitive to rainfall and soil moisture variations. It is apparent from research already undertaken that, utilizing a phenological strategy for mapping at this scale, changes in land cover classes caused by such variations probably will mask changes caused by human activity.

Other areas require further research:

Improvement of the Data Base

- Harmonization of assessment techniques for woody biomass
- Ground sampling that includes destructive sampling to establish the relation between measurable parameters and the weight of woody biomass
- Preservation of permanent plots for long-term monitoring to determine annual growth increment and sustainable yield
- Assessment that includes the herbaceous component of biomass
- Study to relate the age structure of vegetation to its productivity.

Technical Improvement

- Use of remotely sensed imagery having appropriately finer spatial resolution
- Inclusion of *continuous* remotely sensed data, recorded during a span of several contiguous years as opposed to a single year, to produce land cover maps that are more independent of year-to-year climatic variations
- Support of satellite remotely sensed imagery with other forms of assessment, such as aerial photography and ground verification
- Integrated use of three powerful tools: assessments of land cover and woody biomass, environmental data bases for climate, demography, and other factors such as soils, and geographical information systems.

Support Programs

- Early involvement of local, regional, and national organizations
- Technical assistance and support in staffing and the supply of necessary hardware, software, and training programs in developing countries.

Links with Studies of the Demand Side

- Coordination with studies of woodfuel demand in the same land cover classes
- Supporting studies of competing end uses of wood.

The following areas in Sub-Saharan Africa should receive high priority as areas in which to conduct the research just outlined:

- *The Sudanian Woodlands of the Sudano-Sahelian zone.* The work reported here clearly identifies and maps three belts of Sudanian Woodland that are differentiated by the length of their dry seasons and by their moisture regimes. These belts stretch from Senegal to Sudan, covering 1.3 million square kilometers, and are a principal source of woodfuel. This area currently is experiencing a dual threat to the fuelwood supply. First, people are migrating northward into these woodlands from the environmentally stressed areas of the Sahel. Second, the large population centers to the south, along the West African coast, are increasingly relying on woodfuel from these woodlands to meet their domestic energy demands.

  The main factors to be covered by such work are (a) mapping of the land cover classes within the Sudano-Sahelian zone at a finer spatial resolution and (b) a greatly enhanced data base on the woody biomass growing stock and sustainable yield. Such work could be accomplished by using Landsat imagery for interpretation of land cover and by establishing a local network of biomass assessments within the woodlands.

- *The dry-zone corridor of southern and East Africa.* This diverse belt, typified by arid and semiarid vegetation and farming systems, extends from Ethiopia and Somalia southward through east-central Africa to Botswana and Zimbabwe. The zone comprises several areas of ecological stress resulting from scant rainfall and strongly seasonal moisture regimes. This area is important for woodfuel supply because of the great rural population density and large urban centers. Woodfuel supplies have been depleted by clear felling for agriculture. Several local studies of good quality provide benchmark data within this region, and these can be used as a starting point for extending the data base (Chidumayo 1987; Munslow and others 1989; Stomgaard 1985).

- *The mountainous regions of East Africa.* Differentiation of land cover classes in this work was least satisfactory in the Ethiopian Highlands and on the East African mountains. This resulted mainly from a mismatch between the intensive agricultural systems and small landholdings in these areas, which was a consequence of using imagery having a spatial resolution of 8 kilometers. It is possible that the proposed biomass assessment project for Ethiopia, funded by the World Bank as part of the Energy I Project, will help to resolve this problem and will provide more accurate information on land cover types as well as growing stocks and yields.

Following such priority research, the data base for Sub-Saharan Africa can be extended to supplement this and other work like that conducted by the Overseas Development Administration (ODA) in Ghana, Somalia, and elsewhere. Regular updating of the data base would increase its value as a base line for fuelwood supply data.

As noted in the introduction, reliable data are scarce on the growing stock and sustainable yield of biomass for use as fuel within the economic catchments of demand centers (Ryan and Openshaw 1991). This particularly applies to woody biomass (woodfuels), the principal form of energy in many developing countries. Without reliable data, it is difficult to undertake meaningful planning of energy policy and investment in situations that significantly involve biomass fuel. The material presented in this volume is a first step toward removing this deficiency for Sub-Saharan Africa.

References

Every effort has been made to facilitate access to the documents listed here. Some documents, however, lack full bibliographic information because it was unavailable; also, some documents are of limited circulation. The word "processed" describes informally reproduced works that may not be commonly available through libraries.

Andeke Lengui, M. A. 1987. *Formulation du projet aménagement et reboisement forestiers à buts multiples.* FO:NER/85/009 Document de terrain No. 2. Niamey, Niger: Ministère de l'Hydraulique et de l'Environment. FAO.

Baines, A. C. 1980. "Sierra Leone: Assistance to Forestry Development." Freetown. Processed.

Barber, Dennis R. 1992. "The Application of Change Detection Techniques to Rangeland Monitoring in South Australia." *Proceedings, Australasian Remote Sensing Conference (6th: 1992: Wellington, New Zealand),* 2–6 November, Vol. 1, 1.413–1.416.

Bartholomew, W. V., J. Meyer, and H. Landelot. 1953. "Mineral Nutrient Immobilization Under Forest and Grass Fallow in the Yangambi (Belgian Congo) Region." *Série Scientifique* 57 Institut National.

Bianchi, H. 1986. *Assistance au développement forestier, Guinée Bissau: Planification forestière.* TCP/GBS/4506(A). Document de terrain. Rome: FAO.

Black, R. 1989. "Africa's Refugee Crisis." *Geographical Magazine* 61(1):12–18.

Bonkoungou, L., L. Bortoli, and J. M. Oudba. 1988. "Note on Burkina Faso Ecosystems: Woody and Herbaceous Vegetation Characteristics of a Protected Fallow in Degraded Sudanian Zone." Institut de Recherche en Biologie et Ecologie Tropicale (IRBET), Ouagadougou, Burkina Faso. Processed.

Brook, K. D., J. O. Carter, T. J. Danaher, G. M. McKeon, N. R. Flood, and A. Peacock. 1992. "The Use of Spatial Modelling and Remote Sensing for Monitoring and Forecasting of Drought-Related Land Degradation Events in Queensland." *Proceedings,*

Australasian Remote Sensing Conference (6th: 1992: Wellington, New Zealand), 2–6 November, Vol. 1, 1.140–1.149.

Brown, S., and A. E. Lugo. 1984. "Biomass of Tropical Forests: A New Estimate Based on Forest Volumes." *Science* 223:1290–93.

Catinot, R. 1978. "The Forest Ecosystems of Gabon: An Overview." In UNESCO (United Nations Educational, Scientific and Cultural Organization), *Tropical Forest Ecosystems.* Paris: UNESCO.

Chidumayo, E. N. 1987. "Species Structure in Zambian *Miombo* Woodland." *Journal of Tropical Ecology* 3(2): 109–18.

Christiansen, J. H. 1978. "Biomass and Primary Productivity of *Rhizophora Apiculatea* in a Mangrove in Southern Thailand." *Aquatic Botany* 4:43–52.

Clément, J. 1982. "Estimation des volumes et de la productivité des formations mixtes forestières et graminénnes tropicales." *Revue Bois et Forêts des Tropiques,* 198(4):35–58.

Cline-Cole, Reginald, J. A. Falola, H. A. C. Main, Michael E. Montimore, Janet E. Nichol, and Frank D. O. O'Reilly. 1987. *Woodfuel in Kano.* Department of Geography, Bayero University, Kano, Nigeria. Final report of the rural energy research project submitted to the U.N. University Department, Studies Division.

Cracknell, Arthur P., and Ladson D. Hayes. 1991. *Introduction to Remote Sensing.* London: Taylor and Francis.

Danaher, T. J., J. O. Carter, K. D. Brook, A. Peacock, and G. S. Dudgeon. 1992. "Broad-Scale Vegetation Mapping Using NOAA-AVHRR Imagery." *Proceedings, Australasian Remote Sensing Conference (6th: 1992: Wellington, New Zealand),* 2–6 November, Vol. 3, 3.128–3.137.

Daus, Steven J., Mamane Guero, and Lawally Ada. 1986. "A Remote-Sensing-Aided Inventory of Fuelwood Volumes in the Sahel Region of West Africa: A Case Study of Five Urban Zones in the Republic of Niger." Paper presented at Symposium on Remote Sensing for Resources Development and Environmental Management, August 1986, Enschede, Netherlands.

Diallo, O., A. Diouf, N. P. Hanan, A. Ndiaye, and Y. Prevost. 1991. "AVHRR Monitoring of Savanna Primary Production in Senegal, West Africa, 1987–1988." *International Journal of Remote Sensing* 12(6): 1259–80.

Drury, Steven A. 1990. *A Guide to Remote Sensing: Interpreting Images of the Earth.* Oxford: Oxford University Press.

Eidenshink, J. C., and R. H. Haas. 1992. "Analysing Vegetation Dynamics of Land Systems with Satellite Data." *Geocarto International* 7(1):53–61.

ETC (Education and Training Consultants) Foundation. 1987. *Wood Energy Development: Biomass Assessment, a Study of the SADCC Region.* Leusden, Netherlands.

FAO (Food and Agricultural Organization of the United Nations). 1984. "Etudes sur les volumes et la productivité des peuplements forestiers tropicaux. 1. Formation forestières sèches." Etude FAO Forêts 51/1. Rome.

FAO (Food and Agricultural Organization of the United Nations). 1984. *World Food Report.* Rome.

FAO/PNUD (Food and Agricultural Organization of the United Nations/Programme des Nations Unies pour le Développement). 1978. "La conservation des écosystèmes forestières dans la Région des Marts Kouffe." Rome.

Flemons, P. K. J. 1992. "Estimating Pasture Variables Using AVHRR Data. The Effect of Soil Colour and Cover Extent." *Proceedings, Australasian Remote Sensing Conference (6th: 1992: Wellington, New Zealand),* 2–6 November, Vol. 3, 3.199–3.202.

Franklin, J., and P. H. Y. Hiernaux. 1991. "Estimating Foliage and Woody Biomass in Sahelian and Sudanian Woodlands Using a Remote Sensing Model." *International Journal of Remote Sensing* 12(6): 1369–86.

Gatlin, J. A., R. J. Sullivan, and Compton J. Tucker. 1983. "Monitoring Global Vegetation Using NOAA AVHRR Data." *Proceedings of the IGARRS Symposium,* San Francisco, Part I, PF2, 7.1.

Golley, F. B., J. T. McGinnis, R. G. Clements, G. Child, and M. J. Duever. 1969. "The Structure of Tropical Forests in Panama and Colombia." *Bioscience* 19: 693–96.

Golley, F. B., and others. 1971. "La biomasa y la estructura mineral de algunos bosques de Darien, Panama." *Turrialba* 21:189–96.

Goward, Samuel N., Compton J. Tucker, and N. A. Dye. 1985. "North American Vegetation Patterns Observed with NOAA-7 Advanced Very High Resolution Radiometer." *Vegetation* 64:3–14.

Graham, A. 1969. "Man-Water Relations in East Central Sudan." In M. F. Thomas and G. W. Whittington, eds., *Environment and Land Use in Africa.* London: Methuen.

Gray, T. I., and D. G. McCrary. 1981. "Meteorological Satellite Data—a Tool to Describe the Health of the World's Agriculture." AGRISTARS Report EW-NI-04042. NASA/GSC (National Aeronautics and Space Administration/Goddard Space Center), Houston.

Greenland, Denis T., and Jan M. L. Kowal. 1960. "Nutrient Content of Moist Tropical Forest of Ghana." *Plant and Soil* 12(2):154–74.

Guerreiro, M. G. 1966. "A floresta Africana e os factores bioticos. Primeiras observações de um ensaio em Mocambique." Paper in M. J. A. Werger, ed., 1987, *Biogeography and Ecology of Southern Africa.* The Hague: Junk.

Gutmann, Garik G. 1991. "Monitoring Land Ecosystems Using the NOAA Global Vegetation Index Data Set." *Palaeogeography, Palaeoclimatology, Palaeoecology* 90:195–200.

Guy, P. R. 1970. "*Andansonia Digitata* and Its Rate of Growth in Relation to Rainfall in South Central Africa." *Proceedings and Transactions of Rhodesian Science Association.* 54:68–84.

Guy, P. R. 1981. "Changes in the Biomass and Productivity of Woodlands in the Sengwa Wildlife Research Area, Zimbabwe." *Journal of Applied Ecology* 18(2):507–19.

Henderson-Sellers, A., G. Seze, F. Drake, and M. Debois. 1987. "Surface Observed and Satellite-Received Cloudiness Compared for the 1983 ISCCP (International Satellite Cloud Climatology Program) Special Study in Europe." *Journal of Geophysical Research* 22:D4, 4019.

Henrickson, B. L., and J. W. Durkin. 1986. "Growing Period and Drought Early Warning in Africa Using Satellite Data." *International Journal of Remote Sensing* 7(11):1583–1608.

Holben, Brent N. 1986. "Characteristics of Maximum-Value Composite Images from Temporal AVHRR Data." *International Journal of Remote Sensing* 7(11): 1417–34.

Huntley, B. J. 1978. "Characteristics of South African Biomass." In P. de Booysen and N. M. Tainton, eds., *Ecological Effects of Fire in South African Ecosystems.* Ecological Studies 48. Berlin: Springer-Verlag.

Jackson, J. K. 1971. "Productivity of Natural Woodland and Plantations in the Savanna Zones of Nigeria." Savanna Forestry Research Station, Federal Department of Forestry Research, Samuru, Zaria, Nigeria.

Justice, Christopher O., ed. 1986. "Monitoring the Grassland of Semiarid Africa Using NOAA-AVHRR Data." *International Journal of Remote Sensing* 7(11): 1383–1622.

Justice, Christopher O., and P. H. Y. Hiernaux. 1986. "Monitoring the Grasslands of the Sahel Using NOAA

AVHRR Data: Niger." *International Journal of Remote Sensing* 7(11):1475–97.

Kamweti, D. M. 1984. *Fuelwood in Eastern Africa: Present Situation and Future Prospects.* Rome: FAO.

Kennard, D. G., and B. H. Walker. 1973. "Relationships Between Tree Canopy Cover and *Panicum maximum* in the Vicinity of Fort Victoria." *Rhodesian Journal of Agricultural Research* 11:145–53.

Koomanoff, V. A. 1989. "Analysis of Global Vegetation Patterns: A Comparison Between Remotely Sensed Data and a Conventional Map." Biogeography Research Series, Report 890201, Dept. of Geography, University of Maryland.

Laclavère, G. 1980. *Atlas of the United Republic of Cameroon.* Paris: Les Éditions Jeune Afrique.

Lamotte, M., and F. Bourlière. 1978. "Energy Flow and Nutrient Cycling in Tropical Savannas." In F. Bourlière, ed., *Ecosystems of the World 13. Tropical Savannas.* Amsterdam: Elsevier.

Leach, G., and R. Mearns. 1988. *Beyond the Woodfuel Crisis—People, Land, and Trees in Africa.* London: Earthscan.

Lewis, L. A., and L. Berry. 1988. *African Environments and Resources.* Boston: Unwin Hyman.

Lundgren, B., ed. 1975. *Land Use in Kenya and Tanzania.* Stockholm: Royal College of Forestry.

Malaisse, F. P. 1978. "Phenology of the Zambezian Woodland Area with Emphasis on the *Miombo* Ecosystem." In H. Leith, ed., *Phenology and Seasonal Modelling.* Berlin: Springer-Verlag.

Malaisse, F. P., and Binzangi, K. 1985. "Wood as a Source of Fuel in Upper Shaba." *Commonwealth Forestry Review* 64(3):227–39.

Malingreau, J.-P. 1986. "Monitoring Tropical Wetland Rice Production Systems: a Test for Orbital Remote Sensing." In M. J. Eden and J. T. Parry, eds., *Remote Sensing and Tropical Land Management.* Chichester, England: John Wiley.

Malingreau, J.-P., and Compton J. Tucker. 1988. "Large-Scale Deforestation in the Southeastern Amazon Basin of Brazil." *Ambio* 17:49–55.

Malleaux, H. 1980. *The Forest Inventory of Mozambique.* Rome: FAO.

Malo, A. R., and Sharon E. Nicholson. 1990. "A Study of Rainfall and Vegetation Dynamics in the African Sahel Using Normalized Difference Vegetation Index." *Journal of Arid Environments* 19:1–24.

Marsch, H. E. 1978. *Inventaire d'aménagement de forêt de Lama.* Document de terrain No. 3. République Populaire de Bénin. Rome: FAO.

Menaut, J. C., and J. César. 1979. "Structure and Primary Productivity of Lamto Savannas, Ivory Coast." *Ecology* 60(6):1197–1210.

Menaut, J. C., and J. César. 1982. "The Structure and Dynamics of a West African Savanna." In B. J. Huntley and B. H. Walker, eds., *Ecology of Tropical Savannas.* Berlin: Springer-Verlag.

Millington, Andrew C., John R. G. Townshend, Pam A. Kennedy, Richard Saull, Steven D. Prince, and Robert Madams. 1989. *Biomass Assessment. Woody Biomass in the SADCC Region.* London: Earthscan.

Munslow, Barry, Yemi Katerere, Adrian Ferf, and Philip O'Keefe. 1989. *The Fuelwood Trap: A Study of the SADCC Region.* London: Earthscan.

Norwine, J., and D. H. Greegor. 1983. "Vegetation Classification Based on AVHRR Satellite Imagery." *Remote Sensing of Environment* 13(1):69–87.

Nye, P. H. 1961. "Organic Matter and Nutrient Cycles Under Moist Tropical Forest." *Plant and Soil* 13(4): 333–46.

Openshaw, Keith. 1982. *Somalia: The Forest Sector. Problems and Possible Solutions.* Mogadishu, Somalia: World Bank/Energy/Development International/Ministry of Mineral and Water Resources.

Pecha, Albert. 1986. *Développement forestier.* Document de terrain No. 18, République Populaire du Congo. Rome: FAO.

Perry, Charles R., Jr., and Lyle F. Lautenschlager. 1984. "Functional Equivalence of Spectral Vegetation Indices." *Remote Sensing of Environment* 14 (13): 169–182.

Persson, R. 1975. *Forest Resources of Africa, Part I, Country Descriptions.* Stockholm: Royal College of Forestry.

Prince, Steven D. 1986. "Monitoring the Vegetation of Semiarid Tropical Rangelands with the NOAA-7 Advanced Very High Resolution Radiometer." In M. J. Eden and J. T. Parry, eds., *Remote Sensing and Tropical Land Management.* Chichester, England: John Wiley.

Prince, Steven D. 1991. "Satellite Remote Sensing of Primary Production: Comparison of Results for Sahelian Grasslands 1981–1988." *International Journal of Remote Sensing* 12(6):1301–11.

Prince, Steven D., and Compton J. Tucker. 1986. "Satellite Remote Sensing of Rangelands in Botswana II: NOAA AVHRR and Herbaceous Vegetation." *International Journal of Remote Sensing* 7(11):1555–70.

République du Togo. 1987. "Document de terrain—Systèmes d'exploitation rurales et la situation actuelle de la forêt." République Togolaise TOG/83/008. Rome: FAO.

Roderick, M. L., and R. C. G. Smith. 1992. "Use of NOAA Derived Seasonal Vegetation Data Within a GIS for Broad Scale Vegetation Management in Western Australia." *Proceedings, Australasian Remote Sensing Conference (6th: 1992: Wellington, New Zealand),* 2–6 November, Vol. 3, 3.194–3.198.

Rutherford, M. C. 1978. "Primary Production Ecology in Southern Africa." In M. J. A. Werger, ed.,

Biogeography and Ecology of Southern Africa. The Hague: Junk.

Ryan, Paul, and Keith Openshaw. 1991. "Biomass Assessment: A Discussion on its Need and Methodology." Working paper. World Bank, Industry and Energy Department, Washington, D.C.

Sarmiento, G., and M. Monasterio. 1983. "Life Forms and Phenology." In F. Bourlière, ed., *Ecosystems of the World 13. Tropical Savannas*. Amsterdam: Elsevier.

Saull, R. J., Andrew C. Millington, and M. Crosetti. 1991. "Pakistan: A Land Cover Zonation from Multitemporal AVHRR and Environmental Data Using GIS Techniques." Proceedings of the 5th AVHRR Data Users Meeting, EUMETSAT, Tromso, Norway, 25–28 June, 217–21.

Schneider, Stanley R., David F. McGinnis, Jr., and J. A. Gatlin. 1981. "Use of NOAA AVHRR Visible and Near-Infrared Data for Land Remote Sensing." NOAA (National Oceanographic and Atmospheric Administration) Technical Report NESS 84. Washington: USDC.

Schneider, Stanley R., David F. McGinnis, Jr., and George Stephens. 1985. "Monitoring Africa's Lake Chad Basin with Landsat and NOAA Satellite Data." *International Journal of Remote Sensing* 6(1): 59–73.

Sellers, Piers. 1985. "Canopy Reflectance, Photosynthesis, and Transpiration." *International Journal of Remote Sensing* 6(8):1335–72.

Sellers, Piers. 1986. "Canopy Reflectance, Photosynthesis, and Transpiration II. The Role of Biophysics and Their Interdependence." *Remote Sensing of Environment* 18:205.

Stonehouse, Bernard. 1985. *Pocket Guide to the World*. London: George Philip and Co., Ltd.

Stomgaard, Peter. 1985. "Biomass Estimation Equations for *Miombo* Woodland, Zambia." *Agroforestry Systems* 3(1):3–13.

Tarpley, D. R. 1991. "The NOAA Global Vegetation Index Product—A Review." *Palaeogeography, Palaeoclimatology, Palaeoecology* 90:189–94.

Tarpley, J. D., Stanley R. Schneider, and R. L. Money. 1984. "Global Vegetation Indices from NOAA-7 Meteorological Satellite." *Journal of Climate and Applied Meteorology* 23:491.

Townshend, John R. G., Christopher O. Justice, and V. Kalb. 1987. "Characterization and Classification of South American Land Cover Types Using Satellite Data." *International Journal of Remote Sensing* 8(8): 1189–1207.

Townshend, John R. G., Christopher O. Justice, Li Wei, Charlotte Gurney, and Jim McManus. 1991. "Global Land Cover Classification by Remote Sensing: Present Capabilities and Future Possibilities." *Remote Sensing of Environment* 35:243–55.

Townshend, John R. G., and Compton J. Tucker. 1984. "Objective Assessment of AVHRR Data for Land Cover Mapping." *International Journal of Remote Sensing* 5:492.

Tucker, Compton J., J. A. Gatlin, and Stanley R. Schneider. 1984. "Monitoring Vegetation in the Nile Valley with NOAA-6 and NOAA-7 AVHRR." *Photogrammetric Engineering and Remote Sensing* 50(1):53–61.

Tucker, Compton J., Brent N. Holben, and Thomas E. Goff. 1984. "Intensive Forest Clearing in Rondonia, Brazil as Detected by Satellite Remote Sensing." *Remote Sensing of Environment* 15(3):255–261.

Tucker, Compton J., and Piers Sellers. 1986. "Satellite Remote Sensing of Primary Production." *International Journal of Remote Sensing* 7(11):1395–1416.

Tucker, Compton J., John R. G. Townshend, and Thomas E. Goff. 1985. "African Land Cover Classification Using Satellite Data." *Science* 227(4685): 369–75.

Tucker, Compton J., C. L. Vanpraet, M. J. Sharman, and G. Van Ittersum. 1985. "Satellite Remote Sensing of Total Herbaceous Biomass Production in the Senegalese Sahel: 1980–1984." *Remote Sensing of Environment* 17(3):233–49.

van Wilgen, B. W., K. B. Higgins, and Du Bellstedt. 1990. "The Role of Vegetation Structures and Fuel Chemistry in Excluding Fire from Forest Patches in the Fire-Proven *Fynbos* Shrublands of South Africa." *Journal of Ecology* 78(1):210–22.

Werger, M. J. A. 1978. *Biogeography and Ecology of Southern Africa*. The Hague: Junk.

White, F. 1983. "The Vegetation of Africa." *Natural Resources Research Series* 20. Paris: UNESCO/AETFAT/UNSO (United Nations Educational, Scientific, and Cultural Organization/Association pour l'Etude Taxonomique de la Flore de l'Afrique Tropicale/United Nations Sudano-Sahelian Office).

Whitney, J. B. R., C. M. Dufournaud, and B. W. Murck. 1987. "An Examination of Alternatives to Traditional Fuelwood Use in Sudan." *Journal of Environmental Management* (25):319–46.

Woodward, F. Ian. 1987. *Climate and Plant Distribution*. Cambridge, England: Cambridge University Press.

World Resources Institute/International Institute for Environment and Development. 1986. *World Resources 1986*. New York: Basic Books.

PART II

Regional Distribution
of Land Cover Classes

Part II of this book examines the regional distribution of land cover classes (figure 3-5). Broad ecological, environmental, and economic criteria defined our six regions:

- West African Sahel (Mauritania, Senegal, the Gambia, Mali, Burkina Faso, Niger, and Chad)
- West African coast (Guinea-Bissau, Guinea, Sierra Leone, Liberia, Côte d'Ivoire, Ghana, Togo, Benin, and Nigeria)
- The Horn of Africa (Sudan, Ethiopia, Djibouti, and Somalia)
- East Africa (Uganda, Burundi, Rwanda, Kenya, and Tanzania)
- Central Africa (Cameroon, Central African Republic, mainland Equatorial Guinea, Gabon, Congo, and Zaire)
- Southern Africa (Angola, Zambia, Malawi, Mozambique, Zimbabwe, Namibia, Botswana, South Africa, Swaziland, and Lesotho).

In each region, we describe only large classes of land cover (those having an areal extent greater than 1 percent of the region), with the exception of smaller areas that are important to an individual country. We present each class in relation to its geographical distribution within the region, describe the main elements in the natural vegetation and land use, and address the current and future fuelwood resource, as appropriate.

For some important classes in each region, NDVI profiles are plotted to afford an impression of how land cover can be differentiated using the methods described in Chapter 3. At the conclusion of each regional chapter are tables that provide national data on the area, growing stock, and sustainable yield for each land cover class. References are also supplied. At the end of Part II are the four regional maps depicting the land cover classes.

8

The West African Sahel

Richard W. Critchley

This chapter presents a detailed description of the most important land cover classes in this region. Helpful figures in other chapters include figure 3-1 (cloud cover); figures 3-2, 3-3, and 3-4 (NDVI summary land cover profiles); figure 3-5 (regional summary map of land cover classes); figures 7-1 and 7-2 (continental maps of growing stock and sustainable yield); and the "Regional Land Cover Class Map of West Africa" at the end of this volume.

Helpful tables in other chapters include table 3-2 (land cover classes); table 4-1 (data and sources for growing stock and sustainable yield); and table 6-1 (West African Sahel estimated woody biomass by summary class).

Class 0—Desert

Within the region that we designate West African Sahel in this study, the dominant land cover class in relation to surface area is desert, about 50 percent. The countries of Mauritania, Mali, Niger, and Chad extend well into the northern Sahara. In these areas, where precipitation is scant and evaporation is rapid, the flora is restricted to more favorable locations where groundwater may be available, such as dried-up river beds or wadis.

Rainfall in the western half of the Sahel region has been less than average in every rainfall season since 1968 (IIED 1989). Because much of the region is marginal for agriculture, an extended reduction in precipitation is likely to be disastrous. The contribution of this class to the regional growing stock of woody biomass obviously is negligible, although locally it may be significant to nomadic people in the southern portion of the area.

The vegetation is not obvious for most of the year, although significant below-ground biomass may exist, consisting of the extensive rooting systems of drought-resistant geophytic plants. This, of course, cannot be considered a sustainable fuel resource. Other species are ephemerals, surviving by the production of seed during favorable periods. Rainfall in the more northern areas of the region may be extremely rare, as little as one event in 10 years, and may occur as brief rainfall events of 10 to 30 millimeters.

Following a rainfall, a brief explosion of vegetative activity forms a herbaceous carpet, creating a system referred to as *acheb*. The plants that emerge are mostly tropical, although overlap exists with more Mediterranean species in areas of greater relief, such as the Aïr Mountains in Niger and the Tibesti Plateau in Chad. Herbaceous species also are present, such as the tussock-forming grass *Stipagrostis pungens* and the grass *Panicum turgidum*, which responds to water availability by rapid vegetative growth from dormant buds at ground level, providing an important food for antelope. The annual grass *Coelachyrum compressa* can establish itself on the banks of wadis, and the perennial stolon-forming grass, *Eleusine compressa*, also represents the tropical desert flora (Gillet 1986). *Aristida pungens, A. longiflora, Cornulaca monocantha, Cymbopogon monocantha, Eragrostis* sp., *Stipagrostis uniplumis*, and *Tribulus* sp. also are described in these areas and farther south.

The southernmost edge of the class approximately coincides with the 200-millimeter isohyet. In this area, or in depressions farther north where runoff may augment the available water supply, woody species do occur. The commonest is *Acacia tortilis* ss. *raddiana*, which has a restricted but uniform distribution. *Leptadenia pyrotechnica* is restricted to dunes and *Balanites aegyptiaca* and *Commiphora africana* occur locally in the low peneplains (Le Houérou 1980).

The south Saharan plateaus of Aïr (Niger), Adrar des Iforas (Mali), and Ennedi (Chad) provide areas of increased runoff and slightly greater rainfalls. They are characterized by the Semidesert Wooded Grassland Vegetation (Class 21), and in general, woody biomass gradually increases as total precipitation, frequency, and regularity increase toward the south of the zone.

Because of the supervised classification technique used to map land cover (Chapter 3), some small parcels of land in Niger appear to be Class 0, but are surrounded by Class 21. These are believed to represent degraded areas. The causes of this degradation are considered to be of two kinds. The first cause is cyclic climatic variation, leading to an increase in aridity spanning a period of years. In such areas, the vegetation would be expected to recover in years of greater rainfall. The second cause is believed to be anthropogenic activity, and such areas could return to greater production levels with appropriate land management.

Class 12—Hydromorphic Grassland

This class has a very restricted distribution in the West African Sahel, covering less than 1 percent of the region. Its contribution to the regional woody biomass growing stock and sustainable yield is negligible (table 6-1).

The biomass activity produces a relatively small NDVI response (for example, in Chad, figure 8-1) but with a marked seasonality. It is small in March and April at 0.35, but smallest in September at 0.20. Peaks in production are indicated in December and January with values of 0.55, somewhat later than the peak precipitation of the wet season (July and August). This lag relates to the drying out of waterlogged soils.

Hydromorphic Grassland occurs as small patches within the Dry Sudanian Woodland (Class 62) in the western and southeastern portions of Mali and to a lesser extent in eastern Senegal and in Chad, southwest of N'Djamena. Its distribution in Mali is associated with the Manding Plateau and similarly with minor plateaus in Senegal. In Chad it represents a grassland vegetation formed on Quaternary clays in areas susceptible to flooding. White (1983) considers this to be the most widespread kind of edaphic grassland, similar to that forming on seasonally or permanently waterlogged soils in the upper Nile Basin. More extensive areas occur in Sudan and northern Nigeria and complete an east-to-west pattern within Class 62, Dry Sudanian Woodland.

For a more complete description, please refer to the regional reports for the West African coast (Chapter 9) and Horn of Africa (Chapter 10).

Figure 8-1. NDVI **Profile, Hydromorphic Grassland (Class 12)**

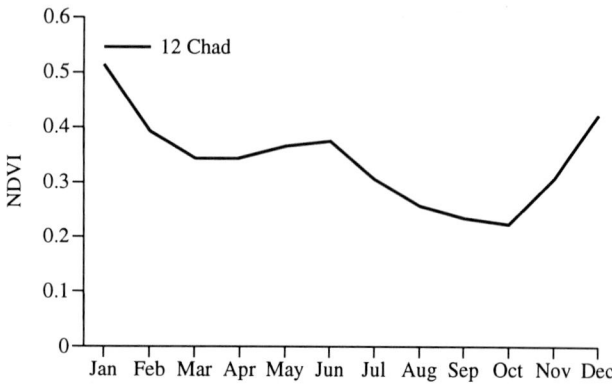

Class 21—Semidesert Wooded Grassland

The Semidesert Wooded Grassland class forms a relatively narrow belt, at its broadest about 400 kilometers, running from Mauritania in the west through central Mali, northern Burkina Faso, southwestern and central Niger, central Chad, and extending into East Africa. It covers about 19 percent of the surface of the West African Sahel region. A more limited distribution occurs north and south of this zone within the Desert (Class 0) to the north and the Sahel-Sudanian *Acacia* Wooded Bushland (Class 44) farther south. The Semidesert Wooded Grassland class corresponds largely to mapping unit 54a of White (1983). The more northerly extensions are associated with the massifs of Adrar des Iforas (Mali), Aïr (Niger), and Ennedi (Chad), but the floral constitution of these areas is different due to the inclusion of Saharan desert species.

Rainfall in the region is about 250 millimeters a year, with a marked seasonality. Rainfall peaks in July and August and coincides with summer temperatures to produce warmth and humidity that can cause rapid and vigorous plant growth. The monthly NDVI curves show a marked seasonality, with peak values of 0.44 in September in response to the increased rainfall, falling to 0.10 in May (figure 8-2). The northern limit of the class coincides approximately with rainfalls of 100 to 150 millimeters a year, whereas the southern limit coincides with the 500-millimeter isohyet. We estimate the class to contribute about 6 percent of the growing stock and 13 percent of the sustainable yield in the region.

The vegetation consists of grassland with a crown cover of less than 10 percent woody species. In the northern extreme of the zone, annual plants are associated with wadis or plateau regions, but are typically tropical rather than Mediterranean in origin. Annual grasses such as *Coelachyrum brevifolium* and perennial

Figure 8-2. NDVI **Profile, Semidesert Wooded Grassland (Class 21)**

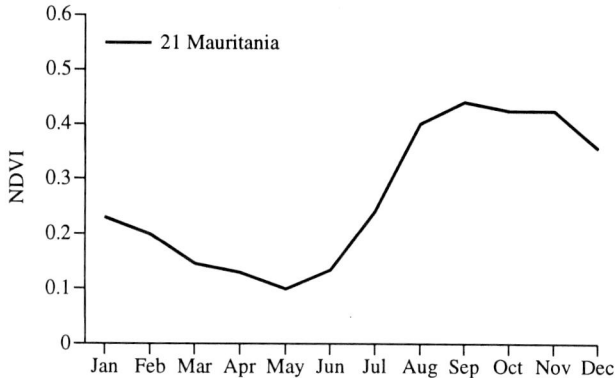

species such as *Eleusine compressa* are cited by Gillet (1986) as representing the basic Sahelian flora.

Where rainfall exceeds 100 millimeters a year and is predictable, a more permanent steppelike vegetation develops. Indeed, on a north-south gradient, a general increase occurs in the proportion of trees. This more southerly region coincides with the second Sahelian zone described by Boudet (1972). Local conditions vary and considerable differences exist in the proportion of woody species, but White's description (1983) of bushes and small bushy trees within grassland seems appropriate, with more favorable areas developing sufficient woody biomass to be described as bush.

The species represented in the herbaceous layer vary considerably from east to west, with local variations related to the proportion of sand and clay within the soil. Generally, however, tussock-forming grasses gradually give way to perennial, steppelike species southward along a north-south transect. Annual grasses such as *Aristida mutabilis*, *A. adscensionis*, *A. stipoides*, *Cenchrus biflorus* (cram-cram), *Schoenefeldia gracilis*, and *Tragus racemosus* are widespread and typical of the northern Sahelian semidesert, whereas perennial grasses such as *A. mutabilis* and *Panicum turgidum* extend quite far south.

On unstable soil systems such as dunes, grasses such as *Stipagrostis pungens* are more common; such species also are associated with lesser rainfall. The instability of tall dunes farther south favors the presence of *Aristida sieberana* and *Panicum turgidum* and their distribution is increased at the expense of woody species by processes that encourage desertification.

Of the woody biomass component, the *Acacia* genus is represented by a number of species. Widespread are *Acacia tortilis* ss. *raddiana*, *A. senegal*, *A. seyal*, and, in the more southerly limits of the zone, *A. nilotica*. In the northern part of the region, White (1983) reports

A. tortilis attaining 5 meters height, with total woody biomass rarely exceeding 3 percent of the aboveground biomass. This species also provides shade within which herbaceous vegetation may thrive and persist long after the end of the dry season. Also described among the more important woody species reported by White (1983) are *Acacia laeta*, *A. ehrenbergiana (flava)*, *Balanites aegyptiaca*, *Boscia senegalensis*, *Commiphora africana*, and *Leptadenia pyrotechnica*.

Anogeissus leiocarpus, an important fuelwood tree, also increases to the south in this class. The size to which trees grow varies, those in the northern Sahel semidesert being stunted and attaining full stature only with increased rainfall in the south of the region. *Calligonum comosum*, a typical Saharan shrub, often is associated with *Panicum turgidum* in areas where rainfall is locally greater.

Human population density is low in the region but generally increases along a north-south transect. The great drought of 1968–73 saw extensive movement southward (Granier 1980), intensifying the pressure on a system that has a low carrying capacity. In this region, the most characteristic activity is pastoralism, associated with seasonal movements of herds and flocks.

Class 44—Sahel-Sudanian *Acacia* Wooded Bushland

This class lies south of the Semidesert Wooded Grassland (Class 21) and forms a parallel belt of similar width across the region, extending into East Africa, with significant representation in all nations in the Sahel except the Gambia. In Mauritania, only the areas along the Senegal River Valley, the escarpments of 'Assâba and Tagant, the Massif of Afollé, and the Hodh Basin in the southeast are included. All other countries in the West African Sahel include significant areas that add to 17 percent of the surface of the region. The class is associated with greater rainfall than Class 21, with annual precipitation between 250 and 500 millimeters occurring largely in July and August. Regionally, this class is important for woody biomass, contributing an estimated 5 percent of the growing stock and 31 percent of the sustainable yield (131 million tonnes and 30.7 million tonnes, respectively).

The NDVI phenologies of this type of land cover exhibit the greatest value during September (figure 8-3). The greater NDVI period has a duration of 2 to 3 months, and throughout the rest of the year NDVI values are approximately 0.08. The growing season is longer in the west, and there herbaceous vegetation persists through a large part of the year.

The herbaceous layer consists mainly of annual grasses, including species represented farther north

Figure 8-3. NDVI **Profiles, Sahel-Sudanian *Acacia* Wooded Bushland (Class 44)**

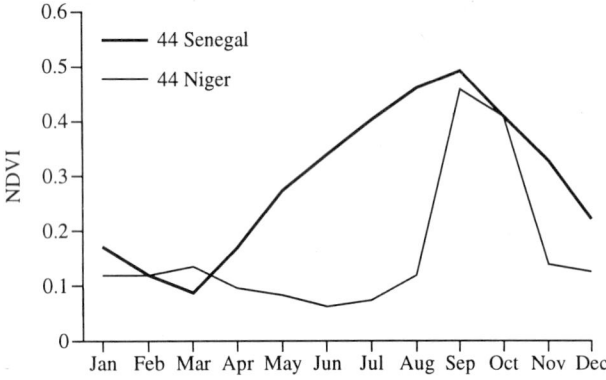

such as *Aristida stipoides*, *A. mutabilis*, *Cenchrus biflorus* (cram-cram), *Chloris prieurii*, and *Schoenefeldia gracilis*. Perennial grass species such as *Andropogon gayanus* occur farther south. The ground vegetation is more persistent than in Class 21 to the north and survives almost to the end of the dry season, attaining heights greater than a meter.

Of the woody species, *Acacia tortilis* ss. *raddiana* is the most common in the north of the zone. According to White (1983), it varies in form throughout the region, being short and bushier farther north with a maximum height of 4 meters, whereas in the south it is a bushy tree of up to 8 meters. Regional descriptions identify differences in species composition across the Sahel. Of widespread occurrence are *Acacia senegal*, *A. seyal*, members of the Combretaceae, especially *Combretum glutinosum* in the Sahelian zone of northern Burkina Faso, and *Ziziphus mauritiania*.

Adansonia digitata (baobab) is locally abundant, particularly around villages or on more saline soils. *Butyrospermum* spp. and *Parkia biglobosa*, more Sudanian species, are particularly important as fuelwood in Burkina Faso. Bush species include *Acacia mellifera*, *Commiphora africana*, and *Euphorbia balsamifera*.

A significant difference exists between vegetation in the north of the region, where the main human land use is nomadic and pastoral, and that in the south, where agricultural activity is more sedentary. This is related largely to the ability to grow crops such as millet in the south. In Niger, this difference is associated with the 350-millimeter isohyet. It may also be related to seasonal transhumance in response to the growth of vegetation in the wet season (Peyre de Fabregues 1980). Granier (1980) describes how people in Mali use indigenous woody and herbaceous species, and shows the intimate relationship between people and plants that must be borne in mind when considering the potential fuelwood resource in this region.

A case study of two traditional systems of consumption in the Sahel-Sudanian zone of Burkina Faso (März 1986) suggests that 75 percent of all wood collected is for fuel and that farmers very carefully select different tree species for different uses. The area required to supply the wood demand of each person is 1.4 to 2.8 hectares from a system producing 0.16 to 0.48 tonne a hectare. In areas seen during field visits in December 1988, a great intensity of land use was noted for many kilometers around both Dakar in Senegal and Bamako in Mali, with agricultural plots and fallows in regular patterns, serving both the cities and numerous villages.

In recent years, in Burkina Faso in particular, large numbers of people have moved from the north to the southwest in response to population increase. The associated agricultural activity has led to degradation of the environment and the effect is apparent on the land cover class maps, where areas of Semidesert Wooded Grassland are surrounded by Sahel-Sudanian *Acacia* Wooded Bushland. Areas of wooded bushland also extend farther south into the Dry Sudanian Woodland (Class 62).

A recent study of the fuelwood resource in Niger (Daus, Guero, and Ada 1986) indicates varying degrees of vegetation degradation associated with areas of great population density in the south and southwest of the country. In study areas of 200 kilometers diameter around five urban zones, the greatest degradation occurred around Tahoua, where the ratio of woody biomass to surface area is smallest. This juxtaposition of two classes, the class associated with lesser woody biomass occurring against a background of greater woody biomass, suggests that extensive vegetation land cover degradation can be mapped from 8-kilometer-resolution AVHRR NDVI imagery.

Class 62—Dry Sudanian Woodland

This class lies south of the Sahel-Sudanian *Acacia* Wooded Bushland (Class 44), forming a narrow belt through southern Senegal, the Gambia, southern and western Mali, southern and western Burkina Faso, the southernmost area of Niger, and southern Chad. Although forming quite extensive areas—about 10 percent of the West African Sahel—it is penetrated by many other vegetation classes from farther north and south, forming complex mosaics in western Mali and Chad in particular.

This class corresponds to the Sudanian Undifferentiated Woodland mapping unit 29a of White (1983) and lies within the Sudanian regional center of endemism (III). In many instances, White considers the area to be a product of agriculture, consisting of land under cultivation or bush fallow, the original vegetation possibly having been dry forest. Certainly in Senegal, Mali, and Burkina Faso it coincides with areas of dense permanent settlement. Regionally, 57 percent of the growing stock and 41 percent of the sustainable yield are estimated to come from this class (1,439 million tonnes and 41.5 million tonnes, respectively).

The area is associated with annual rainfalls of 750 to 1,250 millimeters with a dry season of 7 to 9 months. In areas of greater rainfall, such as lower Casamance in Senegal, the region merges with the Cultivation and Forest Regrowth Mosaic (Class 73). Phenology curves for sites in Burkina Faso and Senegal (figure 8-4) show increases in NDVI beginning in June or July, peaking in August and September at 0.48 in the Burkina Faso site before falling back to baseline values of approximately 0.10. This contrasts with an earlier start to the growing season for the vegetation in Classes 64 and 65, for which growth begins in March or April.

The vegetation is typically open woodland with mainly deciduous trees, 15 to 25 meters in height, with a herbaceous carpet dominated by grasses, although shrubs, bushes, and climbers are present. The perennial herbs include *Andropogon gayanus* on sandy soils; *Cymbopogon giganteus*, *Ctenium newtonii*, and *Hyparrhenia dissoluta* on wetter soils; and *Loudetia simplex* on slopes or rocky outcrops. Among the annual grasses are *Andropogon pseudopricus*, *Eragrostis tremula*, *Pennisetum pedicellatum*, and *Utenium elegans*, the latter considered by Steentoft (1988) to be an indicator of disturbed ground. During the dry season, the herbaceous layer becomes highly combustible, and frequent fires burn when the harmattan blows. The merger with the Cultivation and Forest Regrowth Mosaic (Class 73) in southern Senegal is marked by the appearance of herbaceous species such as *Daniellia oliveri* and *Erythrophleum guineense*.

The woody layer contains a significant number of thorny species typical of the Sahel-Sudanian *Acacia* Wooded Bushland (Class 44) to the north. Large numbers of Combretaceae, including *Anogeissus leiocarpus* and *Combretum glutinosum*, can form distinct woodlands, and *Acacia* spp. are common. Many trees are short-trunked, thick-barked, fire-resistant species, with thin-barked species surviving in areas that escape regular burning or agriculture where they form areas of closed woodland. Farther south, more water-demanding species are in evidence.

Butyrospermum parkii, which traditionally marks the start of the Sudanian zone, and *Parkia biglobosa* are more common, particularly around villages. Both species are used by the local population. They probably represent the remnants of dry forest cover eliminated by human activity. *Butyrospermum* spp. also exhibit good regeneration from a root crown and many of the Combretaceae show cryptogeal germination, enabling them to recover from fire damage.

Other tree species typically present around villages include *Acacia albida*, *Adansonia digitata* (baobab), *Bombax costatum*, *Borassus aethiopum*, *Mangifera indica*, and *Tamarindus indica*. *Khaya senegalensis*, *Parkia biglobosa*, and *Pterocarpus erinaceus* are reported as forming dry woodland in Senegal (Ndiaye 1983). *Boswellia dalzieli*, *Commiphora* spp., and *Ziziphus* spp. also are present. Areas that Steentoft (1988) describes as "savanna regrowth" are vegetated by *Adansonia digitata*, *Acacia albida*, *Balanites aegyptiaca*, *Butyrospermum* sp., *Ceiba* sp., and *Tamarindus* sp.

Bush and shrub species include *Hyphaene thebaica*, a palm used for building, together with members of the Combretaceae, Rubiaceae, and Euphorbiaceae. In moister regions of Senegal such as river valleys, *Borassus flabellifer* (rhun palm) and *Oxytenanthera abyssinica* (a bamboo) occur (Ndiaye 1983). Gallery forests including *Cola laurifolia* and *Mitragyna inermis* exist in Niger (Peyre de Fabregues 1980).

Class 64—Sudanian Woodland

This class occupies 3 percent of the West African Sahel and occurs in southeastern Senegal, western and southwestern Mali, southwestern Burkina Faso, and southernmost Chad. It corresponds approximately to mapping zone 11a of White (1983) as a mosaic of lowland rain forest and secondary grassland, lying in the center of the Sudanian regional center of endemism. It incorporates species from the drier northern part of the Sudanian region, including *Acacia* spp. and members of the Combretaceae, together with species more typical of the humid southern parts such as *Khaya senegalensis* and *Lophira lanceolata*, both Sudanian endemic species.

Class 64 is intimately associated in the south of the region with Dry Sudanian Woodland (Class 62), and is considered in some areas to be a degraded form of dry woodland. It may represent parcels of ground that are less accessible or are forest reserves. (The imagery appears to show areas of Sudanian woodland coinciding with "forêts classées" around Bobo Dioulasso in

Figure 8-4. NDVI **Profiles, Sudanian Woodlands (Classes 62, 64, and 65)**

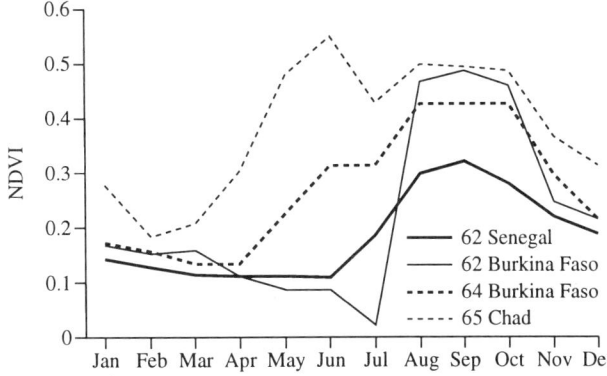

western Burkina Faso). In the region, Sudanian Woodland is estimated to be 15 percent of the growing stock and 6.5 percent of the sustainable yield (378 million tonnes and 6.5 million tonnes, respectively).

Climatically, this zone has a precipitation of at least 1,100 millimeters a year with a dry season of 6 to 7 months from October to May. Monthly rainfall values during the wet season of July and August are 250 to 350 millimeters, with mean daily temperatures of about 28–30°C. This vegetation is clearly the northern limit of the coastal West African region. The factor that limits the growth of woody species is the length of the dry season, and characteristically the trees are deciduous. The canopy is seasonally open and, in combination with the relatively wide spacing of the trees, light is able to penetrate, allowing tussocky grasses such as *Andropogon gayanus, Hyparrhenia cyanescens,* and *Schizachyrium sanguineum* to attain heights greater than 1.5 meters.

Granier (1980), in describing the biogeography of Mali, classifies this area as part of the Sudano-Guinean domain with a savanna-woodland mosaic, subject to regular fires and with gallery forest in the valleys. Tsetse is common and this may explain somewhat the lesser exploitation of this area, compared to more degraded examples of Sudanian Woodland (Class 64). Characteristic species are *Isoberlinia doka,* a fire-resistant species, *Afzelia africana, Daniellia oliveri, Erythrophleum guineense,* and *Uapaca somon.*

Ndiaye (1983) similarly identifies the equivalent region in southern Senegal with a number of dominant woody species, including *Khaya senegalensis* and *Lophira lanceolata* as well as *Parkia biglobosa* and *Pterocarpus erinaceus.* In valleys, the bamboo *Oxytenanthera abyssinica* and the rhun palm *Borassus flabellifer* occur. The rhun palm also occurs as plantations in southwestern Burkina Faso. Sanford and Ischei (1986), in describing "Guinea savanna," identify *Daniellia oliveri, Isoberlinia doka, I. tomentosa,* and *Parkia clappertoniana* as important species, but in the more northern areas the *Isoberlinia* genus is considered by White to be restricted to small pockets on rocky hills.

The term "derived savanna" is used for large parts of this area, but the effect of human activity is much less than in the neighboring Class 62. A gradient of tree cover probably exists, with the cover diminishing as one moves from Sudanian Woodland (Class 64) into areas more properly described as Dry Sudanian Woodland (Class 62).

The phenological curves (figure 8-4) indicate that the maximum NDVI value of 0.43 occurs in August and September with the smallest value of 0.13 in January, February, and March. These indicate the distinct seasonality of the vegetation.

This class in the Sahel is probably a mixture of a number of elements of Sudanian and Sahelian origin.

The class is more fully and typically described in Chapter 9 on the coastal West African region.

Class 65—Moist Sudanian Woodland

Class 65 has a very limited distribution, only 1 percent of the land area in the West African Sahel region, occurring to a significant extent only in southeastern Chad, with isolated small pockets in Burkina Faso. It corresponds to mapping unit 27 of White (1983) or the northern Guinea savanna of Steentoft (1988).

Climatically, the region is similar to that in which Sudanian Woodland (Class 64) develops, having a dry season of 5 to 7 months. The vegetation is similar in structure to that class, but is dominated by *Burkea africana, Daniellia* sp., *Erythrophleum africanum, Isoberlinia doka,* and *Lophira lanceolata.* It represents the most northerly extension of a zone of vegetation occurring in the coastal West African region and spreading eastward to Cameroon, Central African Republic, and Sudan.

Where it occurs in the Sahel, this class forms a mosaic with Sudanian Woodland (Class 64) and Dry Sudanian Woodland (Class 62). The contribution to the regional growing stock is significant, estimated as 4 percent of the growing stock and 1.7 percent of the sustainable yield (100 million tonnes and 1.7 million tonnes, respectively).

Phenologies appear to be similar to those of Classes 62 and 64, but with a maximum NDVI value of 0.55 occurring earlier in the year in May and June and remaining at approximately 0.45 until September (figure 8-4).

Moist Sudanian Woodland is an open woodland savanna, considered by some to be a reduced form of *miombo* because fewer species are present and trees rarely attain heights greater than 15 meters. Typically, it includes *Monotes kerstingii* on dry eroded slopes (Steentoft 1988). In the northernmost part of the area, common species are *Cussonia barteri, Entada africana, Isoberlinia tomentosa, Lannea microcarpa, Terminalia avicennioides,* and *Ximenia americana,* but traces of Sahelian vegetation occur in the form of *Acacia* spp. (particularly *A. albida), Burkea africana,* and *Piliostigma thonningii.*

In areas where the woody biomass is regenerating, species such as *Isoberlinia* sp., *Piliostigma* sp., and *Terminalia avicennioides* exhibit the ability to regrow from underground suckers and hence have an enhanced ability to withstand fire. The appearance of *Isoberlinia* is considered to correspond with the 1,000-millimeter isohyet. An understory of climbers and shrubs may be present, and a herbaceous carpet of perennial grasses occurs, including *Andropogon* spp., *Eragrostis* spp., *Schizachyrium* spp., and *Pennisetum* spp.

Class 65 is largely the product of anthropogenic influences of burning, grazing, and agriculture, with regeneration of vegetation occurring during fallow periods. Consequently, this class is unlikely to offer fuelwood shortages, considering the fairly low human population densities that exist.

Land Cover Class Tables

Tables 8-1 through 8-7, beginning on page 60, present summaries for each land cover class of the area, showing growing stock and sustainable yield for the West African Sahelian nations of Burkina Faso, Chad, the Gambia, Mali, Mauritania, Niger, and Senegal.

References

Every effort has been made to facilitate access to the documents listed here. Some documents, however, lack full bibliographic information because it was unavailable; also, some documents are of limited circulation.

Boudet, G. 1972. "Désertification de l'Afrique tropicale sèche." *Adansonia*, ser. 2, 12(4):505–24.

Daus, Steven J., Mamane Guero, and Lawally Ada. 1986. "A Remote-Sensing-Aided Inventory of Fuelwood Volumes in the Sahel Region of West Africa: A Case Study of Five Urban Zones in the Republic of Niger." Paper presented at Symposium on Remote Sensing for Resources Development and Environmental Management, August, Enschede, Netherlands.

Gillet, H. 1986. In G. W. Lawson, ed., *Plant Ecology in West Africa*. London: John Wiley and Sons, Ltd.

Granier, C. 1980. In M. Traore, ed., *Atlas du Mali*. Paris: Les Editions Jeune Afrique.

IIED (International Institute for Environment and Development). 1989. "Rainfall in the Sahel." Paper No. 10, September, IUCN (International Union for Conservation of Nature) /Sahel Programme.

Le Houérou, H. N. 1980. "The Rangelands of the Sahel." *Journal of Environmental Management* 33(1): 41–46.

März, U. 1986. "Wood Consumption in Traditional Systems in the Sudano-Sahelian Zone of Burkina Faso." *Quarterly Journal of International Agriculture* 25(1):49–58.

Ndiaye, P. 1983. *Atlas du Sénégal*. Paris: Les Editions Jeune Afrique.

Peyre de Fabregues, B. 1980. In E. Berniss and S. A. Hamidou, eds., *Atlas du Niger*. Paris: Les Editions Jeune Afrique.

Sanford, W. W., and A. O. Ischei. 1986. In G. W. Lawson, ed., *Plant Ecology in West Africa*. London: John Wiley and Sons, Ltd.

Steentoft, M. A. 1988. *Flowering Plants in West Africa*. Cambridge, England: Cambridge University Press.

White, F. 1983. "The Vegetation of Africa." *Natural Resources Research Series* 20. Paris: UNESCO/AETFAT/ UNSO (United Nations Educational, Scientific and Cultural Organization/Association pour l'Etude Taxonomique de la Flore de l'Afrique Tropicale/ United Nations Sudano-Sahelian Office).

Table 8-1. Land Cover Classes—Burkina Faso (West African Sahel Region)

Land cover class		Area km²	Area Percent	Growing stock Thousand tonnes	Growing stock Percent	Sustainable yield Thousand tonnes per year	Sustainable yield Percent
0		1,792	0.66	0.00	0.00	0.00	0.00
	12	5,691	2.10	1,291.86	0.29	56.91	0.39
1		5,691	2.10	1,291.86	0.29	56.91	0.39
	21	10,065	3.71	3,321.45	0.74	100.65	0.68
	22	158	0.06	52.14	0.01	1.58	0.01
2		10,223	3.76	3,373.59	0.75	102.23	0.70
	44	93,692	34.50	13,210.57	2.94	3,279.22	22.30
4		93,692	34.50	13,210.57	2.94	3,279.22	22.30
5		211	0.08	453.65	0.10	13.29	0.09
	62	110,555	40.71	295,181.85	65.77	8,512.74	57.89
	64	44,053	16.22	117,621.51	26.21	2,026.44	13.78
	65	5,164	1.90	13,787.88	3.07	247.87	1.69
6		159,772	58.84	426,591.24	95.05	10,787.05	73.36
	81	158	0.06	3,899.44	0.87	465.94	3.17
8		158	0.06	3,899.44	0.87	465.94	3.17
Total		271,539	100.00	448,820.35	100.00	14,704.64	100.00
(Percentage of region)		(5.17)		(17.96)		(14.74)	

Note: In the following tables, details may not add to totals because of rounding.
Source: Authors' calculations from data bases derived from land cover classification and table 4-1.

Table 8-2. Land Cover Classes—Chad (West African Sahel Region)

Land cover class		Area km²	Area Percent	Growing stock Thousand tonnes	Growing stock Percent	Sustainable yield Thousand tonnes per year	Sustainable yield Percent
0		494,756	38.74	0.00	0.00	0.00	0.00
	12	23,344	1.83	5,299.09	0.62	233.44	0.75
1		23,344	1.83	5,299.09	0.62	233.44	0.75
	21	228,592	17.90	75,435.36	8.78	2,285.92	7.31
	22	5,849	0.46	1,930.17	0.22	58.49	0.19
2		234,441	18.36	77,365.53	9.00	2,344.41	7.50
	41	3,899	0.31	5,419.61	0.63	81.88	0.26
	44	230,120	18.02	32,446.92	3.78	8,054.20	25.75
4		234,019	18.33	37,866.53	4.41	8,136.08	26.01
	62	210,465	16.48	561,941.55	65.39	16,205.81	51.81
	63	527	0.04	1,407.09	0.16	23.19	0.07
	64	26,400	2.07	70,488.00	8.20	1,214.40	3.88
	65	29,615	2.32	79,072.05	9.20	1,421.52	4.54
6		267,007	20.91	712,908.69	82.95	18,864.92	60.31
	73	369	0.03	621.40	0.07	7.01	0.02
	74	7,325	0.57	12,335.20	1.44	139.17	0.44
7		7,694	0.60	12,956.70	1.51	146.18	0.47
	81	527	0.04	13,006.36	1.51	1,554.12	4.97
8		527	0.04	13,006.36	1.51	1,554.12	4.97
Lakes		15,229	1.19	0.00	0.00	0.00	0.00
Total		1,277,017	100.00	859,402.89	100.00	31,279.15	100.00
(Percentage of region)		(24.31)		(34.40)		(31.36)	

Source: Authors' calculations from data bases derived from land cover classification and table 4-1.

Table 8-3. Land Cover Classes—The Gambia (West African Sahel Region)

Land cover class	Area km²	Area Percent	Growing stock Thousand tonnes	Growing stock Percent	Sustainable yield Thousand tonnes per year	Sustainable yield Percent
12	53	0.33	12.03	0.02	0.53	0.01
1	53	0.33	12.03	0.02	0.53	0.01
22	580	3.60	191.40	0.35	5.80	0.16
2	580	3.60	191.40	0.35	5.80	0.16
44	2,740	16.99	386.34	0.71	95.90	2.62
4	2,740	16.99	386.34	0.71	95.90	2.62
62	11,856	73.53	31,655.52	58.23	912.91	24.96
6	11,856	73.53	31,655.52	58.23	912.91	24.96
81	896	5.56	22,113.28	40.68	2,642.30	72.24
8	896	5.56	22,113.28	40.68	2,642.30	72.24
Total	16,125	100.00	54,358.57	100.00	3,657.45	100.00
(Percentage of region)	(0.31)		(2.18)		(3.67)	

Source: Authors' calculations from data bases derived from land cover classification and table 4-1.

Table 8-4. Land Cover Classes—Mali (West African Sahel Region)

Land cover class	Area km²	Area Percent	Growing stock Thousand tonnes	Growing stock Percent	Sustainable yield Thousand tonnes per year	Sustainable yield Percent
0	586,024	46.17	0.00	0.00	0.00	0.00
12	7,957	0.63	1,806.24	0.24	79.57	0.30
1	7,957	0.63	1,806.24	0.24	79.57	0.30
21	264,214	20.82	87,190.62	11.79	2,642.14	10.11
22	4,953	0.39	1,634.49	0.22	49.53	0.19
2	269,167	21.21	88,825.11	12.01	2,691.67	10.30
41	685	0.05	952.15	0.13	14.39	0.06
44	178,005	14.02	25,098.70	3.39	6,230.18	23.84
4	178,690	14.07	26,050.85	3.52	6,244.57	23.90
62	146,282	11.52	390,572.94	52.80	11,263.71	43.10
63	8,747	0.69	23,354.49	3.16	384.87	1.47
64	68,662	5.41	183,327.54	24.78	3,158.45	12.09
65	2,582	0.20	6,893.94	0.93	123.94	0.47
6	226,273	17.82	604,148.91	81.67	14,930.97	57.14
74	316	0.02	532.14	0.07	6.00	0.02
76	105	0.01	176.82	0.02	1.99	0.01
7	421	0.03	708.96	0.10	8.00	0.03
81	738	0.06	18,213.84	2.46	2,176.36	8.33
8	738	0.06	18,213.84	2.46	2,176.36	8.33
Total	1,269,270	100.00	739,753.92	100.00	26,131.13	100.00
(Percentage of region)	(24.16)		(29.61)		(26.20)	

Source: Authors' calculations from data bases derived from land cover classification and table 4-1.

Table 8-5. Land Cover Classes—Mauritania (West African Sahel Region)

Land cover class		Area		Growing stock		Sustainable yield	
		km^2	Percent	Thousand tonnes	Percent	Thousand tonnes per year	Percent
0		797,596	77.03	0.00	0.00	0.00	0.00
	21	211,624	20.44	69,835.92	92.85	2,116.24	66.72
	22	580	0.06	191.40	0.25	5.80	0.18
2		212,204	20.50	70,027.32	93.10	2,122.04	66.90
	41	211	0.02	293.29	0.39	4.43	0.14
	44	25,399	2.45	3,581.26	4.76	888.97	28.03
4		25,610	2.47	3,874.55	5.15	893.40	28.17
	81	53	0.01	1,308.04	1.74	156.30	4.93
8		53	0.01	1,308.04	1.74	156.30	4.93
Total		1,035,463	100.00	75,209.91	100.00	3,171.73	100.00
(Percentage of region)		(19.71)		(3.01)		(3.18)	

Source: Authors' calculations from data bases derived from land cover classification and table 4-1.

Table 8-6. Land Cover Classes—Niger (West African Sahel Region)

Land cover class		Area		Growing stock		Sustainable yield	
		km^2	Percent	Thousand tonnes	Percent	Thousand tonnes per year	Percent
0		710,174	59.95	0.00	0.00	0.00	0.00
	12	2,108	0.18	478.52	0.36	21.08	0.19
1		2,108	0.18	478.52	0.36	21.08	0.19
	21	232,966	19.66	76,878.78	58.49	2,329.66	20.57
	22	105	0.01	34.65	0.03	1.05	0.01
2		233,071	19.67	76,913.43	58.52	2,330.71	20.57
	41	632	0.05	878.48	0.67	13.27	0.12
	44	229,119	19.34	32,305.78	24.58	8,019.17	70.79
4		229,751	19.39	33,184.26	25.26	8,032.44	70.91
	62	6,060	0.51	16,180.20	12.31	466.62	4.12
	64	158	0.01	421.86	0.32	7.27	0.06
6		6,218	0.52	16,602.06	12.63	473.89	4.18
	76	211	0.02	355.32	0.27	4.01	0.04
7		211	0.02	355.32	0.27	4.01	0.04
	81	158	0.01	3,899.44	2.97	465.94	4.11
8		158	0.01	3,899.44	2.97	465.94	4.11
Lakes		3,004	0.25	0.00	0.00	0.00	0.00
Total		1,184,695	100.00	131,433.03	100.00	11,328.07	100.00
(Percentage of region)		(22.55)		(5.26)		(11.36)	

Source: Authors' calculations from data bases derived from land cover classification and table 4-1.

Table 8-7. Land Cover Classes—Senegal (West African Sahel Region)

Land cover class		Area km²	Area Percent	Growing stock Thousand tonnes	Growing stock Percent	Sustainable yield Thousand tonnes per year	Sustainable yield Percent
	12	1,844	0.93	418.59	0.22	18.44	0.19
1		1,844	0.93	418.59	0.22	18.44	0.19
	21	14,860	7.48	4,903.80	2.59	148.60	1.57
	22	6,534	3.29	2,156.22	1.14	65.34	0.69
2		21,394	10.77	7,060.02	3.73	213.94	2.26
	44	115,455	58.13	16,279.15	8.59	4,040.93	42.64
4		115,455	58.13	16,279.15	8.59	4,040.93	42.64
	62	54,434	27.41	145,338.78	76.69	4,191.42	44.23
	63	2,213	1.11	5,908.71	3.12	97.37	1.03
	64	3,004	1.51	8,020.68	4.23	138.18	1.46
6		59,651	30.03	159,268.17	84.04	4,426.97	46.72
	81	263	0.13	6,490.84	3.42	775.59	8.18
8		263	0.13	6,490.84	3.42	775.59	8.18
Total		198,607	100.00	189,516.77	100.00	9,475.87	100.00
(Percentage of region)		(3.78)		(7.59)		(9.50)	

Source: Authors' calculations from data bases derived from land cover classification and table 4-1.

9

The West African Coast

Andrew C. Millington

This chapter presents a detailed description of the most important land cover classes in this region. Helpful figures in other chapters include figure 3-1 (cloud cover); figures 3-2, 3-3, and 3-4 (NDVI summary land cover profiles); figure 3-5 (regional summary map of land cover classes); figures 7-1 and 7-2 (continental maps of growing stock and sustainable yield); and the "Regional Land Cover Class Map of West Africa" at the end of this volume.

Helpful tables in other chapters include table 3-2 (land cover classes); table 4-1 (data and sources for growing stock and sustainable yield); and table 6-2 (West African Coast estimated woody biomass by summary class).

Class 12—Hydromorphic Grassland

Only small patches of Hydromorphic Grassland are mapped in this region. They form a discontinuous belt at the foot of the Gambaga Scarp and along the Niger Valley in Benin, and in northern Nigeria. This area extends across the inselberg-dominated landscape of northwestern Nigeria as far as the Jos Plateau. Further small patches occur to the east of the Jos Plateau at a similar latitude. Together these areas account for approximately 1.86 percent of the region (38,415 square kilometers), the largest area by far being in Nigeria (33,461 square kilometers).

These areas mainly occur in the *"Burkea africana Savanna Zone"* as defined by Areola (1982a), although some areas also occur in the northernmost parts of the *"Isoberlinia* Savanna" and "Wooded Savanna." These areas are sparsely populated with an attendant low cultivation density; the grasslands themselves are used mainly for cattle grazing.

The fuelwood demand in these areas is widely variable. On the basis of population density, it is greatest in northeastern Ghana (100 persons per square kilometer) and lowest in Benin and Nigeria (2 persons per square kilometer). In areas of greater population density, fuelwood demand undoubtedly exceeds woody biomass stock, which is relatively low in these dominantly grassland areas.

Class 44—Sahel-Sudanian *Acacia* Wooded Bushland

Acacia wooded grassland and deciduous bushland occurs only in the very north of the region, with small areas in northeastern Ghana and the adjacent parts of Togo and northeastern Benin. The class has its maximum development in northern Nigeria where it forms a broad, continuous belt 50 to 300 kilometers wide across the country, north of 9° N latitude. It accounts for approximately 7.9 percent of the region, or 163,145 square kilometers; the largest proportion is in Nigeria (161,827 square kilometers).

In Nigeria, this area is dominated by three vegetation types—*"Burkea africana* Savanna," "Moist Sudanian Woodland," and "Wooded Savanna." All three types are grass and tree savannas, the latter two classified as Sudanian and the former as Guinean (Areola 1982a). In Togo and Ghana, this class falls into the three areas mapped as (a) "Sudanian Savanna" (Brunel 1981; Brookman-Amissah 1987), (b) the northernmost parts of the "Dry Forest Zone" in Togo (Brunel 1981), and (c) the northernmost parts of the Guinea Savanna Zone in Ghana (Brookman-Amissah 1987). A typical NDVI curve for this class (figure 9-1) shows a strongly developed seasonality with generally lower NDVI levels than the Sudanian Woodlands (Classes 62, 64, and 65) to the south.

The drier areas are dominated by a short, feathery, continuous grass sward which attains a height of only

Figure 9-1. NDVI **Profiles, Sahel-Sudanian** *Acacia* **Wooded Bushland (Class 44)**

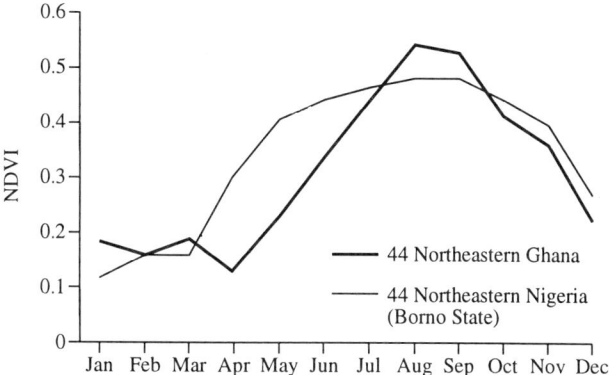

1.5 meters. Interspersed in this grass sward are short savanna trees dominated by *Acacia* spp. Other common trees are *Adansonia digitata, Burkea africana, Butyrospermum parkii, Capparis* spp., *Commiphora africana, Diospyros thespiliformis, Maytenus senegalensis,* and *Tamarindus indica.* The dry woodlands that are mapped within this class are dominated by trees such as *Adansonia digitata* and *Isoberlinia doka.* These latter savanna types covered by this class range from treeless grass savanna to savanna woodland with a more-or-less closed canopy and occur mainly in the south of the mapped area.

Land use in this class is dominated by extensive grazing and intensive cultivation of both food crops (beans, millet, and sorghum) and cash crops (cotton, groundnuts, and tobacco). The population density in parts of this zone can be great—for example, 20 to 39 persons per square kilometer in Circonscription Administrative de Dapadong in northern Togo (Gu-Konu 1981a). Typical of these areas of dense population are the Kano and Sokoto Close-Settled Zones in northern Nigeria. Between the population concentrations are areas with very low densities and consequently a greater frequency of woodlands.

Conversely, population pressure leads to great cultivation density; an example is parts of Bauchi and Kaduna states in Nigeria and the Savanes Region of Togo (Afolayan and Barbour 1982; Gu-Konu 1981b). Population pressure also creates high stock numbers; an example is the more than 80 million cattle in the Savanes Region (Gu-Konu 1981b; National Atlas of Ghana).

As a consequence of these high cultivation and stock densities, the woody biomass resource in this class is under severe pressure in some areas. Land already has been extensively cleared for cultivation and grazing over much of this class in Benin, Ghana, Togo, and in many parts of Nigeria. In areas of low population density, particularly northern Nigeria, the woody biomass condition is much more favorable. Many of the

remaining trees have multiple use, and their potential as fuelwood is limited. This, combined with the relatively low sustainable yield for trees and shrubs in this class (0.35 tonne per hectare per year), means that these areas are suffering from a shortfall in woody biomass.

Class 62—Dry Sudanian Woodland

Dry Sudanian Woodland occurs in a belt covering the northernmost parts of Guinea-Bissau, Guinea, Ghana, Togo, Benin, and northern Nigeria. The main areas in this belt are:

- In Nigeria, a discontinuous belt runs between 7° N and 13° N, attaining its greatest areal extent in Bauchi, Borno, and Sokoto states.
- In Ghana, the class extends southward in a strip along the Black Volta as far as Dochire and over a large area in northeastern Ghana, which extends as far south as Tamale.
- In Togo and Benin, an almost continuous belt runs across the northern part of these countries. In Togo it occurs only north of 10° N, but in Benin it occurs as far south as Banikoara and Kandi, as well as in a strip along the Chaîne de l'Atakora.

Dry Sudanian Woodland covers 192,338 square kilometers (about 9.3 percent of the land area), making it the fifth-largest land cover class in the region (table 9-1).

This is one of the driest savanna woodland classes in the West African coastal region. The only drier class that occurs to any significant extent is Sahel-Sudanian *Acacia* Wooded Bushland (Class 44). Although the latter class is restricted mainly to Nigeria, Dry Sudanian Woodland is far more widespread. Brookman-Amissah (1987) maps Dry Sudanian Woodland as the most northerly of the "Guinea Savanna Woodlands" in Ghana. In the adjacent parts of Togo, Dry Sudanian Woodland falls within the dry northern parts of Brunel's "Forêt Sèche" zone (1981). In Nigeria, it

Table 9-1. Areal Distribution, Dry Sudanian Woodland (62), West African Coast Region

Country	Area (km²)	Class in region (percent)	Class in country (percent)
Nigeria	154,134	80.1	17.2
Ghana	18,233	9.5	7.7
Benin	11,435	5.9	9.9
Togo	4,268	2.2	7.9
Guinea	2,793	1.5	1.2
Guinea-Bissau	1,475	0.8	4.9
Total	192,338	100.0	

Source: Authors' data bases.

corresponds to the northern parts of these types of savanna: "Moist Sudanian Woodland," "Wooded Savanna," and "*Burkea africana* Savanna" (Areola 1982a).

The area is dominated by very open tree and shrub savanna, and it is a drier variant of the closed-canopy savanna woodland to the south. The main trees and shrubs include *Adansonia digitata, Burkea africana, Butyrospermum parkii, Cardenia* spp., *Combretum* spp., *Entada africana, Hyphaene thebaica, Isoberlinia doka, Parkia* spp., and *Terminalia* spp. The understory grasses are dominated by *Andropogon* spp. and *Hyparrhenia* spp.

In Nigeria, land use in this class varies from intensively cultivated and grazed savanna in the east to areas of very low cultivation density and only moderate grazing on the Benin border (Areola 1982b). The intensity of cultivation and grazing reflects population distribution (Afolayan and Barbour 1982). Cropping is dominated by subsistence sorghum cultivation, but cotton and groundnuts are the main cash crops. Similar crop combinations exist in the other parts of the region covered by this class.

Population density in this zone is generally low: <19 persons per square kilometer in Togo (Gu-Konu 1981a) and 2 persons per square kilometer in Nigeria (Afolayan and Barbour 1982). This suggests that fuelwood demand is probably quite low except around the main settlements. Nevertheless, standing stock and sustainable yield in this zone are low and, although they probably meet current demand, any increase in population could reverse this favorable situation.

Class 64—Sudanian Woodland

This is the second most extensive land cover class in the region and occurs in all nine countries except Liberia (table 9-2). It covers 381,040 square kilometers, approximately 18.5 percent of the area, and extends

Table 9-2. Areal Distribution, Sudanian Woodland (64), West African Coast Region

Country	Area (km²)	Class in region (percent)	Class in country (percent)
Nigeria	153,765	40.4	17.2
Guinea	103,283	27.1	43.2
Ghana	47,110	12.4	19.8
Benin	40,839	10.7	35.4
Côte d'Ivoire	16,546	4.3	5.1
Togo	11,224	2.9	20.7
Guinea-Bissau	5,480	1.4	18.2
Sierra Leone	2,793	0.7	4.0
Total	381,040	100.0	

Note: Details may not add to totals because of rounding.
Source: Authors' data bases.

Figure 9-2. NDVI **Profiles, Sudanian Woodland (Class 64)**

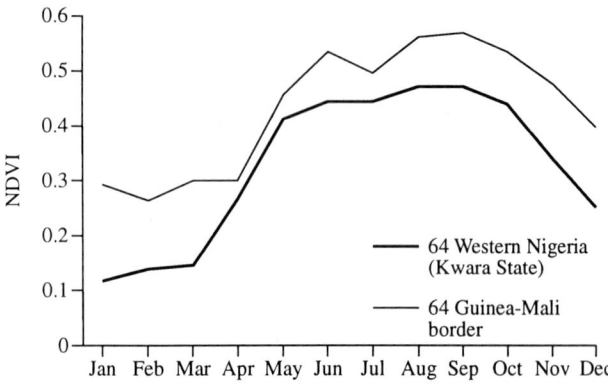

64 Western Nigeria (Kwara State)
64 Guinea-Mali border

from the Gabù Region in eastern Guinea-Bissau to the Nigeria-Cameroon border. Sudanian Woodland is very extensive in Guinea, occurring on the plains between the Sankarani, Niger, and Tinkisso rivers, and on the higher ground along the Mali border. Its occurrence in Côte d'Ivoire is restricted to areas north of 9°30' N where mean annual rainfall is less than 1,300 millimeters.

Sudanian Woodland covers large areas in northwestern Ghana and between Tamale and the Togo Hills. In northern Togo and Benin, it occurs between 9°30' N and 11°30' N. In Nigeria, it covers most of the Nigerian "Middle Belt," demarking a broad tract of high ground from the Benin border in the west to the Jos Plateau and then southward to include the middle Benue River Valley.

White (1983) classifies this and a number of other land cover classes as "Sudanian Woodland with abundant *Isoberlinia*" and a "Mosaic of Lowland Rain Forest and Secondary Grassland." The NDVI curve (figure 9-2) shows a seasonality similar to the Guinean Woodland (Class 74, figure 9-3) but with slightly lower NDVI values during the wet season.

In Guinea-Bissau, this and other land cover classes have been classified as (1) "Dry Forest Types including Tree Savannas" (2) "Forêt Demi-sèche Dense et Claire" and "Forêt Claire Degradée et Savane Boisée" by SCET (Bianchi 1986), and (3) "Savane Boisée" and "Savane Très Claire" by Atlanta Consulting G.m.b.H. (Bianchi 1986). In northern Côte d'Ivoire, Monnier (1983) classifies this land cover type as "Sudanian Savanna." In Ghana, this land cover is classified in conjunction with very large parts of central and northern Ghana as "Guinea Savanna Woodland" (Brookman-Amissah 1987). In Togo, Sudanian Woodland corresponds to the southern part of Brunel's "Forêt Sèche" (1981).

In Nigeria, this class covers much of the "Middle Belt" and represents in part the "*Isoberlinia* Savanna," "Mixed Leguminous Wooded Savanna," "*Afzelia*

Figure 9-3. NDVI **Profiles, Guinean Woodland (Class 74)**

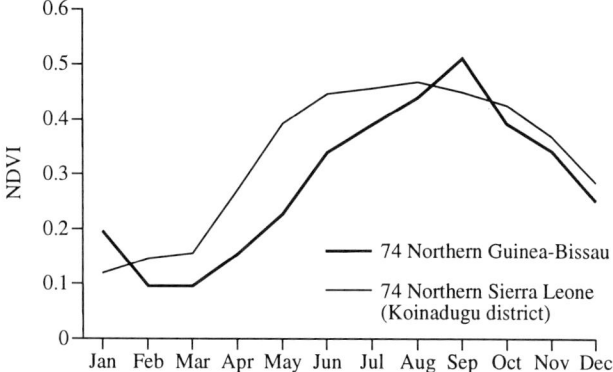

and geomorphological characteristics of the area, which in turn are controlled by the underlying lithology (Stobbs 1963; Millington, Helmisch, and Rhebergen, 1985).

The area is classified as "Riverain Grassland" by Cole (1968) and the "*Lophira* Tree Savanna/Boliland Swamps Complex" by FAO (1981). The "*Lophira* Tree Savanna" is developed on the thin, infertile soils of the intervening high ground and is dominated by stunted, fire-resistant *Lophira lanceolata* in a grass sward of *Chasmopodium caudatum*. The ecological development of this vegetation is controlled by cultivation, annual burning, and intensive coppicing of the trees. The seasonally flooded grass and herb swamps between the interfluves (known as *bolis*) are almost treeless and have been extensively cleared for swamp rice cultivation.

Apart from the Sierra Leonean "bolilands," this class represents mature Sudanian Woodland with low-to-moderate levels of disturbance. Therefore, at present levels of fuelwood demand, wood supplies are probably assured in the short-to-medium term. However, the Sudanian woodlands in the West African coastal region are coming under increasing pressure from charcoal producers. Trees are being cut, converted to charcoal, and then transported by road to the coastal cities such as Abidjan and Accra, where charcoal is an important domestic fuel. The Sierra Leonean "bolilands" represent a worse situation where local demand is relatively great and the locally available woody biomass resource is severely limited.

Class 65—Moist Sudanian Woodland

Moist Sudanian Woodland occurs in scattered patches throughout the Sudanian zone of West Africa. These patches occur westward and eastward from 0° longitude in two groups: (a) a more-or-less continuous belt from eastern Guinea to north-central Ghana and (b) an area from central Togo to the Cameroon-Nigerian border.

The first area includes the northern flanks of the Guinea Highlands (the Dongoroma, Going, Kourandou, and Tourou ranges). It also encompasses much of Côte d'Ivoire north of 8° N, including a number of areas above 400 meters elevation (for example, the Chaîne de Tiemé, Chaîne de Madinani, and the Mont Yévélé-to-Mont Bowé de Kiendi axis) in the Régions du Nord and de l'Est. In the northern region of Ghana, the area extends as far east as Lake Volta.

The second area contains many smaller areas. Two main concentrations of these occur in Benin: (a) a thin strip of land between Tchaourou and Kalale, which includes the towns of Parakou, Péréré, and Nikki, and (b) the densely populated area to the southeast of Chaîne de l'Atakora, which includes the towns of Natitingou, Ouake, Kopargo, Djougou, Pehonko, and Kouandé.

Savanna/Semideciduous Forest," and "Leguminous Wooded Savanna of the Jos Plateau" (Areola 1982a).

The savanna formations in this class vary from closed canopy savanna woodland (for example, "*Afzelia*-Savanna/Semideciduous Woodland" in Nigeria and "Forêt Demi-Sèche Dense" in Guinea-Bissau) to thicket savanna. The spatial variation in savanna formation is caused by variation in both rainfall amount and distribution, and in the moister parts, woodland is well developed. In such woodland, trees and shrubs form a more-or-less closed canopy over grasses.

Moisture availability becomes more of a problem farther north in this land cover class, and in such areas open-canopy savanna woodland and shrub savanna occur. There appears to be an overlap with Dry Sudanian Woodland (Class 62), but the latter class generally represents much drier Sudanian savanna woodland than woodlands in this class. The dominant trees are *Afzelia africana*, *Anogeissus schimperi*, *Daniellia oliveri*, *Isoberlinia doka*, *Monotes kerstingii*, and *Pterocarpus erinaceus*. Dominant grasses include *Andropogon* spp., *Ctenium elegans*, *Hyparrhenia* spp., and *Loudetia superba*.

The area covered by this class varies in population density. But, apart from the area around principal towns, density rarely exceeds 50 persons per square kilometer and is mostly <10 persons per square kilometer. Cultivation density is therefore moderate, as are stock numbers. The main subsistence crops are sorghum and yams. Important cash crops include cotton, groundnuts, and beniseed (Berron and Vennetier 1983; Gu-Konu 1981b; and Nwafor 1982b).

The only large area mapped as this class that does not correspond to the general pattern of Sudanian Woodland is in Sierra Leone. Here, a tract in the northern interior plains is placed in this land cover class. This area corresponds to the "bolilands." This is an extensive area of seasonally flooded hydromorphic grasslands and swamps, with degraded Guinean savanna woodland on the intervening high ground. The vegetation pattern is controlled by the hydrological

Table 9-3.   Areal Distribution, Moist Sudanian Woodland (65), West African Coast Region

Country	Area (km²)	Class in region (percent)	Class in country (percent)
Nigeria	121,252	51.9	13.6
Côte d'Ivoire	43,474	18.6	13.4
Guinea	24,611	10.5	10.3
Ghana	22,975	9.8	9.7
Benin	14,755	6.3	12.8
Togo	6,165	2.6	11.4
Guinea-Bissau	422	0.2	1.4
Sierra Leone	53	> 0.1	> 0.1
Total	233,707	100.0	

Note: Details may not add to totals because of rounding.
Source: Authors' data bases.

Much of central Nigeria also falls into this class. In the south, Moist Sudanian Woodland is limited by the coastal plains, and it does not extend north of 10° N. It is the fourth-largest land cover class in the region, covering 233,707 square kilometers. It accounts for approximately 11.3 percent of the land area and occurs in all countries except Liberia (table 9-3).

Moist Sudanian Woodland is equivalent to a number of vegetation classes mapped by Areola (1982a) in Nigeria—"Mixed Leguminous Wooded Savanna," "*Afzelia*-Savanna/Semideciduous Forest," "Moist Sudanian woodland," and parts of the "Forest- Savanna Mosaic." These are mainly savanna communities, the exception being the "Forest-Savanna Mosaic," which is a transitional class between the rain forest and the savanna woodland to the north.

In Côte d'Ivoire, this class corresponds to the northernmost parts of the "Sub-Sudanian Savanna" (Monnier 1983). These areas are mainly grass and herb savannas with either trees, shrubs, or thicket formations. In Ghana, this area forms part of the "Guinean Savanna Woodland," but is not differentiated from other savanna types by Brookman-Amissah (1987). In Togo, Moist Sudanian Woodland is mainly restricted to hills above 400 meters elevation, and is classified as "Forêt Sèche" (Brunel 1981).

Despite the fact that this zone is named after a single tree, *Isoberlinia doka*, by many workers (including White 1983), it is not a single-dominant community. Other equally important trees occur, such as *Afzelia africana, Anogeissus schimperi, Daniellia oliveri, Khaya senegalensis, Monotes kerstingii, Pterocarpus erinaceus,* and *Uapaca togoensis.* Important shrubs and small trees in drier areas include *Butyrospermum parkii, Cardenia* spp., *Combretum* spp., *Entada africana, Parkia* spp., and *Terminalia* spp.

Cultivation density is much lower in this zone than in the woodland to the south and north. In many areas, extensive tracts of savanna woodland are used only

for grazing. However, around both Enugu and Ilorin in Nigeria, edaphic and derived-savanna grasslands, respectively, are extensively cultivated. This area is one of mixed-grain and root crop agricultural systems, the main subsistence crops being maize, millet, sorghum, and yams (Berron and Vennetier 1983; Gu-Konu 1981b; Nwafor 1982).

In the southern areas of this class, tree and root crops are more common, whereas in the north, cereals predominate. Cash cropping is less common than in many other savanna woodland zones in West Africa, although significant tobacco cultivation occurs in northwestern Côte d'Ivoire (Berron and Vennetier 1983). Grazing is common, although herd sizes generally are much smaller than in the drier savanna woodland zones to the north.

Population density varies from <4 persons per square kilometer in many parts of Côte d'Ivoire (Lecomte and Monnier 1983) to 100 persons per square kilometer in parts of Togo (Gu-Konu 1981a). The spatial variation in both woody biomass and population within the Moist Sudanian Woodland zone means that the firewood supply is extremely variable, ranging from significant surplus to severe shortage.

Class 74—Guinean Woodland

The area covered by Guinean Woodland in this region is very extensive—243,241 square kilometers or about 11.8 percent of the land area. It occurs in all countries of the region except Liberia (table 9-4).

As mapped, this class is probably synonymous with White's "Guineo-Congolian Secondary Grassland and Wooded Grassland" (1983) and occurs in a belt varying between 50 and 950 kilometers wide from the Guinea-Bissau–Senegal border to the Nigeria-Cameroon border. In Guinea-Bissau, Guinean Woodland dominates the higher ground away from the

Table 9-4.   Areal Distribution, Guinean Woodland (74), West African Coast Region

Country	Area (km²)	Class in region (percent)	Class in country (percent)
Guinea	63,024	25.9	26.4
Côte d'Ivoire	59,598	24.5	18.4
Nigeria	53,064	21.8	5.9
Sierra Leone	30,563	12.6	43.4
Guinea-Bissau	14,491	6.0	48.2
Togo	8,958	3.7	16.5
Ghana	7,694	3.2	3.2
Benin	5,849	2.4	5.1
Total	243,241	100.0	

Note: Details may not add to totals because of rounding.
Source: Authors' data bases.

coastal plains (for example, in Bafatá, Gabù, and Oio regions). In Guinea and Sierra Leone, it occurs in three distinct areas:

- From the northern coastal plains of Guinea inland as far as Koundara
- In an area stretching northward from Bo in Sierra Leone to the Fouta Djallon in Guinea and then back southeastward to Kissidougou in Guinea. This area includes the interior plateaus and hill region of Sierra Leone and Guinea
- An area extending from Nzérékoré in Guinea into eastern Côte d'Ivoire.

In Côte d'Ivoire, Guinean Woodland covers most of the ground between 300 and 400 meters above sea level north of 7°40′ N, with the exception of the Mt. Bendi–Mt. Ko axis and some areas in the extreme northeast. The vegetation is poorly developed in Ghana and Benin but is extensive, albeit rather patchy, in Togo and Nigeria. The main occurrence in Nigeria is on land higher than 300 meters elevation on the Abeokuta-Ibadan-Ondo axis, along the Niger Valley as far north as the Jos Plateau, and on the Bamenda and Mandara hills. The NDVI profiles (figure 9-3) show the seasonality of the vegetation and its intermediate position between the forest zone and the drier woodlands.

Rain forest trees within these savannas are either remnants or invasive. The communities themselves are extremely heterogeneous, ranging from grassland through shrubby thicket to well-developed savanna woodland and gallery forest. Three strata can be identified in the savannas, although they are not present in all of the communities:

- A lower stratum of grasses, herbs, young saplings, and shrubs attaining 2 meters in height
- A middle stratum of shrubs and small trees which vary between 2 and 8 meters
- Emergent trees and palms that exceed 8 meters.

At a small scale, the distribution of trees and shrubs is controlled by edaphic factors (Clayton 1958, 1961; Hambler 1964; Jones 1963; Keay 1959; Menaut and César 1979, 1982; Morison, Hoyle, and Hope-Simpson 1948; Ramsay 1964; Ramsay and de Leeuw 1964, 1965; White 1965). Trees and shrubs usually are best developed on the most fertile soils and here the woody biomass is greatest; however, in these areas, competition from grasses is also important and fire is an important factor in savanna formation (Menaut and César 1982). Stands of palms, dominated by *Borassus aethiopum*, are widespread and are tolerant of a wide range of environmental conditions.

Menaut and César (1979, 1982) have reported extensively on Guinean savannas from the Lamto study area in central Côte d'Ivoire (the UNESCO Man and Biosphere or MAB Programme). At Lamto, a number of vegetation communities have been identified that typify the potential diversity of woody biomass stocks in this and similar land cover classes (table 9-5). These variations are mainly related to topographic and soil conditions.

On the highest ground at Lamto, freely draining ferralsols are dominated by "Andropogoneae Savannas," of which five types are recognized. These are all dominated by *Andropogon* spp. grasses. They are classified by woody biomass stock and, in order of decreasing stock, they are:

- Savanna Woodland, in which the trees form a continuous canopy between 2 and 6 meters, although palms often emerge to heights of 10 to 12 meters. The trees are underlain by a dense grass sward.
- Dense Shrub Savanna, which has a lower overall tree and shrub cover varying between 20 and 25 percent, with shrubs attaining a maximum height of 8 meters.
- Open Shrub Savanna, which occurs in areas with better soil condition than the transitional shrub and grass savannas. Consequently, the shrub species

Table 9-5. Woody Biomass Stocks and Growth Rates, Guinean Woodland Communities, Lamto, Côte d'Ivoire

Parameter	Intermediate shrub savanna	Open shrub savanna	Dense shrub savanna	Savanna woodland
Individuals	120	160	300	800
Canopy cover (percent)	7	15	20	45
Leaf area index	0.1	0.2	0.4	1
Woody biomass				
Above ground (kg/ha)	7,400	21,900	32,600	54,200
Below ground (kg/ha)	3,600	9,200	1,430	26,600
Woody biomass growth rate				
Above ground (kg/ha/yr)	120	330	420	760
Below ground (kg/ha/yr)	50	130	230	370

Source: After Menaut and César 1982.

are better developed, attaining ground covers of 15 to 20 percent.

- Intermediate Shrub Savanna, which has a very low woody cover (less than 10 percent), with individual shrubs ranging from 3 to 6 meters in height.
- Grass Savanna, which occurs where shallow ferricretes restrict tree and shrub growth, consisting exclusively of grasses up to 2.5 meters in height.

On the hydromorphic gleysols, which characterize the slopes down to the rivers, *Loudetia simplex* grass-dominated savannas occur. These have a very poor woody element (mainly palms and shrubs on termite mounds) because of various edaphic constraints related to waterlogging in the gleysols.

Menaut and César (1979, 1982) also have investigated the role of burning on the evolution of these savannas. On the one hand, *Loudetia* savannas have a sharp boundary between savanna and forest, especially in burned areas. On the other hand, Andropogoneae savannas have a gradual increase in woody component from open grassy areas to savanna woodland. The forest-savanna boundary is dependent on the date, duration, and intensity of fires and there is no one definite regional succession.

In some areas, forest trees establish themselves in swamps, on termite mounds, or on rocky outcrops. These communities then act as precursors for forest invasion, once burning ceases. In other areas, gallery forest trees may invade the adjacent savannas after herbs have invaded. As the tree canopy closes, the shading effect restricts grass growth, less fuel for the annual fires is produced, and fires therefore do not penetrate the invasive forest stands.

Cole (1968) recognizes a moist savanna woodland in northeastern Sierra Leone which he terms the "South Guinean Savanna." It grades from moist woodland on the wetter soils to a shrub or grass savanna on mountain slopes. Moist woodland and open tree savanna are the dominant ecological structures. In the moist woodland and tree savanna, the trees are gnarled, stunted, fire-resistant *Daniellia oliveri* and *Lophira lanceolata,* attaining heights of 10 to 17 meters, growing in a dense grass understory. On thinner soils with less capacity for holding moisture, shrub and tall grass savannas dominate. In the "South Guinean Savanna" zone, gallery forest mixes savanna trees (for example, *Cussonia longissima, Parkia biglobosa,* and *Terminalia glaucescens*) with rain forest trees (especially *Chlorophora regia, Erythrophleum guineense, Parinari excelsa,* and *Uapaca guineensis*).

Cole argues that moist semideciduous woodland is the climax community on the best soils, but that degradation of these woodlands in northern Sierra Leone has been caused by dry-season burning to promote better grazing, by settlement schemes for nomadic

cattle owners (between 1951 and 1963), and by dry-season farming promoted by small irrigation schemes. The best examples of moist closed-canopy woodland now exist around villages, where they are reserved for a variety of reasons.

In the "Southern Guinean Savanna" of Sierra Leone and adjacent areas of Guinea, land use is dominated by Fula pastoralism, and the grass understory is extensively grazed. In northwestern Sierra Leone, the unusual "*Lophira* Tree Savanna" occurs in this class (Cole 1968). Here, a combination of infertile thin soils, intensive cultivation, and fierce annual burning have created an ecological community completely dominated by *Lophira lanceolata* in a grass understory of *Chasmopodium caudatum*. The trees have been intensively coppiced, both during agricultural clearance and for fuelwood. Only shrubby regrowth remains in most areas. In slightly less disturbed areas, however, the trees attain 7.5 to 10 meters in height, and at the edge of the "*Lophira* Tree Savanna" other savanna trees are invasive.

Many parts of the Guinean Woodland are sparsely populated, and thus have a low cultivation density. Other areas have a high cultivation density, mainly for upland (rain-fed) rice in the wetter areas and sorghum and millet in the drier areas. Vegetation is cut, dried, and burned to create fields that are commonly cultivated for up to 2 years. The woodland is then left to regenerate as a woody fallow.

Although older woodland regrowth resembles the climax savanna communities in structure and floristic relations, the derived savannas are characterized by the frequent occurrence of the economically important fire-resistant oil palm (*Elaeis guineensis*) and a greater proportion of fire-resistant trees because of the frequent burning. Fire-resistant and fire-tolerant successional trees (for example, *Lophira alata, Parkia biglobosa, Piliostigma thonningii, Pterocarpus erinaceus,* and *Syzygium guineense*) first occur in the younger regrowth stages and become subdominant in older successions. They are always an important component in the Guinean Woodland communities, however, because of the annual burning of the understory grasses to create better grazing for cattle in the woodlands.

The Guinean Woodland class is dominated mainly by well-developed closed-canopy woodland with a moderately high sustainable yield. This, combined with the variable but generally low population density (2 persons per square kilometer), suggests that few woody biomass supply problems exist, although localized problems undoubtedly occur (for example, in the "*Lophira* Tree Savanna" in Sierra Leone). In many West African states, this situation could change with a switch in charcoal production for coastal towns from Sudanian Woodland (64) to Guinean Woodland. The high-density wood that is characteristic of trees in the Guinean Woodland would produce high-quality charcoal if the supply from Sudanian Woodland became scarce.

Overview: West African High Woody Biomass Mosaics

The humid tropical forest zone in West Africa is densely populated, and only in a few areas do extensive tracts of primary or mature secondary forest remain. Much of the area is cultivated, the two main agricultural systems being bush fallowing and plantation cropping. Consequently, the land cover of the region is a complex mosaic of primary forest and mature secondary forest reserves, various stages of forest regrowth, cultivated fields, and plantations. The fragmentary nature of land use is such that individual elements cannot be routinely identified and mapped using 8-kilometer resolution data.

After extensive discussion with foresters and land use planners from West Africa, we mapped the area as a high woody biomass "Mixed Agriculture and Forest Fallow Mosaic." We divided the mosaic into two phases, based on overall level of productivity:

- High-Productivity West African Cultivation and Forest Mosaic (Class 75), corresponding roughly to the evergreen or ombrophilous humid forest zone
- Medium-Productivity West African Cultivation and Forest Mosaic (Class 76), corresponding roughly to the semideciduous or mesophilous forest zone.

Class 75—High-Productivity West African Cultivation and Forest Mosaic

This mosaic occurs along the coastal belt of West Africa in two main areas: (a) between the Gola Mountains on the Sierra Leone–Liberia border and central Ghana, and (b) in southern Nigeria, mainly in Bendel State and the Cross River Basin. Small patches also occur in eastern Guinea, central Côte d'Ivoire, and in the Atacora Mountains on the Ghana-Togo border (table 9-6a). The class is mainly restricted to the lowland plains and plateaus of the coastal belt. For example, in Côte d'Ivoire, it occurs mainly within 150 kilometers of the coast at elevations below 200 meters, although it also exists on Mont Bowé de Kiendi and Mont Nangbion. Other occurrences on higher ground are, for example, in the Loma-Man Dorsale.

The class covers 101,438 square kilometers, about 4.9 percent of the total area. Its greatest extent is in Côte d'Ivoire (47,584 square kilometers), followed by Liberia (24,925 square kilometers) and Ghana (18,812 square kilometers).

Rainfall is great over most of the area mapped in this class, ranging from about 1,400 to more than 4,000 millimeters a year in most areas. In Ghana, however, the class occurs in areas with as little as 1,000 millimeters annual rainfall. To a large extent, the greater productivity of vegetation in this class is due to rainfall. But evidence also exists for a minor depression in photosynthetic activity in the wet season, approximately in August, in Nigeria, Togo, and Ghana, a period known as the "little dry season." In the more westerly areas covered by this class, rainfall is restricted to a single wet season with a dry season ranging from 4 to 6 months. The pattern of vegetation response to rainfall is reflected in the NDVI profiles (figure 9-4).

Much of the humid tropical forest that once covered this area has been cleared. White (1983) reported three types of wet lowland rain forest in this area: "Hygro-

---

Table 9-6a. Areal Distribution, High-Productivity Cultivation and Forest Mosaic (75), West African Coast Region

Country	Area (km²)	Class in region (percent)	Class in country (percent)
Côte d'Ivoire	47,584	46.9	14.7
Liberia	24,925	24.6	26.5
Ghana	18,812	18.5	8.4
Nigeria	7,641	7.5	0.9
Togo	1,107	1.1	2.0
Guinea	1,001	1.0	0.4
Sierra Leone	369	0.4	0.5
Benin	0	0.0	0.0
Guinea-Bissau	0	0.0	0.0
Total	101,439	100.0	

Note: Details may not add to totals because of rounding.
Source: Authors' data bases.

Table 9-6b. Areal Distribution, Medium-Productivity Cultivation and Forest Mosaic (76), West African Coast Region

Country	Area (km²)	Class in region (percent)	Class in country (percent)
Côte d'Ivoire	150,023	26.8	46.3
Nigeria	132,739	23.7	15.2
Ghana	99,383	17.7	44.6
Liberia	60,494	10.8	64.3
Benin	38,046	6.8	33.3
Guinea	31,512	5.6	13.2
Sierra Leone	27,981	5.0	39.8
Togo	18,865	3.4	34.9
Guinea-Bissau	1,739	0.3	5.8
Total	560,782	100.0	

Note: Details may not add to totals because of rounding.
Source: Authors' data bases.

Figure 9-4. NDVI **Profiles, High-Productivity West African Cultivation and Forest Mosaic (Class 75)**

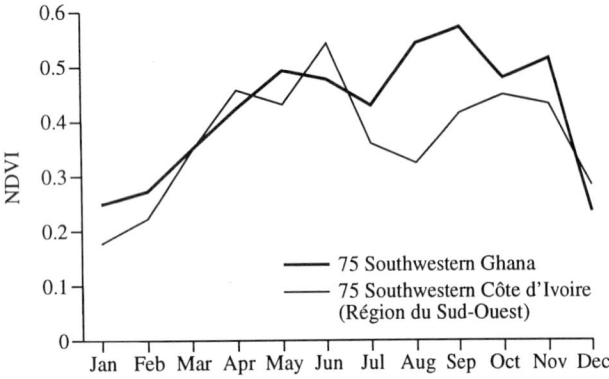

philous Coastal Evergreen Guineo-Congolian Rain Forest," "Mixed Moist Semi-Evergreen Guineo-Congolian Rain Forest," and "Single-Dominant Moist Evergreen and Semi-Evergreen Rain Forest." Little forest now occurs in the region because of extensive cultivation of food and plantation crops. Forest still occurs, however, in less densely populated areas and in forest reservations.

"Hygrophilous Coastal Evergreen Guineo-Congolian Rain Forest"

Of the three forest types, this was the most common in West Africa. Hall and Swaine (1976) describe "Hygrophilous Coastal Evergreen Rain Forest" as shorter than "Mixed Moist Semi-Evergreen Rain Forest," with an upper canopy attaining 30 meters and a few emergents attaining 40 meters.

The evergreen characteristic of the forest is caused by most species shedding their leaves intermittently, although in areas with a pronounced dry season (for example, Liberia and Sierra Leone), many species shed their leaves simultaneously and are immediately replaced with new ones. This type of forest is equivalent to "Forêt Hyperombrophile" in Côte d'Ivoire (Monnier 1983), which occurs only in areas with a mean annual precipitation exceeding 1,800 millimeters and a reduced dry season.

This type undoubtedly occurs in Liberia, but Jansen (1972) does not differentiate it from other evergreen humid forest types. He does, however, record "Wet Coastal Rain Forest" vegetation dominated by *Tetraberlinia tubmaniana* in a belt that stretches inland about 80 kilometers between River Cess and Greenville in Liberia. The canopy dominants of this forest are *Crudia gabonensis, Didelotia unifoliolata, Gilbertiodendron preussii, Gluma ivorensis, Lophira alata, Mapania* spp., *Tarietia utilis, Terminalia utilis, Tieghemella (Dumoria) africana,* and *Sacoglottis gabonensis.*

"Mixed Moist Semi-Evergreen Guineo-Congolian Rain Forest"

This type is less developed in West Africa than "Hygrophilous Coastal Evergreen Guineo-Congolian Rain Forest." White (1983) attributes this to the abrupt transition from very wet coastal conditions to the dry interior. Areola (1982a) describes "Moist Lowland Forest" in southern Nigeria as typically having the three layers associated with tropical rain forest: the upper layer of emergents 40 to 50 meters high, a middle layer of trees between 16 and 40 meters, and a lower layer of trees varying between 10 and 16 meters. The upper and middle layers form one continuous canopy and the lower layer forms a second continuous canopy. Underneath are shrub and herb layers with saplings of the taller trees.

In this class, the mean deciduousness in Sierra Leonean forest reserves was 23.5 percent (Cole 1968), although other forests classed as evergreen by Cole had deciduousness values from 16.5 to 26.2 percent.

This type of forest corresponds to Monnier's "Forêt Ombrophile" (1983), which occurs south of the 1,600- millimeter isohyet in Côte d'Ivoire. It is characterized by emergents such as *Lophira alata,* which attain a height of 50 meters. In Sierra Leone, this type of forest occurs in the Gola Mountains on the Liberian border and spills across the border into Lofa County in Liberia. Other interior evergreen humid tropical forests in Liberia (in Montserrado, Grand Bassa, Grand Gedeh, and Sinoe counties) probably fall into White's "Mixed Moist Semi-Evergreen Forest" class. Cole (1968) notes that when the Gola Forest was first enumerated in 1923, ". . . most of it was virgin primary rain forest and even areas which had been disturbed by man, had secondary forests of considerable age."

Dominant canopy trees are *Anthonotha fragrans, Combretodendron* spp., *Cynometra leonensis, Chrysophyllum* spp., *Diospyros* spp., *Entandrophragma* spp., *Guarea* spp., *Lophira alata, Nauclea diderrichii, Oldfieldia africana,* and *Tarietia utilis.*

"Single-Dominant Moist Evergreen and Semi-Evergreen Guineo-Congolian Rain Forest"

This type is interspersed within the other two types. Single-dominant forest is much rarer in West Africa than in the Zaire Basin and is mainly restricted to the wetter evergreen forest along the coast. White (1983) notes that only two of the dominant trees in the Zaire Basin extend into southern Nigeria: *Gilbertiodendron dewevrei* dominates some fringing forests or swamp forests, and *Julbernardia seretii* occurs within the coastal evergreen forest. Single-dominant forest is

Figure 9-5. Côte d'Ivoire Wood Production, 1980–84

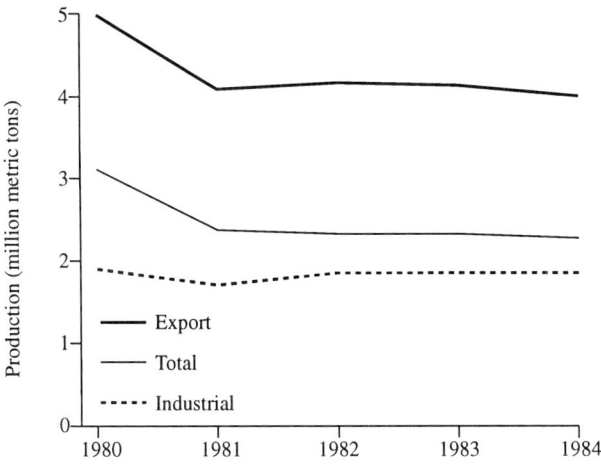

dense, typically 35 to 45 meters high, and is composed almost entirely of one species.

Some areas in the high-productivity zone still have extensive mature secondary evergreen humid tropical forest. These areas always occur where population density is low. For instance, in southwestern Côte d'Ivoire (Région du Sud-Ouest), population density is <10 persons per square kilometer, and most of these people live in a few coastal towns or along roads. Consequently, cultivation density is low and livestock herds are small in these areas.

In Sierra Leone, Liberia, Côte d'Ivoire, and Ghana, population densities are much greater and forest clearance for arable and tree crop cultivation is common. The main crops are rice and cocoa, although some extensive rubber plantations in Liberia are mapped in this class. The low population density and high woody

biomass reserve mean that fuelwood supply problems are, at worst, only localized.

The main problem facing the woody biomass resource in this area is exploitation of hardwood timber by local contractors and under concession to foreign companies. Valuable hardwoods such as *Oldfieldia africana*, *Khaya* spp., *Entandrophragma* spp., *Turraeanthus africana*, *Chlorophora excelsa*, and *Tieghemella hockelii* all are harvested from these forests (Cole 1968; Arnaud 1983). The trade in timber is of great importance in Ghana, Côte d'Ivoire, and Liberia (for an example in Côte d'Ivoire, see figure 9-5). Comparison of the quality of timber felled in Côte d'Ivoire in 1980 and 1984 shows a slight deterioration in quality (figures 9-6 and 9-7) that might be expected with overexploitation.

Class 76—Medium-Productivity West African Cultivation and Forest Mosaic

This mosaic occurs in all nine countries in the region (table 9-6b), extending from the Tombali Region in Guinea-Bissau to the Nigeria-Cameroon border. In Guinea-Bissau, it is restricted to the Tombali Region. An extensive area in Guinea runs from the Fouta Djallon foothills to the Forecariah Plains. The coastal and interior plains of southern Sierra Leone and Liberia form another large contiguous area, including the forested southern interior plateau, parts of the Gola, Gori, and Nimini hills, and the Upper Moa Basin of Sierra Leone.

The mosaic continues across much of the southern half of Côte d'Ivoire into Ghana, where three main areas exist: (a) on high ground to the west of Sunyani, (b) around Lake Volta in the center of the country, and (c) on the Accra Coastal Plains. This

Figure 9-6. Côte d'Ivoire Wood Production by Class, 1980

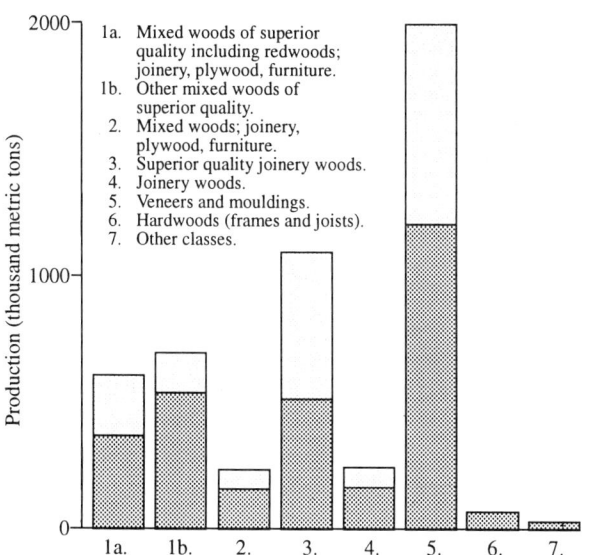

Figure 9-7. Côte d'Ivoire Wood Production by Class, 1984

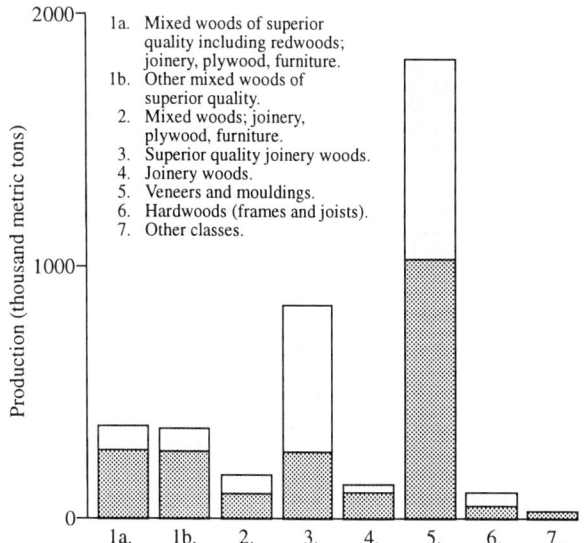

last area extends from slightly west of Accra to southern Nigeria, covering similar land uses in southern Togo and Benin. In southern Nigeria, the mosaic covers the western coastal plains, eastern parts of the Lower Niger Valley (Anambra, Benue, Imo, Kwara, Lagos, Ogun, Ondo, and Oyo states), as well as other parts of the region.

The mosaic of agriculture and forest fallow that has developed in this region is known by a variety of local names (table 9-7). As the local names suggest, this land cover class represents a mosaic of forest areas that have been cleared for cultivation or are in different stages of regrowth after farms have been abandoned. The broad land cover pattern is therefore a consequence of the bush-fallowing and plantation agriculture typical of the West African forest zone.

This area originally was covered mainly by "Semi-Deciduous Guineo-Congolian Rain Forest," but, like the wet evergreen forest types, little remains. Most has been cleared for timber and to provide land for agriculture. The remaining forest differs from the evergreen humid tropical forest in the level of deciduousness. Because of the longer dry season and associated moisture stress, the majority of trees lose their leaves during some part of the dry season. Cole (1968) examined the degree of deciduousness in five Sierra Leonean forest reserves. The only semideciduous forest he examined had 67.5 percent deciduous tree species compared with much lower values for four evergreen forests (table 9-8).

Table 9-7. Equivalent Names for Cultivation and Forest Mosaic in West African Coast Region

Country	Equivalent name
Guinea-Bissau	"Derrubadas Pelas Cultures" "Forêt Sub-humide Degradée"; "Forêt Sub-humide de Transition" (SCET International in Bianchi 1986)
Sierra Leone	"Farm Bush" (Clarke 1966) "Forest Regrowth and Farmland" (FAO 1981; Cole 1968)
Liberia	"Farm and Secondary Forest" (Jansen 1972)
Côte d'Ivoire	"Secteur Préforestier" (Monnier 1983)
Ghana	"Coastal Thicket and Grassland" (Brookman-Amissah 1987) "Coastal Thicket of Tree Savanna" (Lawson 1968)
Togo	"La terre de barre" and "Zone de transition" (Brunel 1981)
Nigeria	"Secondary Forest" and "Forest Savanna Mosaic" (Areola 1982a)

Source: Authors' data bases.

Table 9-8. Degree of Deciduousness, Reserved Moist Forests, Sierra Leone

Reserve	Forest type	Number of species sampled	Degree of deciduousness (percent)
Tama-Tonkolili	Semi-deciduous	30	67.5
Nimini-South	Evergreen	102	24.4
Dodo Hills	Evergreen	88	25.9
Kambui Hills and Dambaya Valley	Evergreen	37	16.5
Gola North	Evergreen	156	26.2

Source: After Cole 1968.

The seasonal difference between high-productivity and medium-productivity phases of the West African cultivation and forest mosaics shows distinctly in the NDVI curves for the two classes (figure 9-8).

"Semideciduous Guineo-Congolian Rain Forest" has a similar three-layered structure to the moist evergreen forest types, although greater diversity exists in the trees, both in the canopy and understory layers, because of the slightly greater penetration of sunlight. The main canopy species are *Afzelia africana, Aningeria* spp., *Canarium schweinfurthii, Celtis* spp., *Chlorophora* spp., *Chrysophyllum perpulchrum, Cola* spp., *Mansonia altissima, Parinari excelsa, Terminalia* spp., and *Triplochiton scleroxylon.*

Numerous small swamps and fluvial grasslands occur in the forest zone and, as a consequence, special types of swamp and gallery forest develop. Under favorable ecological conditions, "Humid Tropical Swamp Forest" resembles mature rain forest in structure and height, with the tallest trees attaining 45 meters in height. More typically, however, the canopy is more open than undisturbed rain forest. Cole (1968) calls such forests "Freshwater Inland Swamp Forests" in Sierra Leone. Their canopies range from closed to moderately open, with the trees ranging from 12.5 to 20 meters in height.

The dominant trees are usually *Mitragyna stipulosa* and *Raphia* spp., but in some areas forest trees (commonly *Carapa procera, Mitragyna ciliata, Nauclea* spp., *Pandanus candelabrum, Phoenix reclinata, Spondianthus preussii, Symphonia globulifera, Uapaca guineensis,* and *U. heudelotii*) are interspersed with *Raphia* spp. and the large and small rattan palms (*Ancistrophyllum secundiflorum* and *Calamus deeratus*).

The Accra coastal plains fall into this land cover class and were mapped as "West African Coastal Mosaic" by White (1983). The vegetation is dominated by sparse, short (<80 centimeters high) grassland, or grass savanna that is controlled by shallow, poorly drained soils (Jenik and Hall 1976). These pedogenic factors restrict the root development of trees and

Figure 9-8. NDVI Profiles, Medium-Productivity West African Cultivation and Forest Mosaic (Class 76)

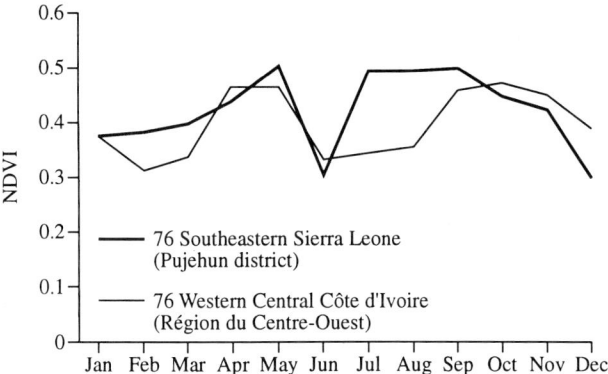

shrubs. However, within the grasslands, shrubby thickets occur on old termite mounds. These thickets form a dense, closed canopy of *Capparis erythrocarpos*, *Flacourtia indica (flavescens)*, *Grewia carpinifolia*, *Securinega virosa*, *Uvaria chamae*, and *Zanthoxylum xanthoxyloides* about 5 meters high with emergents up to 10 meters of *Elaeophorbia drupifera* and *Diospyros mespiliformis*.

On inselbergs, patches of evergreen and semi-evergreen coastal forest have developed since the cessation of farming (Swaine, Lieberman, and Hall 1990). These are multilayered like the rain forest, although they rarely exceed 10 to 12 meters in height. The dominant canopy trees are *Cynometra megalophylla*, *Diospyros abyssinica*, *D. mespiliformis*, *Manilkara obovata*, and *Millettia thonningii*. The lower layer is chiefly composed of *Drypetes floribunda*, *D. parvifolia*, and *Vepris heterophylla*. Emergents such as *Antiaris toxicaria*, *Ceiba pentandra*, *Celtis mildbraedii*, and *Nesogordonia papaverifera* also exist.

The ecology of a forest on an inselberg differs from that of the rain forest because inselberg forest is shorter, lacks evergreen tree species, and has a poor regeneration rate from seed. The agricultural pattern is similar to that elsewhere in the land cover class.

Grassland

"Guineo-Congolian Edaphic Grassland," which occurs on hydromorphic soils throughout the region, is also mapped within this land cover class. White (1983) suggests that most of these grasslands represent transitions between aquatic vegetation and forest, the grassland structure being maintained by annual burning. Menaut (1983) considers these to be "Coastal Grass Savannas" in Côte d'Ivoire and suggests that many are fire-proclimax communities derived from forests or wooded savannas (Adjanohoun 1962).

Nevertheless, truly hydromorphic grasslands recognized in Côte d'Ivoire ("Savanes de Basse Côte" or "Savanes Lagunaires") occupy small tracts along a

narrow coastal strip at Tabou, Néro-Mer, Lagune Taio, Lagune Ebrie, and Lagune Ehi. These savannas are dominated by a dense grass cover of *Brachiaria brachylopha*, *Hyparrhenia chrysargyrea*, and *Loudetia phragmitoides*. Trees occur in the grasslands and, in a few places, are remnants of large forests with typical rain forest trees such as *Bridelia ferruginea* and *Ficus capensis*, although Adjanohoun (1962) ascribes their origin to wooded savannas.

Two hydromorphic grassland communities have been recognized in Sierra Leone: "Riverain Grasslands" and "Grass-Herb Swamps" (Cole 1968). Riverain grasslands occur extensively along the lower courses of the Sewa and Waanje rivers and are dominated by grasses, sedges, and herbs, the only commonly occurring shrubs being *Anthostema senegalense*, *Clappertonia ficifolia*, *Croton scarciesii*, *Dissotis cornifolia*, and *Hyptis spicigera*. The wetter parts of the coastal savanna zone of Liberia around Greenville, Harper, and River Cress are also mapped in this class. These are assumed to be areas of derived savanna (Voorhoeve 1968) or hydromorphic grassland.

Origin

In its northernmost occurrences, this land cover class forms one of the main ecotones between the humid tropical forest and the wetter savanna woodlands to the north. Characteristically, this area contains both rain forest and savanna elements. The former occur as small areas of semideciduous rain forest and gallery forest along rivers that are set in large tracts of wooded savanna.

One body of researchers says that much of the savanna is maintained by fire and terms it "derived savanna." They argue that, if burning ceased, the rain forest would invade these areas, and they cite evidence of remnant humid tropical forest areas in the savanna and the existence of fire-tolerant and fire-resistant trees to support this hypothesis.

Another body of researchers questions this interpretation on the basis of the sharpness of the forest-savanna boundary and the lack of penetration of fires into moist forests. This lends credence to the hypothesis that, before cultivation and burning in these areas, a continuum of vegetation existed from rain forest to dry savanna (Hopkins 1974; Keay 1952; Swaine, Hall, and Lock 1976).

Population and Agriculture

Population density varies in the area covered by this class. Where it is low, forest and woodland areas dominate, but in areas of greater population density, forests are cleared extensively. Many large cities and towns in the region are situated in this class, and around them the anthropogenic impact is severe. Due to the continued forest clearance and fuelwood demand within the zone, this mosaic is a principal source of fuelwood in the region. Not only is local demand from within the class great, but the land cover class is probably a net exporter of fuelwood to the cities and wood-deficit areas of the region as a whole.

Extensive subsistence and cash crop agriculture occurs in the area covered by this class. Subsistence agriculture is dominated by upland and swamp rice cultivation in Sierra Leone, Liberia, and the south of Côte d'Ivoire. In Guinea and the central areas of Côte d'Ivoire, the main subsistence crop is rice, although significant areas of bananas, plantains, cassava, and yams are grown. These latter four crops are also typical of the mosaic in Ghana, Togo, and Nigeria.

The main cash crops in these forest zones are cocoa, coffee, oil palm, and rubber, often grown on extensive plantations owned by companies or wealthy farmers. However, extensive peasant cultivation of these crops also occurs at a much smaller scale. In Sierra Leone, many riverain grasslands are important rice-growing areas, whereas in Côte d'Ivoire and Liberia, important agricultural areas are fewer but produce a broader range of crops.

The main control on the distribution of forest regrowth vegetation adjacent to and within the forest zone is primarily the continual clearance and burning during the dry season, which is part of the local farming systems. Clearance and burning are carried out for three reasons: to prepare fields, to promote grazing, and to flush game.

In Sierra Leone, such areas are mapped as "Forest Regrowth and Farmland" (Cole 1968; FAO 1981; Clarke 1966). This area was formerly closed forest but was opened to timber exploitation in the nineteenth century. Since then it has been extensively cleared and burned to provide land for upland rice, coffee, cocoa, and oil palm cultivation (Millington 1987).

The ecological communities within this class range from recently cleared farms to relatively mature secondary forest regrowth. Secondary forest and bush regrowth is cleared between January and April, and the wood is dried and burned just before the seasonal rains start in May. The dominant crop, upland rice, is grown in a mixed cropping system and harvested from October to December.

The fields are sometimes used for a second year, but often are allowed to revert to bush after a single year's cultivation. After 2 to 3 years, a dense, low thicket forms, composed of herbs, razor grass (*Scleria barteri*), climbers, and coppicing stumps left during clearance (particularly *Canthium glabrifolium*, *Craterispermum laurinum*, and *Musanga cecropioides*). Trees such as *Albizia*

zygia, *Dichrostachys glomerata*, *Harungana madagascariensis*, *Trema guineensis*, and *Xylopia quintasii* grow very fast in the early successional stages in response to large amounts of sunlight. Other common forest trees at this stage are *Ceiba pentandra*, *Chlorophora regia*, *Cola nitida*, *Elaeis guineensis*, *Mangifera indica*, and *Sterculia tragacantha*.

Similar agricultural systems to that described for Sierra Leone exist throughout this mosaic, and can be termed bush-fallowing agricultural systems. In these, the forest regenerates in the fallowing period, allowing soil nutrient levels and structure to recover. For instance, in Côte d'Ivoire, the "Secteur Préforestier" (Monnier 1983) is part of this class; it falls entirely into the "Zone de l'igname." In Togo, this area is very extensive, stretching from the coastal plains to 8°30' N, and falls into four agricultural zones mapped by Gu-Konu (1981): (a) the southern parts of the maize-sorghum-yam zone, (b) the bean-maize-yam zone, (c) the cassava-maize-yam zone, and (d) the cassava-maize zone.

Similar subsistence cropping patterns exist in Benin and Nigeria, where the main cash crops are cocoa (Liberia, Nigeria, Togo, and Sierra Leone); coffee (Côte d'Ivoire, Liberia, Togo, and Sierra Leone); cotton (Côte d'Ivoire, Liberia, Togo, and Sierra Leone); oil palm (Liberia, Nigeria, and Sierra Leone); rice (Côte d'Ivoire, Liberia, and Sierra Leone); sugarcane (Liberia and Togo); and tobacco (Côte d'Ivoire) (Berron and Vennetier 1983; Gu-Konu 1981; Nwafor 1982; von Gnielinski 1972).

Class 81—Mangrove

Mangrove swamps are common along the West African coast, but are particularly important in these areas:

- From the Casamance River in Senegal to the Rio Nunez in Guinea, a coastal strip that includes the extensive mangrove of the Bijagós Islands and the Cacine, Cacheu, Corubal, and Gêba rivers in Guinea-Bissau and the Komponi river in Guinea
- Around Conakry in Guinea
- In Sierra Leone, especially in the estuaries of the Scarcies, Rokel, and Jong rivers, and around Sherbro Island
- As a series of small isolated occurrences in Liberia and Côte d'Ivoire
- In the Volta Delta in Ghana
- Along the Nigerian coast, especially in the Niger and Cross deltas.

These areas account for 63,814 square kilometers, about 3.1 percent of the region. The largest areas are in Nigeria (35,200 square kilometers), Liberia (5,533 square kilometers), Guinea (5,480 square kilometers), and Ghana (5,111 square kilometers).

Mangrove swamps occur only in tidal and brackish water along the coast in places where adequate shelter from storm waves usually exists. They commonly occur fringing wide estuaries, in shallow creeks, and behind islands. Mangrove is swamp forest in which a dense network of tree trunks, stilt roots, and pneumatophores usually form an impenetrable thicket. The trees root directly in the mud of the tidal flats. The canopy cover is high and mostly evergreen, generally with no ground cover.

The mangrove forests of West Africa are dominated by three genera: *Avicennia, Laguncularia,* and *Rhizophora.* Spatial relations among the genera, however, are not always clear. Mangrove trees of the genus *Avicennia* generally occur inland of *Rhizophora.* However, along the Gambia River, for instance, no obvious zonation occurs (Giglioli and Thornton 1965), and the Niger Delta is dominated by a mixture of *Rhizophora harrisonii, R. mangle, R. racemosa,* and *Avicennia germinans.*

In Sierra Leone, mangrove swamp zonation is thought to be controlled by soil conditions, topography, and water salinity (Cole 1968). In Sierra Leone, *Rhizophora racemosa* is the pioneer species on silty, fibrous soil, while on sandy soils the pioneer species are *Avicennia africana, R. mangle,* and *R. harrisonii.* Trees at the mouths of rivers attain maximum heights of 7.5 to 10 meters, while upriver, at the limit of tidal incursion, they grow to 20 meters.

Inland from the area of pioneer trees, the open woodland changes to a dense, thicketlike swamp forest often dominated by *Conocarpus erectus* and *Laguncularia racemosa.* In these areas, grasses and sedges characteristic of freshwater swamps invade to form an undergrowth, whereas other areas display a mixture of mangrove and freshwater swamp forest tree and shrub species.

Mangrove productivity and biomass are very closely correlated with mean annual rainfall. The greatest productivity exists in humid tropical forest areas, although they show only moderate seasonality (figure 9-9). This relation can be illustrated using tree height as an indicator of the growing stock. In the Niger Delta, where the mean annual rainfall is more than 4,000 millimeters, *Rhizophora mangle* stands attain 45 meters in height (Rosevear 1947). But at the northern and southern limits of mangrove on the west coast of Africa, tree heights are much lower. At the northern limit on Ile Tidra, Mauritania, 19°50′ N, where mean annual rainfall is 100 millimeters, *Avicennia* spp. trees are only a few meters high (Chapman 1977). At the southern limit in Benguela, Angola, 12°30′ S, where mean annual rainfall is approxi-

Figure 9-9. NDVI **Profile, Mangrove (Class 81)**

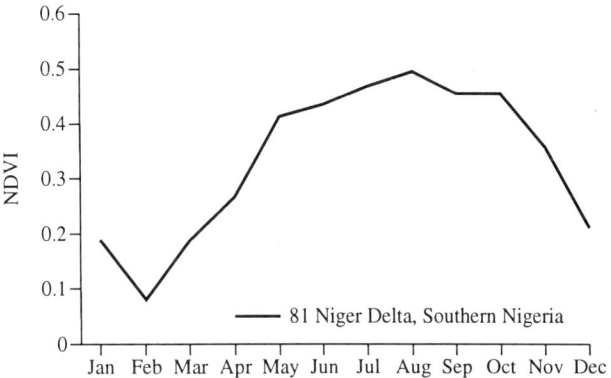

mately 150 millimeters, mangrove again is quite stunted (Barbosa 1970).

Many of the mangrove swamps in the region have been cleared for swamp rice cultivation. This clearance has had an important impact on woody biomass stock, as is well illustrated by Sierra Leone. Although Temne farmers cultivated swamp rice in northern Sierra Leone estuaries in the nineteenth century, the main impetus to clear mangrove swamp vegetation for cultivation came with twentieth century colonial penetration of the interior (Richards 1985; Millington 1987). Clearance of mangrove swamps leads to a buildup of sulfuric acid to toxic levels in the thionic fluvisols of the swamps. This acidity initially restricted rice cultivation, but research by the West African Rice Development Association (WARDA) in Sierra Leone on soil management techniques and the introduction of specific rice varieties has led to the widespread use of mangrove swamps for rice farming.

Due to the concentration of population along the West African coast, pressure on mangrove swamps can locally be very great. Pressure arises from land clearance for rice cultivation, from fuelwood exploitation, from the preferential use of mangrove wood for smoking fish and baking bread, and from its use in building construction. Areas of mangrove adjacent to towns suffer severe exploitation, and in these areas fuelwood supply problems undoubtedly exist, accentuated by the value placed on mangrove wood because of its high calorific value.

Land Cover Class Tables

Tables 9-9 through 9-17, beginning on page 80, present summaries for each land cover class of the area, showing growing stock and sustainable yield for the West African coast nations of Benin, Ghana, Guinea-Bissau, Guinea, Côte d'Ivoire, Liberia, Nigeria, Sierra Leone, and Togo.

References

Every effort has been made to facilitate access to the documents listed here. Some documents, however, lack full bibliographic information because it was unavailable; also, some documents are of limited circulation.

Adjanohoun, E. 1962. "Etude phytosociologique des savanes de basses Cote d'Ivoire (savanes lagunaires)." *Vegetatio* 11:1–38.

Afolayan, A., and K. M. Barbour. 1982. "Population Distribution and Density." In K. M. Barbour, J. S. Oguntoyinbo, J. O. C. Onyemelukwe, and J. C. Nwafor. *Nigeria in Maps*. London: Hodder and Stoughton.

Areola, O. 1982a. "Vegetation." In K. M. Barbour, J. S. Oguntoyinbo, J. O. C. Onyemelukwe, and J. C. Nwafor. *Nigeria in Maps*. London: Hodder and Stoughton.

Areola, O. 1982b. "Land Use." In K. M. Barbour, J. S. Oguntoyinbo, J. O. C. Onyemelukwe, and J. C. Nwafor. *Nigeria in Maps*. London: Hodder and Stoughton.

Arnaud, J.-C. 1983. "Economie du bois." *Atlas de la Côte d'Ivoire*. Paris: Les Editions Jeune Afrique.

Barbosa, L. A. 1970. *Carta fitogeográfica de Angola*. Luanda: Inst. Invest. Cient. Angola.

Berron, H., and P. Vennetier. 1983. "Agriculture." *Atlas de la Côte d'Ivoire*. Paris: Les Editions Jeune Afrique.

Bianchi, H. 1986. *Assistance au développement forestier, Guinée-Bissau: Planification forestière*. TCP/GBS/4506(A). Rome: FAO.

Brookman-Amissah, J. 1987. *Multipurpose Management of Woody Vegetations in the Northern Regions of Ghana*. Kumasi.

Brunel, J. F. 1981. "Végétation." In Y. E. Gu-Konu and G. Laclavère. *Togo*. Paris: Les Editions Jeune Afrique.

Chapman, V. J. (ed.) 1977. *Wet Coastal Ecosystems*. Amsterdam: Elsevier Scientific.

Clarke, J. I. 1966. "Vegetation." *Atlas of Sierra Leone*. London: Hodder & Stoughton.

Clayton, W. D. 1958. "Secondary Vegetation and the Transition to Savanna Near Ibadan, Nigeria." *Journal of Ecology* 46(2): 217–38.

Clayton, W. D. 1961. "Derived Savanna in Kabba Province, Nigeria." *Journal of Ecology* 49(3): 595–604.

Cole, N. H. A. 1968. *The Vegetation of Sierra Leone*. Njala, Sierra Leone: Njala University College Press.

FAO (Food and Agriculture Organization of the United Nations). 1981. *Land Systems of Sierra Leone*. FAO-Government of Sierra Leone Land Resources Survey Project, Report 1, Freetown.

Giglioli, M. E. C., and I. Thornton. 1965. "The Mangrove Swamps of Keneba, Lower Gambia River Basin. I. Descriptive Notes on the Climate, the Mangrove Swamps and the Physical Composition of Their Soils." *Journal of Applied Ecology* 2(1): 81–103.

Gu-Konu, Y. E. 1981a. "Population." In Y. E. Gu-Konu and G. Laclavère. *Togo*. Paris: Les Editions Jeune Afrique.

Gu-Konu, Y. E. 1981b. "Agriculture." In Y. E. Gu-Konu and G. Laclavère. *Togo*. Paris: Les Editions Jeune Afrique.

Gu-Konu, Y. E., and G. Laclavère. 1981. *Togo*. Paris: Les Editions Jeune Afrique.

Hall, J. B., and M. D. Swaine. 1976. "Classification and Ecology of Closed-Canopy Forest in Ghana." *Journal of Ecology* 64:913–51.

Hambler, D. J. 1964. "The Vegetation of Granite Outcrops in Western Nigeria." *Journal of Ecology* 52(3): 573–94.

Hopkins, B. J. 1974. *Forest and Savanna*. 2d ed. London: Heinemann.

Jansen, J. W. A. 1972. "Vegetation." In S. von Gnielinski. *Liberia in Maps*. London: University of London Press.

Jenik, J., and J. B. Hall. 1976. "Plant Communities of the Accra Plains, Ghana." *Folia Geobotanica et Phytotaxonomica* 11:163–212.

Jones, E. W. 1963. "The Forest Outliers in the Guinea Zone of Northern Nigeria." *Journal of Ecology* 51(2): 415–34.

Keay, R. W. J. 1952. "*Isoberlinia* Woodlands in Nigeria and Their Flora." *Lejeunia* 16: 17–26.

Keay, R. W. J. 1959. "Derived Savanna—Derived from What?" *Bulletin d'Institut Français d'Afrique Nord*, Série A 21: 427–38.

Lawson, G. W. 1968. "Ghana." *Acta Phytogeographica Suecica* 54:74–76.

Lecomte, G., and N. Monnier. 1983. "Population." *Atlas de la Côte d'Ivoire*. Paris: Les Editions Jeune Afrique.

Menaut, J. C. 1983. "The Vegetation of African Savannas." In F. Boulière, ed., *Tropical Savannas*. Amsterdam: Elsevier.

Menaut, J. C., and J. César. 1979. "Structure and Primary Productivity of Lamto Savannas, Ivory Coast." *Ecology* 60(6):1197–1210.

Menaut, J. C., and J. César. 1982. "The Structure and Dynamics of a West African Savanna." In B. J. Huntley and B. H. Walker, eds., *Ecology of Tropical Savannas*. Berlin: Springer-Verlag.

Millington, Andrew C. 1987. "Environmental Degradation, Soil Conservation and Agricultural Policies in Sierra Leone, 1895–1984." In David D. Anderson and Richard Grove, eds., *Conservation in Africa: People,*

Policies, and Practice. Cambridge: Cambridge University Press.

Millington, Andrew C., Felix Helmisch, and Gaest Rhebergen. 1985. "Inland Valley Swamps and Bolis in Sierra Leone: Hydrological and Pedological Considerations for Agricultural Development." *Zeitschrift für Geomorphologie*, Supplement Band 52, 201–22.

Monnier, Y. 1983. "Végétation." In *Atlas de la Côte d'Ivoire*. Paris: Les Editions Jeune Afrique.

Morison, C. G. T., A. C. Hoyle, and J. F. Hope-Simpson. 1948. "Tropical Soil-Vegetation Catenas and Mosaics." *Journal of Ecology* 36(1): 1–84.

Nwafor, J. C. 1982a. "Agricultural Zones." In K. M. Barbour, J. S. Oguntoyinbo, J. O. C. Onyemelukwe, and J. C. Nwafor. *Nigeria in Maps*. London: Hodder and Stoughton.

Nwafor, J. C. 1982b. "Major Cash Crops and Plantations." In K. M. Barbour, J. S. Oguntoyinbo, J. O. C. Onyemelukwe, and J. C. Nwafor. *Nigeria in Maps*. London: Hodder and Stoughton.

Swaine, M. D., J. B. Hall, and J. M. Lock. 1976. "The Forest-Savanna Boundary in West-Central Ghana." *Ghana Journal of Science* 16:35–52.

Swaine, M. D., D. Lieberman, and J. B. Hall. 1990. "Structure and Dynamics of a Tropical Dry Forest in Ghana." *Vegetatio* 88:31–51.

von Gnielinski, S. 1972. *Liberia in Maps*. London: University of London Press.

Voorhoeve, A. G. 1968. "Liberia." *Acta Phytogeographica Suecica* 54:74–76.

White, F. 1965. "The Savanna Woodlands of the Zambezian and Sudanian Domains. An Ecological and Phytogeographical Comparison." *Webbia* 19: 651–81.

White, F. 1983. "The Vegetation of Africa." *Natural Resources Research Series* 20. Paris: UNESCO/AETFAT/UNSO (United Nations Educational, Scientific and Cultural Organization/Association pour l'Etude Taxonomique de la Flore de l'Afrique Tropicale/United Nations Sudano-Sahelian Office).

Table 9-9. Land Cover Classes—Benin (West African Coast Region)

Land cover class		Area km²	Area Percent	Growing stock Thousand tonnes	Growing stock Percent	Sustainable yield Thousand tonnes per year	Sustainable yield Percent
	12	2,108	1.84	478.52	0.18	21.08	0.32
1		2,108	1.84	478.52	0.18	21.08	0.32
	22	211	0.18	69.63	0.03	2.11	0.03
2		211	0.18	69.63	0.03	2.11	0.03
	43	211	0.18	357.01	0.13	4.43	0.07
4		211	0.18	357.01	0.13	4.43	0.07
	52	211	0.18	453.65	0.17	13.29	0.20
5		211	0.18	453.65	0.17	13.29	0.20
	62	11,435	10.00	30,531.45	11.21	880.50	13.51
	64	40,839	35.70	109,040.13	40.02	1,878.59	28.82
	65	14,755	12.90	39,395.85	14.46	708.24	10.86
6		67,029	58.60	178,967.43	65.69	3,467.33	53.19
	74	5,849	5.11	9,849.72	3.62	111.13	1.70
	76	38,046	33.26	64,069.46	23.52	722.87	11.09
7		43,895	38.37	73,919.18	27.14	834.00	12.79
	81	738	0.65	18,213.84	6.68	2,176.36	33.39
8		738	0.65	18,213.84	6.68	2,176.36	33.39
Total		114,403	100.00	272,459.26	100.00	6,518.61	100.00
(Percentage of region)		(5.60)		(5.08)		(2.53)	

Note: In the following tables, details may not add to totals because of rounding.
Source: Authors' calculations from data bases derived from land cover classification and table 4-1.

Table 9-10. Land Cover Classes—Ghana (West African Coast Region)

Land cover class		Area km²	Area Percent	Growing stock Thousand tonnes	Growing stock Percent	Sustainable yield Thousand tonnes per year	Sustainable yield Percent
	12	1,054	0.45	239.26	0.04	10.54	0.05
1		1,054	0.45	239.26	0.04	10.54	0.05
	22	685	0.29	226.05	0.04	6.85	0.03
2		685	0.29	226.05	0.04	6.85	0.03
	41	263	0.11	365.57	0.06	5.52	0.02
	43	158	0.07	267.34	0.05	3.32	0.01
	44	316	0.13	44.56	0.01	11.06	0.05
4		737	0.31	677.46	0.12	19.90	0.08
	52	1,001	0.42	2,152.15	0.37	63.06	0.28
5		1,001	0.42	2,152.15	0.37	63.06	0.28
	62	18,233	7.70	48,682.11	8.43	1,403.94	6.31
	64	47,110	19.89	125,783.70	21.77	2,167.06	9.74
	65	22,975	9.70	61,343.25	10.62	1,102.80	4.95
6		88,318	37.29	235,809.06	40.81	4,673.80	21.00
	74	7,694	3.25	12,956.70	2.24	146.19	0.66
	75	18,812	7.94	31,679.41	5.48	357.43	1.61
	76	99,383	41.97	167,360.97	28.97	1,888.28	8.48
7		125,889	53.16	211,997.08	36.69	2,391.90	10.75
	81	5,111	2.16	126,139.48	21.83	15,072.34	67.71
	84	53	0.02	524.70	0.09	21.04	0.09
8		5,164	2.18	126,664.18	21.92	15,093.38	67.80
Lakes		13,964	5.90	0.00	0.00	0.00	0.00
Total		236,812	100.00	577,765.24	100.00	22,259.43	100.00
(Percentage of region)		(11.59)		(10.78)		(8.66)	

Source: Authors' calculations from data bases derived from land cover classification and table 4-1.

Table 9-11.   Land Cover Classes—Guinea-Bissau (West African Coast Region)

Land cover class	Area *km²*	*Percent*	Growing stock *Thousand tonnes*	*Percent*	Sustainable yield *Thousand tonnes per year*	*Percent*
22	3,425	11.38	1,130.25	0.96	34.25	0.38
2	3,425	11.38	1,130.25	0.96	34.25	0.38
44	263	0.87	37.08	0.03	9.21	0.10
4	263	0.87	37.08	0.03	9.21	0.10
62	1,475	4.90	3,938.25	3.36	113.58	1.27
64	5,480	18.21	14,631.60	12.49	252.08	2.81
65	422	1.40	1,126.74	0.96	20.26	0.23
6	7,377	24.52	19,696.59	16.81	385.92	4.31
74	14,491	48.16	24,402.84	20.83	275.33	3.07
76	1,739	5.78	2,928.48	2.50	33.04	0.37
7	16,230	53.94	27,331.32	23.33	308.37	3.44
81	2,793	9.28	68,931.24	58.85	8,236.56	91.78
8	2,793	9.28	68,931.24	58.85	8,236.56	91.78
Total	30,088	100.00	117,126.48	100.00	8,974.29	100.00
(Percentage of region)	(1.47)		(2.18)		(3.49)	

Source: Authors' calculations from data bases derived from land cover classification and table 4-1.

Table 9-12.   Land Cover Classes—Guinea (West African Coast Region)

Land cover class	Area *km²*	*Percent*	Growing stock *Thousand tonnes*	*Percent*	Sustainable yield *Thousand tonnes per year*	*Percent*
0	53	0.02	0.00	0.00	0.00	0.00
22	5,006	2.09	1,651.98	0.25	50.06	0.21
2	5,006	2.09	1,651.98	0.25	50.06	0.21
43	632	0.26	1,069.34	0.16	13.27	0.05
44	0	0.00	0.00	0.00	0.00	0.00
45	0	0.00	0.00	0.00	0.00	0.00
4	632	0.26	1,069.34	0.16	13.27	0.05
52	1,581	0.66	3,399.15	0.52	99.60	0.41
5	1,581	0.66	3,399.15	0.52	99.60	0.41
62	2,793	1.17	7,457.31	1.14	215.06	0.89
63	53	0.02	141.51	0.02	2.33	0.01
64	103,283	43.20	275,765.61	42.33	4,751.02	19.56
65	24,661	10.32	65,844.87	10.11	1,183.73	4.87
6	130,790	54.71	349,209.30	53.60	6,152.14	25.33
74	63,024	26.36	106,132.42	16.29	1,197.46	4.93
75	1,001	0.42	1,685.68	0.26	19.02	0.08
76	31,512	13.18	53,066.21	8.15	598.73	2.46
7	95,537	39.96	160,884.31	24.70	1,815.21	7.47
81	5,480	2.29	135,246.40	20.76	16,160.52	66.53
8	5,480	2.29	135,246.40	20.76	16,160.52	66.53
Total	239,079	100.00	651,460.48	100.00	24,290.80	100.00
(Percentage of region)	(11.71)		(12.15)		(9.45)	

Source: Authors' calculations from data bases derived from land cover classification and table 4-1.

Table 9-13. Land Cover Classes—Côte d'Ivoire (West African Coast Region)

Land cover class	Area km²	Area Percent	Growing stock Thousand tonnes	Growing stock Percent	Sustainable yield Thousand tonnes per year	Sustainable yield Percent
22	949	0.29	313.17	0.05	9.49	0.05
23	685	0.21	226.05	0.03	6.85	0.04
2	1,634	0.50	539.22	0.08	16.34	0.09
41	105	0.03	145.95	0.02	2.21	0.01
4	105	0.03	145.95	0.02	2.21	0.01
52	1,475	0.45	3,171.25	0.46	92.92	0.49
5	1,475	0.45	3,171.25	0.46	92.92	0.49
64	16,546	5.10	44,177.82	6.41	761.12	4.03
65	43,474	13.41	116,075.58	16.83	2,086.75	11.05
6	60,020	18.51	160,253.40	23.24	2,847.87	15.08
74	59,598	18.38	100,363.03	14.55	1,132.36	6.00
75	47,584	14.68	80,131.46	11.62	904.10	4.79
76	150,023	46.28	252,638.73	36.64	2,850.44	15.10
7	257,205	79.34	433,133.22	62.81	4,886.90	25.89
81	3,741	1.15	92,327.88	13.39	11,023.21	58.44
8	3,741	1.15	92,327.88	13.39	11,023.21	58.44
Total	324,180	100.00	689,570.92	100.00	18,878.44	100.00
(Percentage of region)	(15.87)		(12.86)		(7.34)	

Source: Authors' calculations from data bases derived from land cover classification and table 4-1.

Table 9-14. Land Cover Classes—Liberia (West African Coast Region)

Land cover class	Area km²	Area Percent	Growing stock Thousand tonnes	Growing stock Percent	Sustainable yield Thousand tonnes per year	Sustainable yield Percent
22	158	0.17	52.14	0.02	1.58	0.01
2	158	0.17	52.14	0.02	1.58	0.01
43	263	0.28	445.00	0.15	5.52	0.03
4	263	0.28	445.00	0.15	5.52	0.03
52	2,266	2.41	4,871.90	1.68	142.76	0.78
5	2,266	2.41	4,871.90	1.68	142.76	0.78
75	24,925	26.50	41,973.70	14.48	473.57	2.59
76	60,494	64.31	101,871.90	35.13	1,149.39	6.30
7	85,419	90.81	143,845.60	49.61	1,622.96	8.89
81	5,533	5.88	136,554.44	47.10	16,316.82	89.37
84	422	0.45	4,177.80	1.44	167.53	0.92
8	5,955	6.33	140,732.24	48.54	16,484.35	90.29
Total	94,061	100.00	289,946.87	100.00	18,257.17	100.00
(Percentage of region)	(4.61)		(5.41)		(7.10)	

Source: Authors' calculations from data bases derived from land cover classification and table 4-1.

Table 9-15. Land Cover Classes—Nigeria (West African Coast Region)

Land cover class		Area		Growing stock		Sustainable yield	
		km^2	Percent	Thousand tonnes	Percent	Thousand tonnes per year	Percent
0		422	0.05	0.00	0.00	0.00	0.00
	12	33,461	3.80	7595.65	0.31	334.61	0.24
1		33,461	3.80	7595.65	0.31	334.61	0.24
	22	1,792	0.20	591.36	0.02	17.92	0.01
	23	4,005	0.46	1,321.65	0.05	40.05	0.03
2		5,797	0.66	1,913.01	0.07	57.97	0.04
	41	105	0.01	145.95	0.01	2.21	0.00
	43	949	0.11	1,605.71	0.07	19.93	0.01
	44	161,827	18.40	22,817.61	0.95	5,663.95	4.07
4		162,881	18.52	24,569.27	1.02	5,686.08	4.08
	52	11,909	1.35	25,604.35	1.06	750.27	0.54
5		11,909	1.35	25,604.35	1.06	750.27	0.54
	62	154,134	17.53	411,537.78	17.07	11,868.32	8.52
	63	2,793	0.32	7,457.31	0.31	122.89	0.09
	64	153,765	17.48	410,552.55	17.03	7,073.19	5.08
	65	121,252	13.79	323,742.84	13.43	5,820.10	4.18
6		431,944	49.12	1,153,290.48	47.84	24,884.50	17.87
	73	790	0.09	1,330.36	0.06	15.01	0.01
	74	53,064	6.03	89,359.78	3.71	1,008.22	0.72
	75	7,641	0.87	12,867.44	0.53	145.18	0.10
	76	132,739	15.09	223,532.48	9.27	2,522.04	1.81
7		194,234	22.09	327,090.06	13.57	3,690.45	2.64
	81	35,200	4.00	868,736.00	36.03	103,804.80	74.51
	84	263	0.03	2,603.70	0.11	104.41	0.07
8		35,463	4.03	871,339.70	36.13	103,909.21	74.58
Lakes		3,320	0.38	0.00	0.00	0.00	0.00
Total		879,431	100.00	2,411,402.51	100.00	139,313.09	100.00
(Percentage of region)		(43.06)		(44.98)		(54.17)	

Source: Authors' calculations from data bases derived from land cover classification and table 4-1.

Table 9-16. Land Cover Classes—Sierra Leone (West African Coast Region)

Land cover class		Area		Growing stock		Sustainable yield	
		km^2	Percent	Thousand tonnes	Percent	Thousand tonnes per year	Percent
	22	474	0.67	156.42	0.07	4.74	0.03
2		474	0.67	156.42	0.07	4.74	0.03
	43	2,002	2.85	3,387.38	1.49	42.04	0.28
4		2,002	2.85	3,387.38	1.49	42.04	0.28
	52	1,475	2.10	3,171.25	1.40	92.93	0.62
5		1,475	2.10	3,171.25	1.40	92.93	0.62
	64	2,793	3.97	7,457.31	3.29	128.48	0.86
	65	53	0.08	4.00	0.00	2.54	0.02
6		2,846	4.05	7,461.31	3.29	131.02	0.88
	74	30,563	43.47	51,468.09	22.70	580.70	3.89
	75	369	0.52	621.40	0.27	7.01	0.05
	76	27,981	39.80	47,120.00	20.78	531.64	3.56
7		58,913	83.79	99,209.49	43.75	1,119.35	7.50
	81	4,592	6.53	113,330.56	49.98	13,541.81	90.68
	84	4	0.01	39.60	0.02	1.59	0.01
8		4,596	6.54	113,370.16	50.00	13,543.40	90.69
Total		70,306	100.00	226,760.02	100.00	14,933.47	100.00
(Percentage of region)		(3.44)		(4.23)		(5.81)	

Source: Authors' calculations from data bases derived from land cover classification and table 4-1.

Table 9-17. Land Cover Classes—Togo (West African Coast Region)

Land cover class	Area		Growing stock		Sustainable yield	
	km²	*Percent*	*Thousand tonnes*	*Percent*	*Thousand tonnes per year*	*Percent*
12	1,792	3.31	406.78	0.33	17.92	0.48
1	1,792	3.31	406.78	0.33	17.92	0.48
22	316	0.58	104.28	0.08	3.16	0.08
2	316	0.58	104.28	0.08	3.16	0.08
44	738	1.36	104.06	0.08	25.83	0.69
4	738	1.36	104.06	0.08	25.83	0.69
62	4,268	7.89	11,395.56	9.19	328.64	8.75
64	11,224	20.74	29,968.08	24.16	516.30	13.74
65	6,165	11.39	16,460.55	13.27	295.92	7.88
6	21,657	40.02	57,824.19	46.62	1,140.86	30.37
74	8,958	16.55	15,085.27	12.16	170.20	4.53
75	1,107	2.05	1,864.19	1.50	21.03	0.56
76	18,865	34.86	31,768.66	25.61	358.44	9.54
7	28,930	53.46	48,718.12	39.27	549.67	14.63
81	685	1.27	16,905.80	13.63	2,020.07	53.76
8	685	1.27	16,905.80	13.63	2,020.07	53.76
Total	54,118	100.00	124,063.23	100.00	3,757.51	100.00
(Percentage of region)	(2.65)		(2.31)		(1.46)	

Source: Authors' calculations from data bases derived from land cover classification and table 4-1.

10

The Horn of Africa

John Kirkby

This chapter presents a detailed description of the most important land cover classes in this region. Helpful figures in other chapters include figure 3-1 (cloud cover); figures 3-2, 3-3, and 3-4 (NDVI summary land cover profiles); figure 3-5 (regional summary map of land cover classes); figures 7-1 and 7-2 (continental maps of growing stock and sustainable yield); and the "Regional Land Cover Class Map of East Africa" at the end of this volume.

Helpful tables in other chapters include table 3-2 (land cover classes); table 4-1 (data and sources for growing stock and sustainable yield); and table 6-3 (Horn of Africa estimated woody biomass by summary class).

Class 0—Desert

Occupying one-quarter of this region, with more than 1 million square kilometers, Desert is the largest of all the land cover classes. Desert covers a significant area of all four countries in the region—Sudan, Ethiopia, Djibouti, and Somalia. The northern third of Sudan is almost entirely desert, and this continues into northern Eritrea and the Danakil of Ethiopia. Virtually all of Djibouti is desert, and this area extends into the northern coastal strip of Somalia. Extensive areas of desert occur on the eastern coast of northern Somalia, particularly in the Bari and Mudug regions. In Ethiopia, areas of desert occur within Ogaden and further patches exist on the borders with Kenya and Somalia. Estimated growing stock and sustainable yield are both 0 percent of the regional resource (table 6-3).

Most of the desert is below 300 meters altitude; indeed, extensive areas of the Danakil are below sea level. Annual precipitation is less than 150 millimeters and is extremely variable from year to year. In parts of the Libyan Desert of Sudan, rainfall is a very rare

event. Soils are at best skeletal and completely absent from the two large areas of erg (mobile sand dunes) in the southwestern Libyan Desert and on part of the Egyptian frontier in the Nubian Desert. Extensive areas of bare rock and rock debris without any soil formation (reg) exist in the Libyan and Nubian deserts, the Red Sea Hills of Sudan, inland areas of Eritrea, the Danakil Alps, and on the northwestern coast of Somalia. Saline soils occur in the Danakil Depression of Ethiopia and on parts of the northeastern coast of Somalia.

Phenological curves for the Danakil and Nubian deserts show expected low rates of activity, with figures for the Danakil being less than those for the southern Nubian Desert. Minima of –0.02 (Danakil) and +0.01 (Nubian) occur from July to September, with maxima of +0.02 and +0.01 in March and November (Danakil), and +0.06 in April and December (Nubian).

Vegetation is virtually absent from much of the area, particularly the Libyan Desert; nevertheless, plant cover does exist, if only temporarily, in some situations. After rare rain episodes, and even after long droughts, grasses (*gizzu*) may briefly flourish in normally unvegetated areas. More substantial plants that are adapted to desert conditions may survive in a dormant state until rainfall occurs.

White (1983) identifies semidesert conditions in the northeastern desert of Sudan. Hemming (1961) finds more advanced vegetation than is generally implied by the term "desert" in the coastal zone of northern Eritrea. Mangrove occurs in a discontinuous strip of coastal swamp along parts of the Red Sea coast; for example, a community of *Avicennia* and *Rhizophora*, the latter particularly valued as fuelwood, exists for about 12 kilometers along the coast of Djibouti.

Farther north, in the Red Sea coastal desert of Sudan, particularly to the north of Port Sudan, vegetation occurs on the coast. Salt marshes have an almost complete vegetation cover. Saline areas of wadis support *Juncus arabicus* and *Tamarix mannifera* and nonsaline areas of wadis have *Acacia tortilis* and *Zilla spinosa* (White 1983).

Inland, small areas of trees exist along the larger desert wadis along terraces of the River Nile in northern Sudan and on sites where surface water concentrates. *Acacia,* usually of low stature, occurs on such sites. In higher areas of the Red Sea Hills, such as the Karora Hills, small relict areas of *Juniperus procera* forest occur. In the Libyan Desert too, more substantial vegetation is associated with moister areas of higher land. Elsewhere, at lower elevation, a scrubby vegetation dominated by *Acacia glaucophylla* (*circummarginata*) and *A. etbaica* occurs (Bari 1968). Along the River Nile, and at a number of places near the Red Sea coast in Sudan, irrigation schemes are associated with much greater NDVI values (see Class 22).

Fuelwood is very scarce and threatened by stock browsing, particularly during sustained drought (Moghraby, Ali, and Seed 1987), when drought stress, overgrazing, and cutting may destroy the limited wood stock. Where sand dunes migrate toward oases or toward the desert margin, as in parts of Kordofan, woody species may be overwhelmed by blowing sand. Because herdspeople must use oases for watering animals, the vegetation in these areas is at particular risk from overgrazing and firewood cutting. In the Danakil Desert, an indication of the severe shortage of woodfuel is the widespread use of dung as fuel (Kamweti 1984).

Class 12—Hydromorphic Grassland

This class occupies 2 percent of the region, and exists mainly in Sudan, where it constitutes more than 3 percent of the area. Principal areas of this class lie in southeastern Sudan, particularly in Jonglei and eastern Sobat, but with discontinuous areas in El Bukeyral and near the Blue Nile on the Ethiopian border, mainly below 500 meters.

Almost all hydromorphic grassland in Sudan is clearly within the flood region of the upper Nile, an area where the annual flood affects the flat terrain to different extents, depending on small variations of altitude. Terrain is everywhere extremely flat, with slopes of less than 1° on the infilled Quaternary lake basin, which consists of dark cracking clays associated with hydromorphic soils (vertisols). The relative imperviousness of the soils when moist contributes to the severity of flooding. Precipitation increases from 400 millimeters a year in the north to 1,000 millimeters in the south, and at more than 700 millimeters the soil is unable to absorb all the rainfall.

Although the area is described as grassland, patches of trees, woodland, and thicket exist on drier soils, particularly toward the margin of the class, on the infrequently flooded zone. Bari (1968) identifies three flooded zones, based on frequency and duration of flooding (seasonal, frequent, and permanent). The area of seasonal flooding is mainly grassland. Swamp species such as *Phragmites communis* and *Hyparrhenia rufa* dominate the frequently or permanently flooded areas. Trees are mainly *Acacia* species; *Acacia mellifera* is common at up to 570 millimeters of precipitation and *Acacia seyal* occurs where precipitation is greater than 570 millimeters. Where flooding is deeper or more frequent, trees are absent. Grasses occurring with the *Acacia* trees include *Cymbopogon nervatus,* *Hyparrhenia anthistirioides,* *Schoenefeldia gracilis,* *Schuma ischaemoides,* and *Sorghum purpureo-sericeum* (White 1983). *Hyparrhenia rufa* and *Setaria incrassata* are, however, the dominant grasses (Bari 1968).

It is probable that grassland and thicket have a cyclic relation through time. Cultivation and burning of the grass have combined to reduce the tree cover of the Sudanese Hydromorphic Grasslands, and because grassland is by far the largest component of the ecosystem, fuelwood is limited. The estimated growing stock and sustainable yield are both 0 percent of the regional resource.

Class 13—Ethiopian Montane Steppe

This class occupies 3 percent of Ethiopia, including all land above 3,500 meters on both the eastern and western plateaus. This class is notable in the northern area of the western plateau in the Simien Region and Choke Mountains, and in the mountains overlooking the east-facing escarpments of the Welo, Tigre, and Shewa regions. In the eastern plateau, the main areas are in the Balē Mountains and Ārsī Mountains. The area is coincident with Von Breitenbach's "Mountain Steppe" and part of his "Mountain Savanna" classes (1963).

The terrain of these areas is high mountain and plateau, formed on volcanic rock, with poorly developed skeletal soils and, in many cases, with steep, unstable slopes. Strong winds, cold temperatures, and frost are limiting factors in the development of vegetation (White 1983). Environmental variables such as soil depth, protection from wind, slope, and precipitation are reflected in vegetation cover, but generally the biomass and the height of vegetation decrease with altitude. The lower limit of the Ethiopian Montane Steppe has remnants of a formerly more extensive tree cover, including *Acacia xiphocarpa* (*abyssinica*), *Apodytes* spp., *Hagenia abyssinica, Juniperus procera,* and *Olea africana*. These may form elfin thicket, particularly in less-accessible places such as gorges. Most of the landscape, however, is covered by Cyperaceae and Graminieae.

It is not clear to what extent the grass is primary or the result of repeated burning. At the highest levels, widely spaced tussocks of *Agrostis sclerophylla, Carex monostachya, Deschampsia caespitosa,* and *Limosella africana* are able to withstand high winds and cold temperatures. In protected sites, scattered shrubs such as *Erica arborea* and *Lobelia rhynchopetalum* add diversity to the vegetation, but are susceptible to burning. Browsing, burning, grazing, and fuelwood collection have all contributed to loss of woody biomass from this region, where the rate of tree growth is very slow and virtually no fuelwood remains. Consequently, estimated growing stock and sustainable yield are negligible.

Class 21—Semidesert Wooded Grassland

This class is the second most extensive in the region, occupying 15 percent of the area. Semidesert Wooded Grassland, which we will refer to generally as semidesert, is present in all four countries of this region, and is particularly significant in Sudan and Somalia.

In Sudan, where this class makes up 14 percent of the area, an irregular wedge extends from Kassala on the Ethiopian frontier, broadening to the west in Northern Darfur. It has significant extensions northward into the Libyan Desert, along larger wadis such as El Milk and Hiwa, with outliers such as the El'Atrun Oasis and on higher ground that receives more precipitation. Small outliers also occur in the winter rain areas of the Red Sea Hills, for example on the Erkowit Hills south of Port Sudan and in the Karora Hills. White (1983) and Sudan Survey Department (1954) indicate that all of the Red Sea Hills and much of Kassala Province is "*Acacia tortilis–Maerua crassifolia* Desert Scrub," which must be transitional to "Sahelian Savanna." Wickens, in Bari (1968), suggests that the Red Sea Hills were rapidly deteriorating at that time (probably as a result of grazing pressure), and that subsequent droughts would further stress the vegetation.

Most nondesert areas of northern Somalia and extensive areas of the Wadi Shebelle Valley of central Somalia are Sahelian Semidesert. Class 21 covers one-third of Somalia.

The nondesert area of Djibouti also is in this class. It is particularly important north of the Gulf of Tadjoura. Here, within the savanna and associated with greater rainfall on Mount Goda between 1,300 and 1,800 meters, is a small area (870 hectares) of *Juniperus procera* and *Buxus* forest in the Forêt du Day (Hamrouni 1984). Surrounding the forest are various forms of woody savanna, with *Acacia arabica, A. flava, A. tortilis,* and *Balanites racemosa.* Near the lakes of western Djibouti a small area is dominated by dom palm, *Hyphaene thebaica.*

In Ethiopia, Class 21 extends between the Danakil Desert and the lower plateau steppes through the areas of Keren, Āsmera, Adīgrat, Korem, Desē, Ankober, and northern Hārergē.

Transition from desert may be gradual or sharp, depending on edaphic conditions. Where water is available near the surface, and in topographic situations that encourage water-gathering, such as wadis, well-developed Semidesert Wooded Grassland may occur in nearly pure desert conditions. Similarly, the dry limit of this class may be as little as 80 millimeters precipitation (White 1983) on sands, although the desert may extend as far as the 150-millimeter isohyet. Generally, Semidesert Wooded Grassland gives way to denser woodland at about 250 millimeters.

The rainfall regime varies between 1 and 3 months with significant rainfall. The phenology shows a limited seasonal variation, with two NDVI maxima of approximately 0.11 and 0.12 in November-December and May-June and NDVI minima of 0.06 and 0.07 in April and September, reflecting the dual precipitation maxima.

Most of the class is on land below 600 meters, on soils that are thin and poorly developed. On the edge of the Danakil Desert, they are saline, and in parts of northern Somalia, they have gypsum crusts. In western Sudan, extensive areas are developed on fixed sand dunes, although incursions of mobile dunes from the Libyan Desert are widespread and destroy vegetation.

Much of the class has long been used for grazing and this has strongly influenced the grasses, herbs, and trees. Because of browsing and grazing, many of the more palatable *Acacia* species have been replaced by unpalatable ones such as *Acacia nubica, Cassia acutifolia,* and *Calotropis procera.* The vegetation is subject to brief fluctuations of precipitation and occasional prolonged droughts. Although the plants are resilient, the combined effect of drought and overgrazing has reduced the tree cover, even of valued species such as *Acacia senegal* that normally would be protected as a source of gum arabic.

Crown cover is usually less than 10 percent (White 1983) and continuous areas of taller trees are almost entirely limited to rocky outcrops and water-receiving sites. Where rainfall is about 100 millimeters, woody plant cover rarely exceeds 3 percent. Annual grasses dominate and trees tend to have thin stems and to be of limited height. In Somalia, for example, the drier scrub is rarely more than 3 meters high. Woody species such as *Acacia ehrenbergiana, A. laeta, A. tortilis, Balanites aegyptiaca, Boscia senegalensis, Commiphora africana*, and *Leptadenia pyrotechnica* (White 1983) are more common and better developed toward the wetter margin of the class, but are usually less than 5 meters in height. Sudan Survey Department (1954) and Bari (1968) found that *Acacia tortilis* and *Maerua crassifolia* dominate in the east of Sudan and *Acacia mellifera* dominates in the west. Grasses such as *Panicum turgidum* dominate the more mobile, drier areas of dune crests, whereas *Cenchrus biflorus* covers lower, more stable sands.

The estimated growing stock is 2 percent of the regional total (139 million tonnes), and estimated sustainable yield is 5 percent of the regional total (5.3 million tonnes). Woody biomass is already grossly overused in this class. Growth rates are low, and the social consequences of overuse are severe for the pastoralists of this region. Overgrazing may lead to dune destabilization and to desertification. The class also is susceptible to deterioration during drought years, for example 1972–73 and 1984–85.

Desertification in this area has been examined in Djibouti by Hamrouni (1984); in Somalia by Okafo (1987); in Sudan by Lewis and Berry (1988), by Moghraby, Ali, and Seed (1987), and by Booth (1984); and in Ethiopia by Kebbede and Jacob (1988). Human population may be very dense in this class, considering the limited resource base. In Somalia, for example, this class supports 40 percent of the population in an environment in which cultivation is not possible, so that grazing puts great stress on the trees (Kamweti 1984). Hamilton and Maizels (1989) predict that, with increasing population, the clearance of woodland in northern Ethiopia in this class will continue. Moghraby, Ali, and Seed (1987) attribute severe deterioration to the northward spread of cultivation during the late 1970s.

Class 22—*Acacia* Wooded Grassland

This class occupies approximately 2 percent of the region. It is present in small patches on the lower slopes to the north of the limestone escarpment in northern Somalia. In Ethiopia, it occurs as numerous discontinuous areas north of the western plateau in Welega, Gojam, and Begemder, with further small patches on the lower eastern slopes of the Eritrean hills near Āsmera. A string of small patches exists along the lower Awash River, almost to the border with Djibouti.

The main area of 68,000 square kilometers is in Sudan, where it occurs in many environmentally contrasting areas of the country. Significant occurrences are along the Nile, conspicuously in the Libyan Desert and in the flood area of the south. Other stretches are along the White Nile, Atbara, and Khor Abu Habi tributaries of the Nile. Two areas occur on the coast,

south of Port Sudan, including the Tokar Delta. The Gash Delta, Khashim El Girba, and Gezira-Managli irrigated areas clearly are included in the class. A large complex also is associated with the edaphic grasslands of north Sobat and an isolated occurrence on the slopes of Jebel Gurgei in Darfur.

Ecologically, the class covers a wide range of situations, and in some areas it reflects strong human influence on the landscape. In fact, in the northern half of Sudan, almost every occurrence of Class 22 is directly related to irrigation and crop production. Monoculture of cotton dominates much of this area, and trees may be virtually absent. In contrast, trees form a conspicuous element of the Class 22 landscape occurring in Sobat and western Ethiopia.

White (1983) associates "*Acacia* Wooded Grassland" with precipitation between 250 and 500 millimeters, but in Sudan, Class 22 also occurs in drier and wetter areas. In the Nubian Desert, for example, pumped-water irrigation allows the development of this class in extremely dry areas, whereas in the humid area of Sobat, the greater rainfall is effectively increased by Nile floodwater. Much of the western plateau of Ethiopia receives up to 1,000 millimeters of precipitation.

A variety of species of *Acacia* is associated with Class 22. In Ethiopia, Von Breitenbach (1963) identifies this area as "Lowland Woodland" with an upper story of trees of 5 to 12 meters, forming a more-or-less closed canopy that is sparse enough to admit light to a thicketlike lower story of shrubs of 1 to 3 meters. Characteristic trees are *Acacia amythetophylla*, *A. asak*, *A. mellifera*, *A. senegal*, *A. sieberana*, *A. tortilis*, *A. venosa*, *Ficus acrocarpa*, *Cassia singueana*, *Combretum molle*, and *Ziziphus abyssinica*. Toward 2,000 meters and in drier conditions, *Euphorbia* thicket 8 to 12 meters high, dominated by *Euphorbia abyssinica*, replaces *Acacia* trees.

A range of *Acacia* spp. also is characteristic of Class 22 along the Nile and Atbara; these include *Acacia nilotica*, *A. senegal*, *A. seyal*, and *A. tortilis*. But in those areas where gravity irrigation, pump irrigation, or flush irrigation is possible (Grove 1978), the dominant vegetation is cropland, particularly cotton. In such areas, woodland may be completely absent.

Acacia seyal is associated with a mosaic of grassland and trees in the flood region of Sobat. Locally, *Acacia* spp. form almost impenetrable thicket, separated by areas of grassland on the annually flooded dark cracking clays. A cyclic relation appears to exist between the thicket and grassland of *Cymbopogon nervatus*, *Hyparrhenia anthistirioides*, and *Sorghum purpureo-sericeum* (White 1983). Areas of shallow flooding are associated with *A. seyal* communities; areas of deeper flooding are associated with open grassland, particularly *Setaria incrassata*. Extensive areas of swamp vegetation such as *Cyperus papyrus* cover severely flooded areas.

Bushland and thicket 2 to 3 meters high covers rocky outcrops and water-receiving sites on the lower slopes of Jebel Gurgei. Elsewhere on the pediment of Jebel Gurgei, *Acacia mellifera, Boscia senegalensis, Combretum africanum,* and *Euphorbia candelabrum* form the principal elements of the plant cover.

The estimated growing stock is 0.5 percent of the regional total (20 million tonnes), and the estimated sustainable yield is 0.5 percent of the regional total (770,000 tonnes). Burning, clearance for cultivation, and grazing pressure have severely reduced woody biomass in much of the class in Ethiopia, Sudan, and Somalia. In central areas of eastern Sudan, agriculture dominates the class, particularly on the cracking clays where Graham (1969) notes that removal of trees is associated with clearance for commercial agriculture. A further pressure on trees in this zone is the demand for fuelwood for the three towns of Khartoum, Khartoum North, and Omdurman (Lewis and Berry 1988). Kassas (1970) notes the progressive removal of tree cover southward from Khartoum (see Class 44). The extension of agriculture and need for fuelwood around refugee settlements has significantly increased the rate of tree removal in Sudan since 1970.

Class 41—Dry *Acacia-Commiphora* Bushland and Thicket

Though isolated areas of this class occur within the Semidesert Wooded Grassland (Class 21) of central Sudan, the main occurrence is in the southern and southeastern parts of Hārergē Province, southern Balē and Gamo-Gofa in Ethiopia, extending into the larger portion of southern Somalia. Several patches occur on the coast and slopes of the northern Somalian hills, extending as a discontinuous line into the lowlands of north Hārergē and the Rift Valley of Ethiopia. Small areas follow the lower escarpment in Eritrea and Welo, and a large section occurs in the drier area of Eastern Equatoria Province of Sudan. The class occupies one-quarter of the area of Ethiopia and one-third of the area of Somalia, or 12 percent of the entire region.

Most of the class is below 1,000 meters and grades into semidesert at its drier limits. Soils are skeletal or weakly developed. Although precipitation may be as great as 500 millimeters in southeastern Sudan and southern Somalia, it is as slight as 200 millimeters in the Ogaden of Ethiopia and below 100 millimeters in northern Somalia. Evaporation is rapid and patches of true desert occur within this class in parts of the Ogaden. Precipitation varies greatly from year to year, and prolonged drought lasting several years is characteristic of this zone.

Phenological curves show some variation in this class, with generally small values of approximately 0.10, but according to the length and precipitation during the wet season, marked increases can occur. Figure 10-1 shows an example from the Ogaden in Ethiopia that demonstrates consistently small values.

Deciduous bushland and thicket covers much of this class, normally in the form of dense bushland 3 to 5 meters high, but with scattered emergent trees to 9 meters. Across extensive areas it is virtually impenetrable because of the density of trunks and branches and thorniness of many plants. White (1983) believes that it is more appropriate to describe the class as thicket than as savanna. In rocky areas, or where more water is available, emergent trees may form a more-or-less full cover with touching crowns. Toward the west of Ogaden, the class is less thicketlike and *Acacia seyal* is more dominant. Along wadis and rivers, riverine forest may occur, and along the course of the Juba, a gallery forest about 200 meters wide once was present over considerable stretches (Mooney 1959). Bally (1968) notes that *Hyphaene* is present along both the Juba River and Wadi Shebclle.

Again, a wide range of *Acacia* spp. is present; White (1983) notes thirty *Acacia* spp. and sixty *Commiphora* spp. as endemic. Shrubs are short-stemmed, multibranched, and underlie scattered umbrella-like trees such as *Acacia nubica, A. mellifera, A. senegal, A. seyal,* and *Commiphora boiviniana.* The thicket-forming scrub is short-stemmed and multibranched (Von Brietenbach 1963). Perennial grasses occur in tussocks between shrubs and trees, and after rainfall, bare ground is covered by annual grasses. Von Breitenbach (1963) describes the class as "Shrub Steppe."

Most of this class is used by migrant pastoralists, many of whom travel in an annual cycle between Somalia and the Ogaden. *Boswellia* spp. is used for frankincense, and *Commiphora* spp. for myrrh, and because individual trees are owned by families, they tend to be carefully managed and not overused. Plantations of *Conocarpus* spp. near Berbera, Somalia are used for urban firewood (Finlayson, Child, and Van

Figure 10-1. NDVI **Profiles, Bushland and Thicket (Classes 41, 44, and 45)**

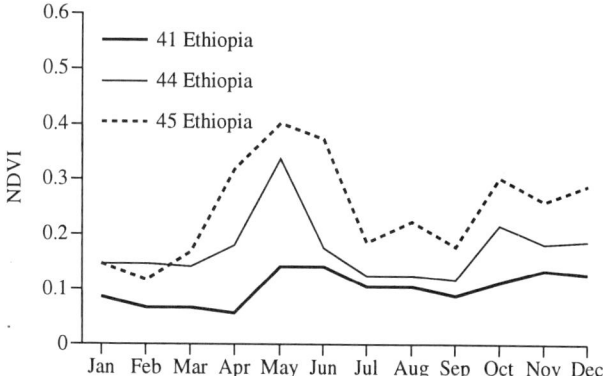

Rensburg 1972) and charcoal from this region is exported to other parts of Ethiopia and exported from northern Somalia to Aden. These researchers found that trees had been planted near villages and were being carefully tended, and Mooney (1959) reported a range of planted trees near Margherita, Somalia.

As might be expected, the pressure of grazing has led to some deterioration of trees. Finlayson, Child, and Van Rensburg (1972) found that *Acacia* spp. and *Commiphora* spp. were degraded near Kisimaya, although in the same area dunes were being fixed by planting of *Commiphora* spp. This class has suffered extreme degradation through overpopulation, partly because of the movement of refugees, and partly due to overgrazing. Hemming (1966) states that overgrazing has occurred since the nineteenth century. The most severe overgrazing is near deep boreholes, around which desertification occurs. Demand for fuelwood increases the rate of depletion. The estimated growing stock is 15 percent of the regional total (729 million tonnes). The estimated sustainable yield is 8 percent of the regional total (11 million tonnes). Kamweti (1984) concluded that fuelwood shortages in Somalia are extreme.

Class 43—Moist *Acacia-Commiphora* Bushland and Thicket

This class covers 2 percent of the region, mainly toward the coast in south Somalia (Shabeellaha Dhexe, Shabeellaha Hoose, Jubbada Dhexe, and Jubbada Hoose) at elevations of up to 200 meters, and includes the valleys of the lower Juba and Wadi Shebelle. About 10 percent of Somalia is in this class. Precipitation varies between 300 and 600 millimeters a year, with two wet seasons. Considerable fluctuation in precipitation occurs from year to year, however.

Class 43 in south Somalia is almost completely surrounded by Dry *Acacia-Commiphora* Bushland and Thicket (Class 41), with a gradual transition between the two. A number of patches also exist in southern Ethiopia at altitudes up to 1,500 meters. The largest area is near Ginir in the Shebelle Valley and in the upper Geneale Valley. Other small areas are on the slopes of the southern Rift Valley and in the lower Omo Valley. Precipitation in Ethiopia may attain 1,000 millimeters, with two rainfall maxima.

Compared with Class 41, NDVI values are generally greater, although June, the month with the greatest value of 0.45, is almost the same as values from Class 41 in inland southern Somalia. Minimum NDVI values of 0.18 and 0.19 for March and April are, however, considerably greater than those for Class 41.

In Somalia, the class is similar to but denser than the Dry *Acacia-Commiphora* Bushland and Thicket (Class 41). Generally, it forms a two-story dense bushland 3 to 5 meters tall with scattered emergent trees up to 9 meters (White 1983), although Mooney (1959) describes the vegetation as scrub less than 3 meters high. *Acacia* and *Commiphora* are spiny, so impenetrable thickets may form, particularly because many species are multistemmed and some species of *Commiphora* have a radial prostrate form. In these cases a single tree may be 12 meters across, although less than 5 meters high.

A few species have well-defined trunks and form emergents, occasionally reaching 10 meters high. *Acacia tortilis*, *Adansonia digitata*, and *Euphorbia robecchii* are in this category, although the baobab here rarely exceeds 8 meters in height. Emergent trees are more common in the wetter areas, and may even form a type of woodland. In the Holawajir Forest near the Kenyan border, emergent trees reach 30 meters, with a middle story higher than 10 to 15 meters. *Acacia bussei* dominates large areas, with few other species under its spreading crown (Bird and Shepherd 1988).

Most species are deciduous, although a small percentage are evergreen, and succulents are widespread. Grasses form an insignificant part of the aboveground biomass, although Bally (1968) notes that where stock are excluded, a much greater profusion of vegetation quickly develops. *Acacia* spp., *Capparidaceae* spp., *Commiphora* spp., and *Grewia* spp. are very common. Toward the south of Somalia, *Euphorbia robecchii*, which is unusual in having a straight bole, may also form a conspicuous and taller element, although it has been used extensively in box-making because it is one of the better timber trees (Mooney 1959), and numbers at that time were much reduced.

In a narrow zone near the Juba River, a number of taller trees that form a type of gallery forest are reported by Finlayson, Child, and Van Rensburg (1972). This forest, however, is being encroached on for both permanent and shifting cultivation and has been cut for commercial timber (Luchini 1986). The largest of the lower Juba forests was about 150 square kilometers in extent in 1948, but had been reduced to less than 2,000 hectares in 1987, and less than 700 hectares remain in the lower Shebelle (Douthwaite 1987). Dominant trees in the riparian Juba forests include *Acacia seyal*, *Afzelia quanzensis*, *Antidesma venosum*, *Mimusops fruticosa*, *Sorindeia madagascariensis*, *Spirostachys venifera*, *Thespesia danis*, and *Trichilia emetica* (Luchini 1986). Douthwaite (1987) identifies about 50,000 hectares of forest in the Ltolawajir Depression bordering Kenya, but predicts that at the present rate of clearance most will have been removed by the year 2000.

Species characteristic of the main canopy of Class 43 in Somalia are *Acacia bussei*, *A. horrida*, *A. mellifera*, *A. nilotica*, *A. reficiens*, *A. thomasii*, *Balanites orbicularis*, *Boscia coriacea*, *Boswellia neglecta*, *Commiphora africana*, *C. boiviniana*, *C. campestris*, *Cadaba* spp., *Dobera* spp.,

Euphorbia spp., *Salvadora* spp., *Sterculia* spp., and *Terminalia* spp. (White 1983; Mooney 1959). Smaller bushes include *Bauhinia taitensis*, *Bridelia taitensis*, *Caesalpinia trothae*, *Combretum aculeatum*, *Ecbolium* spp., *Grewia* spp., *Maerua* spp., and *Sericocomopsis* spp. (White 1983; Mooney 1959; Bird and Shepherd 1988).

The small areas of this class in the lower Omo Valley and the Rift Valley of Ethiopia are similar to the Somalian area described previously, but much of the Ethiopian occurrence of this class is at higher altitudes, above 1,500 meters. Dry *Acacia-Commiphora* Bushland and Thicket (Class 41) adjoins this class at lower altitudes, and Cultivation and Forest Regrowth Mosaic (Class 73) lies at the same altitude to the south. Beals (1968) describes Class 43 as "Deciduous Woodland and Savanna," dominated by *Acacia* woodland with *A. senegal*, *A. seyal*, and *A. tortilis*, locally dominated by other *Acacia* spp. At higher elevations, *Acacia etbaica* is more common.

Von Breitenbach (1963) considers the area to be part of the lowland savanna woodland with a closed canopy at 5 to 12 meters, but allowing sufficient light to penetrate and allow development of shrubs 1 to 3 meters high. This forms *Acacia* thicket with *Euphorbia* thicket at higher elevations.

Species listed by Von Breitenbach include *Acacia amythetophylla*, *A. asak*, *A. mellifera*, *A. seyal*, *A. sieberana*, *A. tortilis*, *A. venosa*, *Albizia amara*, *Cassia singueana*, *Combretum molle*, *Entadopsis abyssinica*, *Euphorbia* spp., *Ficus* spp., *Grewia* spp., *Maerua angolensis*, *Pterolobium stellatum*, and *Ziziphus abyssinica*. Parasites, including Loranthus and Viscum, are common in tree crowns.

Kamweti (1984) identifies the coastal region of Somalia as a region of fuelwood deficit. White (1983) describes the class as having suffered extreme degradation through overgrazing, particularly intensified by the provision of boreholes used as watering sites, and by the increase in cattle allowed by the improvement of veterinary services. Finlayson, Child, and Van Rensburg (1972) found that shifting cultivation and commercial woodcutting were reducing the remaining areas of good-quality riparian woodland and that sedentary agriculture in the Juba and Wadi Shebelle valleys and near Mogadishu was causing further reduction in vegetation cover. It is almost certain that the effects of grazing and agriculture are significantly reducing the quality and quantity of tree cover in Class 43.

Kamweti (1984) identifies fuelwood demand from Mogadishu as a cause of tree loss, extending for a considerable distance (some 200 kilometers along rivers) to supply the 250,000 tonnes needed each year. Bird and Shepherd (1988) found that *Acacia bussei* was extensively exploited for charcoal used in towns, particularly Mogadishu, but that *A. senegal* was becoming important as *A. bussei* was being overexploited, and

that even poor sources such as *Commiphora* were being used for charcoal. Much of the tree canopy in the Bay area east of the Juba River has been lost, causing soil erosion and land degradation and greatly reducing the grazing potential.

Cultivation also is increasing in the Juba Valley. Douthwaite (1987) stresses the significance of farming, logging, mining, and water resource development in reduction of the lowland forest resource of southern Somalia. Uses of timber in the remaining Shebelle gallery forest (approximately 700 hectares) include timber for beehives, charcoal, fuelwood, building timber, agricultural implements, and goat browse. The estimated growing stock and sustainable yield are about 2 percent of the regional total (118 million tonnes and 1.5 million tonnes, respectively).

Class 44—Sahel-Sudanian *Acacia* Wooded Bushland

This class, which occupies 10 percent of the region, occurs in Sudan and Ethiopia. It is particularly extensive in Sudan, where it covers 16 percent of the country. It lies between the Semidesert Wooded Grassland (Class 21) to the north and Dry Sudanian Woodland (Class 62) to the south. It forms a band about 500 kilometers wide, stretching from southern Darfur through central Kordofan, White Nile, and Blue Nile provinces, and continuing into the northern part of Eritrea. Small areas of Sahel-Sudanian *Acacia* Wooded Bushland occur on the Red Sea coast of Sudan, and isolated areas exist along the Nile and in small areas of the Libyan Desert. More substantial outliers are scattered to the south of the main belt in both Sudan and Ethiopia.

Values of NDVI for a site in northern Ethiopia confirm that growth activity is greater than in the Dry *Acacia-Commiphora* Bushland and Thicket (Class 41). These values also reflect the double wet season with peaks in May and October (figure 10-1).

Most of the class lies on plains below 500 meters, although in Ethiopia it extends to more than 1,500 meters. Precipitation ranges between 250 and 500 millimeters, and most of the area covered by this class experiences 2 months with at least 100 millimeters.

Soils vary significantly to the west of the Nile, where the class is best developed. Quaternary sand dunes up to 40 meters high and up to 100 kilometers long, now mainly stabilized by vegetation, overlie kaolinitic clays. These form the *qoz* landscape, which is well developed near El Obeid (Booth 1984). These sandy soils cover about 65,000 square kilometers of Sudan (White 1983). *Acacia senegal* is characteristic of this substrate, and may form almost pure stands. Because it is used for gum arabic production, it has been protected and incorporated in the agricultural system, but in the past two decades it has been extensively cut for fuelwood, and pressure on the land has led to its omission from the cropping cycle.

To the east of the Nile, the central clay plains are dominated by soils with significant levels of montmorillonite and illite (Graham 1969). These clays are impermeable, becoming dry, dusty, and deeply cracked in the dry season, but muddy when wet. Because of the impermeability of the cracking clays, water supply for vegetation is less than in the *qoz* areas. For example, *Acacia senegal* needs 280 to 400 millimeters of precipitation on sandy soils, but 500 millimeters on the cracking clays. *Acacia seyal* forms almost pure stands in areas of 500 millimeters precipitation on flat areas of clay, but in depressions and along streams it occurs with much less precipitation (Smith 1953). *Acacia tortilis* ss. *raddiana* extends along streams into areas with as little as 150 millimeters precipitation, but on flat clays needs about 450 millimeters.

Throughout the clay plains, trees and shrubs are more common along drainage channels. In this class, on favorable sites, tree crowns may be almost touching, but usually are several crown diameters apart. The main tree species include *Acacia albida, A. astringens, A. laeta, A. nilotica, A. senegal, A. seyal, A. tomentosa, A. tortilis, Balanites aegyptiaca, Boscia senegalensis, Combretum cordofanum, Commiphora africana, Dalbergia* spp., *Maerua crassifolia, Leptadenia pyrotechnica, Terminalia* spp., and *Anogeissus schimperi* (White 1983; Bari 1968). In general, trees are of small stature, on the order of 7 to 10 meters. Grass species are mainly annuals such as *Cenchrus biflorus*, but perennials become more common in wetter areas.

The zone is identified by Berry (1983) as suffering severe deforestation because of both direct and indirect physical effects of severe drought, such as that between 1968 and 1973. Drought in itself stresses the trees, and, when other vegetation is lacking, herders cut much more fodder from the trees than normal. Fuelwood collection for urban areas, particularly for the three towns of Khartoum, Omdurman, and Khartoum North, had extended to the Ethiopian frontier by 1960. By 1980, fuelwood was being transported from more than 600 kilometers south. Much of this fuelwood is from the Sahelian-Sudanian zone.

Kassas (1970) records that *Acacia* spp. trees were common near Khartoum in 1955, but that by 1972 the nearest trees were 90 kilometers south of the city. Graham (1969) identifies agricultural and pastoral intensification as an important cause of wood loss in this zone. On the clay areas, mechanized farming has entailed complete removal of trees. In the *qoz* sands, digging of high-yield tube wells since the early 1960s has concentrated grazing pressure around them. By 1962, little vegetation remained within 2 kilometers of these wells—a piosphere—and vegetation had been severely depleted within 5 kilometers of wells (Lewis and Berry 1988). A similar effect has been observed in Somalia in Dry *Acacia-Commiphora* Bushland and Thicket (Class 41).

Desertification, reflecting both climatic fluctuations and anthropogenic factors, has led to the deterioration of extensive areas of Class 44. Grass burning and selective browsing have very considerably altered the tree species present (Bari 1968). Moghraby, Ali, and Seed (1987) attribute deterioration of biomass in this zone to the southward migration of livestock-based communities.

The fuelwood potential of this class has been severely depleted by the processes described, and if deterioration is to be halted, remedial action is essential. In the Ethiopian areas of Class 44, few trees now remain. Booth (1984) investigated restocking with *Acacia senegal* in the area of El Obeid to prevent desertification.

The estimated growing stock is just 1.2 percent of the regional total (59 million tonnes), but the estimated sustainable yield is 11 percent of the regional total (15 million tonnes).

Class 45—Escarpment Wooded Thicket

Class 45 is most extensive in Ethiopia, particularly in a long arc from Eritrea, following the eastern edge of the western plateau through Tigre, Welo, and Shewa, along part of the Rift Valley floor, along the upper escarpment of Ch'erch'er and the Hārer Hills. Small relict patches exist in Shewa, Begemder, and Gojam and in small areas in Bora. The Hārer escarpment section continues into northern Somalia on the high north-facing escarpment, including Daloh Forest. In Djibouti, a small patch of 870 hectares persists in the Forêt du Day. Sudan has areas in the Boma, Didinga, and Dongotona hills of Equatoria, and a small relict area in the Karora hills on the Red Sea coast.

Almost all of the areas covered are relatively high; for example, the Daloh Forest at 2,000 meters is atop a 500-meter vertical limestone scarp. In Ethiopia on the east-facing scarp, *Juniperus* woodland occurs to 3,200 meters, although in Djibouti the Forêt du Day on Mount Goda is only between 1,300 and 1,800 meters (Hamrouni 1984). Precipitation is approximately 1,000 to 1,500 millimeters in Ethiopia, although examples of *Juniperus* exist as emergents in scrub forest with only 650 millimeters of rainfall (White 1983). In the Daloh and Day forests, precipitation is much less, about 700 millimeters in the former and 350 millimeters in the latter. Both examples owe their existence to occult precipitation that creates high humidities for a considerable part of the year (Hamrouni 1984; Hemming 1966). Values of NDVI, however, generally reflect the double rainy season and, in the case of an Ethiopian example from the rift escarpment, exhibit a peak value greater than 0.4 in May (figure 10-1).

Though the class is identified as Woodland Thicket in its climax state, above 2,000 meters much of it originally was forest. *Juniperus procera* formed impressive, almost pure stands up to 3,200 meters (Rochetti 1961), whereas *Podocarpus gracilior* occurred in damper areas at about 2,000 meters. In such forests, both *Juniperus* spp. and *Podocarpus* spp. form an upper evergreen story 40 to 50 meters high, with an understory of *Apodytes dimidiata* var. *acutifolia, Cussonia* spp., and *Ekebergia* spp. at 10 to 20 meters, plus a lower shrub layer. Epiphytes and lianes may be present. *Juniperus procera* is light-demanding, and White (1983) suggests that, because it does not regenerate in its own shade, it probably depends on fire (either natural or anthropogenic) for its survival.

In the Forêt du Day, *Torchonanthus camphoratus* in more open forest follows the degradation of *Juniperus* spp., and *Acacia etbaica, A. seyal,* and *A. tortilis* succeed as more open woodland. In Daloh Forest, *Juniperus procera* forms a relatively open forest with widely branched crowns. Associated with *Juniperus is* a bush vegetation of *Buxus hildebrandtii, Cadia purpurea,* and *Dodonaea viscosa.* Inland, *Juniperus* forest is replaced by *Acacia* scrub (Hemming 1968). On the lower part of the escarpment, *Buxus* spp., *Boswellia carteri, Dracaena schizantha,* and *Superbia grandis* form an evergreen scrub, and lower still is a zone of *Acacia-Commiphora scrub.* In Somalia, Hemming (1966) identifies three types of juniper forest—closed, open degraded, and climatic relicts.

Throughout the class, the *Juniperus* and *Podocarpus* woodland is rapidly declining, partly through commercial exploitation of timber, partly through overgrazing, and partly through a failure to regenerate. On the scarp between Dirē Dawa and Hārer, considerable area had been cleared for agriculture by 1988. Much of the central Ethiopian plateau may have had this type of woodland; for example, it was common near Addis Ababa at the end of the nineteenth century, but very little remains, apart from that which is protected.

Much of the deforestation occurred in the nineteenth century, when timber was used for building and as fuel for towns on the plateau. The Emperor Menelik encouraged the planting of *Eucalyptus* spp., particularly *Eucalyptus globus* as a replacement, so that *Eucalyptus* spp. are probably the most important trees in Ethiopia today. Approximately 15,000 hectares of *Eucalyptus* spp. plantation exist around Addis Ababa, but many small clumps also exist in Shewa, and the tree has been planted widely as a road margin plant for use as fuel and poles. Scattered blocks of *Eucalyptus* spp. elsewhere cover 20,000 to 25,000 hectares (FAO 1963).

The estimated growing stock is 1.4 percent of the regional total (69 million tonnes) and the estimated sustainable yield is 0.6 percent of the regional total (0.7 million tonnes). It seems unlikely that accessible *Juniperus* spp. and *Podocarpus* spp. will survive degradation by grazing, agriculture, lumbering, and cutting for fuelwood, but *Acacia* spp. too are suffering from the same pressures in this class. Our impression is that the rate of loss significantly exceeds sustainable yield. The natural regeneration of *Juniperus procera* appears to need particular conditions that are unlikely to be met in extensively used forest, so the present generation of trees may be the last to survive.

Class 52—East African Low Woody Biomass Mosaic

Covering 3 percent of the region is this class. It is particularly extensive in Ethiopia, where it covers about 5 percent of the country. Scattered patches occur in southern Somalia, but the main areas of this class are in the hills of northern Hārergē and on hilly areas of the Gamo-Gofa and Sidamo provinces in southwestern Ethiopia. Other areas exist on the edge of the lava plateau of Welo Province. A discontinuous zone occurs on the lower slope of the mountains of Eastern Equatoria and in the northern area of Western Equatoria between Dry Sudanian Woodland (Class 62) and Moist Sudanian Woodland (Class 65).

In Sudan, the class occurs at about 600 meters altitude, but in Ethiopia it occurs above 2,000 meters. Soils associated with this class are extremely varied, including those of the lava plateaus and the lateritic soils of southern Sudan. Rainfall exceeds 500 millimeters, approaching 1,000 millimeters in western Hārergē, and the dry season lasts between 5 and 7 months. The pattern of NDVI increases to 0.27 in May and June, with a second slight increase to 0.15 in October, but for the remaining 9 months of the year, values are as small as 0.11 and 0.14.

The East African Low Woody Biomass Mosaic is an intermediate zone between *Acacia-Commiphora* Bushland and Thicket (Classes 41 and 43) and Escarpment Wooded Thicket (Class 45). It is far from uniform in structure or constituents, and may be either floristically rich or relatively poor. Certain species are almost always present (White 1983): *Acokanthera* spp., *Carissa edulis, Dodonaea viscosa, Euclea* spp., *Olea africana, Sansevieria* spp., *Tarchonanthus camphoratus,* and *Teclea* spp. Succulent species of *Aloe* and *Euphorbia* are normally present. The canopy is fairly low, between 3 and 7 meters being common, but emergents such as *Euphorbia candelabrum* may rise to 10 meters, and toward the upper limit tall trees such as *Juniperus procera* start to appear.

In Somalia, dominant species include *Acokanthera schimperi, Buxus hildebrandtii* (which may form scrub forest), *Cadia purpurea,* and *Dodonaea viscosa.* Other important species include *Aloe eminens, Barbeya oleoides,*

Cussonia holstii, Dracaena schizantha, Euphorbia grandis, and *Sideroxylon* spp. (White 1983). A narrow zone of this class lies below the *Juniperus* forest of the escarpment in northern Somalia (Hemming 1966).

Grazing, fuelwood collection, charcoal production, and clearance for agriculture have reduced the fuelwood potential of this class. In the Hārer region, clearance for agriculture has been the main influence on woody biomass, and extensive areas now have a low tree density and are almost entirely cropped. Roadside trees are common; these are often *Eucalyptus* spp., which also occur in small plantations and farms. In Ethiopia, the class covers a complex mosaic of land uses, with, for example, extensive coffee production and remnants of *Podocarpus* forest. The estimated growing stock is 4 percent of the regional total (230 million tonnes), and the estimated sustainable yield is 5 percent of the regional total (6 million tonnes).

Class 62—Dry Sudanian Woodland

Of the total area of the region, 9 percent is in this class. It is particularly important in Sudan, with more than 330,000 square kilometers. The class forms a belt up to 500 kilometers wide in south-central Sudan, south of about 8° N and between Sahel-Sudanian *Acacia* Wooded Bushland (Class 44) and Sudanian *Isoberlinia* savanna and Sudanian Woodland (Class 64). Southern Darfur, Southern Kordofan, and parts of Jonglei also are included in this zone.

Dry Sudanian Woodland almost surrounds the Nile flood zone, but part of the class is within the southeastern part of the flood zone. This class covers 15 percent of the surface of Sudan. Extensive areas of the northern section of the Ethiopian plateau, including much of Begemder, western Tigre, Gojam, and western Welo are within this class, and significant expanses exist in the Rift Valley.

In Sudan, precipitation is between 500 and 800 millimeters; in Ethiopia, up to 1,000 millimeters. Generally, areas of this class in Sudan lack pronounced relief and lie below 500 meters, but in Jebel Marra the land rises to 3,057 meters. In Ethiopia, much of the class occurs on land exceeding 2,500 meters, and the topography is more pronounced. The Precambrian Basement Complex and Nubian (Cretaceous) sedimentary rocks underlie much of the Sudanese section, although Jebel Marra, Jebel Gurgei, and part of the Ethiopian frontier section are on Tertiary basic lavas. Much of this, however, is covered by the *qoz* (stabilized Quaternary aeolian sands) in the west, or by deep, dark, impermeable cracking clays (described in the section on Sahel-Sudanian *Acacia* Wooded Bushland, Class 44). These sands and clays are vastly different in their properties and have a profound influence on drainage conditions and soils, and thus on vegetation.

Toward the west, the dominant cover is "*Combretum cordofanum—Dalbergia* spp., *Albizia sericocephala* Woodland" and "*Anogeissus-Combretum hartmannianum* Woodland." In the east, "*Acacia mellifera* Thornland" and "*Acacia seyal—Balanites* Savanna" are the most important formations, the former on clays, the latter on sands. In both formations, woodland alternates with patches of grassland (Sudan Survey Department 1954). *Anogeissus leiocarpus* may occur in pure stands on thicker soils in Darfur, whereas stony ground is dominated by *Boswellia papyrifera* (White 1983). On flooded soils near the White and Blue Niles, *Acacia astringens, A. nilotica, A. subalata, A. seyal,* and *A. tomentosa* are common.

Much of the woodland that had continuously covered extensive areas has been cleared for agriculture, particularly in central, eastern, and western Sudan and in Ethiopia, because the trees were easily removed with simple equipment. Pastoralism is widespread in the south and west, which are free of tsetse; overgrazing and browsing has almost everywhere led to thinning of the tree cover. Extensive smallholder cultivation occurs in Darfur and throughout Ethiopia (Berry, Taurus, and Ford 1980).

In the past, vegetation was allowed to recover as part of a bush-fallow cycle. More recently, cultivation has become almost continuous. The effect of agriculture and grazing has been to degrade the woodland, which on the slopes of Jebel Marra, for example, is replaced by shrub or thicket. Above 1,800 meters, the upper plains and peaks are now covered by montane grassland, swept annually by fire. This is almost certainly a fire-climax vegetation, although relict trees remain. Riparian forest remains in inaccessible gorges as an indication of widespread former conditions (White 1983).

Firewood collection for Khartoum has caused considerable reduction in forest cover in Eastern Kordofan and Blue Nile (Lewis and Berry 1988). Fuelwood collection has severely depleted woodland in Ethiopia around all towns, but particularly Āsmera (deforestation and overgrazing are probably more important in this part of Ethiopia than elsewhere in the country).

As a result of these pressures, the landscape has changed from mainly woodland with some grassland to parkland or very thinly wooded farmland, and in some places even to extensive treeless steppe. Rocky areas, or those where water for human use is scarce, are the only areas where relatively intact Dry Sudanian Woodland survives. Consequently, the estimated growing stock is 23 percent of the regional total (1,098 million tonnes), and the estimated sustainable yield is 23 percent of the regional total (31.7 million tonnes).

It is almost certain that rates of wood depletion now exceed the sustainable yield of this class and that fuelwood shortage will become an even more serious cause of environmental deterioration and of human suffering.

Class 63—Sudan-Ethiopian Woodland and Thicket

This class, which is equally extensive in Sudan and Ethiopia, occupies 2 percent of the region. Most of it is on the edge of the central Sudan plain and the foothills and west-facing slopes of the Ethiopian plateau. Extensions occur along river valleys, particularly the Blue Nile and Atbara and their tributaries in Gojam and western Welega provinces. Discontinuous extensions exist into the Blue Nile and Sobat provinces of Sudan. A small area in Bahr al Ghazāl adjoins Undifferentiated Dry Sudanian Woodland, and other areas occur in less-frequently flooded areas of the Sudd.

In Sudan, most of the class lies below 600 meters, but in Ethiopia large areas exist above 1,000 meters with some above 1,500 meters. Precipitation in Sudan is about 600 to 800 millimeters and in Ethiopia about 1,000 millimeters. Both countries experience 5 or 6 months with more than 100 millimeters of rain. This is reflected in the phenology for this class, which shows a marked seasonality, with a major contrast between the small NDVI values of 0.08 to 0.15 between December and May and values of 0.55 and 0.56 for August and September. Values of NDVI exceed 0.27 from June through November. The class is developed on dark cracking clays in Sudan and on complex Precambrian basement rocks and Tertiary lavas in Ethiopia.

Von Breitenbach (1963) includes the class in his "Lowland Woodland" and "Lowland Savanna" classes. It forms the central section of White's Class 25b (1983), which extends along virtually the whole border between Sudan and Ethiopia. White identifies among the principal constituents: *Anogeissus leiocarpus, Balanites aegyptiaca, Boswellia papyrifera, Combretum collinum, C. hartmannianum, Commiphora africana, Dalbergia melanoxylon, Erythrina abyssinica,* and *Terminalia brownii.* Von Breitenbach (1963) identifies an upper story at 5 to 12 meters of mainly deciduous trees, forming an almost completely closed canopy, but allowing sufficient light for a shrub layer of 1 to 3 meters height.

Rochetti (1961) records a zone similar to White's, extending from Eritrea to southern Ethiopia "le bois à feuilles caduques"—with an almost identical list of tree species. Like Von Breitenbach, however, he also finds extensive areas of bamboo thicket, particularly *Oxytenanthera abyssinica* and *O. borzii* in the higher areas. He describes the trees as quite small with thick trunks. Von Breitenbach (1963) subdivides the class into "*Combretum* Thicket," "*Croton* Thicket," and "*Balanites* Thicket."

As in much of the Sudanese and Ethiopian woodland, clearance for agriculture and collection of firewood has considerably reduced the woody biomass stock of this class. Currently, the estimated growing stock is 5 percent of the regional total (259 million tonnes), and the estimated sustainable yield is 3 percent of the regional total (4.3 million tonnes). FAO (1963) comments that these woodlands are of value for fuelwood and charcoal, but reports aerial surveys which suggest that bamboo stands may be more limited than was previously believed. Western Sudanian areas have been severely thinned, and in Ethiopia, permanent cultivation dominates the area covered by this class.

Class 64—Sudanian Woodland

This class covers 2 percent of the region. In Sudan, a large stretch of Sudanian Woodland extends parallel to the southwestern border through El Buheyrat into Equatoria to the south of the flood region of the upper Nile. Other areas exist on mountains in Eastern Equatoria. In Ethiopia, numerous patches occur on the east of the plateau above the east-facing escarpment. The largest area lies to the north of Addis Ababa. Isolated areas occur near Aksum and Sek'ot'a, and a small area occurs on the Hārer Hills. The class occurs at altitudes below the Moist Sudanian Woodland (Class 65) in Sudan; in Ethiopia, it is above the Escarpment Wooded Thicket (Class 45).

Sudanian Woodland lies mainly above 2,000 meters, with considerable area above 3,000 meters in Ethiopia. In Sudan, however, the class mainly occurs below 500 meters. Soils differ too, the Ethiopian soils being mainly formed on basic volcanic lavas, whereas those in Sudan are formed on the edge of the ironstone plateau and are lateritic. For more than 6 months of the year, the Sudanian area receives in excess of 100 millimeters of precipitation each month. But in Ethiopia, the dry season is shorter (4 months in Addis Ababa and Desē). In both areas, annual rainfall exceeds 1,000 millimeters.

This class is intermediate between Moist Sudanian Woodland (Class 65) and Dry Sudanian Woodland (Class 62). It is best described as "*Anogeissus-Khaya-Isoberlinia* Deciduous Woodland" in Sudan (Bari 1963), because it has fewer *Isoberlinia* spp. than the "Sudanian *Isoberlinia* Savanna" proper. Trees are rarely more than 15 meters high and species in the south of Sudan include *Acacia dudgeonii, A. gourmaensis, Antidesma venosum, Faurea saligna, Lophira lanceolata, Maprounea africana, Maranthes polyandra, Monotes kerstingii, Ochna afzelii, O. schweinfurthiana, Protea madiensis,* and *Terminalia glaucescens* (White 1983).

In its natural state, tree cover is extensive in this class, but cultivation has profoundly modified much of Class 64 through bush fallowing in less densely populated areas and by permanent cultivation in more densely populated areas. In some areas, trees may be completely eliminated (White 1983). This is particularly true in Ethiopia, where the term "woodland" is appropriate only in a historical sense. In Ethiopia, the woodfuel scarcity is comparable to that in Class 77 (Highland Cultivation Mosaic), and the estimated growing stock is only 4 percent of the regional total (259 million tonnes). Further, the estimated sustainable yield is only 3 percent of the regional total (4.3 million tonnes).

Class 65—Moist Sudanian Woodland

In this region, 4 percent falls within this category, which covers 6 percent of Sudan. A broad band up to 300 kilometers wide follows the southwestern frontier of Sudan. Isolated discontinuous patches occur farther east and in El Buheyrat as "islands" in the Sudd. In Ethiopia, numerous patches occur on the central and southwestern plateaus of Gonder and Shewa, in tributary valleys of the Blue Nile and Atbara, and to the north of Addis Ababa. Larger areas occur toward the western boundary of the rift, and a series of linear areas follows the foot of the Hārer Hills near Dirē Dawa. Another area lies on the upper dipslope of the Hārer Hills and numerous small areas exist in the valleys of Ĭlubabor and Gamo-Gofa.

The main extension of this class in Sudan is at about 500 meters above sea level on the lateritic soils developed on the ironstone plateau. In Ethiopia, most of the class is at considerably higher altitude, more than 2,500 meters in places, and mainly occurs on volcanic rocks having a great variety of soils. Precipitation in both cases is 1,000 millimeters or greater, with a wet season of 4 months in Sudan and 7 months in Ethiopia. This is reflected in a phenology that reveals marked seasonality and NDVI values that exceed 0.45 at the maximum (figure 10-2).

Intact Moist Sudanian Woodland rarely exceeds 15 meters in height. It is taller than Dry Sudanian Woodland (Class 62), but it is identifiable mainly through the presence of *Isoberlinia* spp. which may locally dominate this class. It is restricted, however, in the drier northern woodlands, where it may exist on rocky hills (White 1983). Other species present include *Acacia dudgeonii, A. gourmaensis, Antidesma venosum, Faurea saligna, Lophira lanceolata, Maprounea africana, Maranthes polyandra,* and *Monotes kerstingii.*

White (1983) argues that "*Isoberlinia* Sudanian Savanna" is significantly different in stature and floristics to enable its separation from *miombo* woodland. In Ethiopia, much of this class falls within Von Breiten-

Figure 10-2. NDVI Profile, Sudanian Woodland (Class 65)

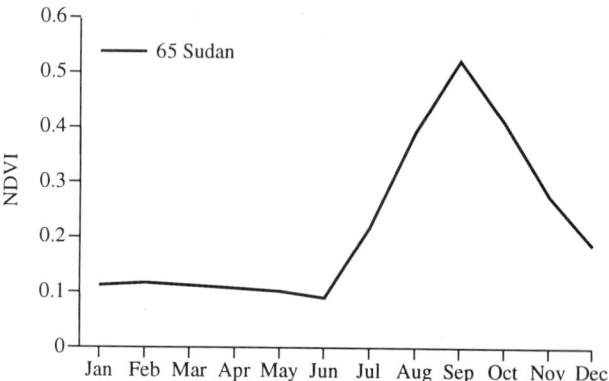

bach's "Mountain Savanna" category (1963), although he classifies it as "Lowland Savanna" in parts of northern Hārergē, the Rift Valley, and the southwest. Von Breitenbach identifies *Acacia abyssinica, Agauria salicifolia, Buddleja polystachya,* and *Olinia usambarensis* as characteristic of the more humid parts of the "Mountain Savanna."

Near Awash Station, field investigations in 1988 showed a mosaic of cultivation and grazing with some evidence of burning and a range of densities of crown cover of *Acacia* spp. varying from approximately 20 percent cover to large areas with relatively few trees. Near the railway, trees had been cut for fuel. Over extensive areas, *Opuntia* spp., or prickly pear, formed the understory or was the main component of vegetation.

As in the case of Dry Sudanian Woodland (Class 62), human activity has severely altered climax Moist Sudanian Woodland, particularly in the highland areas of Ethiopia. In Sudan, both pastoralism and smallholder farming have reduced tree cover in the west, whereas to the southeast smallholder farming has had a major influence (Berry, Taurus, and Ford 1980). Bush fallowing has been an important feature of cultivation, and has reduced biomass, favored some species, and probably altered soil conditions. When the fallow season is shortened, however, significant ecological changes occur, and tree cover is much reduced.

Sudan Survey Department (1954) shows a considerable area of Western Equatoria, which is now within this category, as having been recently derived from rain forest. Fuelwood demand in this area of Sudan is less than it is farther north because of the lack of urban settlements, and the productivity of the woodland is greater. Consequently, more possibility of fuelwood production from this class exists than from other types of woodland. However, in Ethiopia, on the western plateau, the class is severely depleted. Field investigations in 1988 suggested, however, that a reasonable reserve exists in the Awash Station area.

The estimated growing stock is 10 percent of the regional total (468 million tonnes), and the estimated sustainable yield is 6 percent of the regional total (8 million tonnes).

Class 73—Cultivation and Forest Regrowth Mosaic

Approximately 3 percent of the region is in this class. In Ethiopia, where most of this class lies, and in Sudan, this class occurs in two distinctly different environments. Most of the Sudanese area is in a strip along the flood zone of the River Nile south of 12°N. In Ethiopia, the class is widespread on the lower areas of the plateau, between 1,000 and 2,000 meters above sea level. The largest single area, however, about 1,500 square kilometers, is on the dipslope of the eastern plateau, on the slopes of mountains in Balē and Sīdamo provinces. Another series of irregular patches of Cultivation and Forest Regrowth Mosaic lies to the west of the southern end of the Rift Valley. Other significant areas occur toward the west of southern Gojam and Welega. In Somalia, small areas exist in the lower Juba Valley.

Geologically, the areas are disparate. In Sudan, the class occurs on the plains, dominated by cracking clays of Quaternary age and overlying Tertiary sedimentary strata. In Ethiopia, it exists on a variety of extrusive volcanic rocks. Precipitation in Ethiopia is about 1,000 millimeters a year, with 6 or 7 dry months, whereas in Sudan precipitation is significantly less at 600 to 800 millimeters. In the latter area, however, considerable soil moisture is available because of the water-holding properties of the clays and the flooding of the Nile.

The phenological curve for this class shows greater NDVI values from April to November, exceeding 0.43 for 8 months with the exception of September, when the value drops to 0.36. Maximum values of 0.57 and 0.58 occur in May and June. During the 4-month dry season, the NDVI value falls to 0.17.

Most areas of Cultivation and Forest Regrowth Mosaic would fall within Von Breitenbach's "Lowland Woodland" class (1963), but parts of it, particularly near the Kenyan border and in Gamo-Gofa, fall into his "Lowland Savanna" class. In Beal's description (1968), most of this class occurs in the "*Acacia* Woodland-Savanna" zone, with *Acacia senegal*, *A. seyal*, and *A. tortilis* being widespread. On stonier soils, particularly in western Ethiopia, combretaceous woodland is more common, with *Combretum molle* or *Terminalia brownii* the dominant species. Also in western Ethiopia, *Oxytenanthera abyssinica* bamboo may be locally important.

The grasses *Themeda triandra* and *Hyparrhenia* spp. are widespread below the tree layer. Von Breitenbach (1963) recognizes a number of associations in this class, where the upper story is composed of trees 5 to 12 meters high, with a thicketlike lower story of shrubs 1 to 3 meters high. Characteristic associations include "*Acacia* Thicket," "*Combretum* Thicket," "*Balanites* Thicket," and "*Croton* Thicket," with "*Euphorbia* Thicket" in drier areas.

Much of this class has, however, been profoundly modified by grazing, burning, and clearance for agriculture, so that cleared sections around farmed areas may be effectively grasslands. Less-disturbed bushland occurs at a greater distance from settlements. Field investigations in 1988 in the area between Dirē Dawa and Awash Station showed that agriculture, using irrigation water from streams originating in the Hārergē Hills, has locally modified the land cover pattern. Significantly larger trees with trunk diameters up to 1 meter occur in these better-watered areas. Prickly pear (*Opuntia* spp.) forms a common component of the understory in this area, where tree density varies considerably.

Consequently, the class describes a mosaic of different types of agriculture with various stages of regrowth vegetation. Woody biomass stock varies considerably spatially, but it is estimated that the growing stock is 3 percent of the regional total (149 million tonnes) and the sustainable yield is estimated to be about 1 percent of the regional total (1.6 million tonnes).

Class 74—Guinean Woodland

In this region, 2 percent lies within this class. It is particularly extensive in Ethiopia, where it covers 3 percent of the country. A considerable area in southern Equatoria Province of Sudan also is within this class, and scattered patches exist within the flood zone of the upper Nile, including some along the river itself. Larger areas cover parts of western and highland Ethiopia between 2,000 and 3,000 meters above sea level.

The class is particularly important on the western plateau, in Gojam, Welega, Shewa, and Kefa, and a small area occurs in Welo. An almost continuous strip follows the highest edge of the escarpment of the eastern plateau from the Hārer Mountains, through the Gogu Mountains, to the lower sections of the Ārsī Hills. This area also includes an outlier as far south as the hills overlooking Karsa Dek.

Precipitation regimes in Ethiopia vary; rainfall is almost everywhere greater than 700 millimeters, but approaches 1,400 millimeters on the edges of the rain forest on the southwestern plateau. The wet season lasts between 6 and 8 months. In Sudan, elevations are lower, but precipitation is still between 700 and 1,000 millimeters for almost all of the area of this class. A sample NDVI curve for southern Sudan shows the marked seasonality of vegetation growth in this class (figure 10-3).

This class does not fit easily into existing classifications such as those of White (1983) or Von Breitenbach (1963) in Ethiopia, although it accords with Bari's "High Rainfall Savanna Woodland" in Sudan (1968), which is characterized by *Anogeissus-Khaya senegalensis* and *Isoberlinia* spp. Much of this class is on the ferricrete ironstone plateau of Equatoria, which is elsewhere occupied by the remains of *Isoberlinia* woodland. Sudan Survey Department (1954), however, shows that much of the area of this class in Equatoria was recently cleared rain forest, where *Cola cordifolia*, *Erythrophleum guineense*, *Khaya grandifolia*, *Mitragyna stipulosa*, and *Syzygium guineense* were common species. This interpretation also may be valid for some areas along the upper Nile. In Ethiopia, however, it is unlikely that the class originally would have been rain forest, into which it grades on the southwestern plateau.

Elsewhere, because of precipitation and altitude on both the western and eastern plateaus, much of the class consists of degraded and largely cleared *Juniperus procera* and *Podocarpus gracilior* woodland. This has been reduced to a complex mosaic of farmed land with discontinuous areas of secondary woodland and thicket. The amount of woodland remaining reflects the pressure of cultivation and grazing.

On the Ethiopian plateau, more extensively wooded areas are common on steeper slopes, although some hills are almost completely denuded of trees and have been replaced by grassland. Field investigations on the plateau to the west of Addis Ababa in 1988 showed a complex pattern: within a short distance occurred a landscape of extensively wooded escarpments, partially wooded and treeless hills, parkland savanna, and small copses. Indeed, every variation between full woodland and treeless grassland was observed. Tree species varied greatly and included *Acacia* spp., *Euca-*

lyptus spp., *Euphorbia* spp., and *Ficus sycomorus*, and in higher areas, grassland.

In such a varied environment, fuelwood varies in availability across a short distance. Reserves of wood remain, and in some cases are obviously protected, but woodfuel has become a marketed good rather than a free good. This, together with lopped and trimmed trees, is certain evidence of woodfuel shortage. As population increases, the remaining wood will be even more stressed, because it will be even more extensively used for fuel and because it is the only place where cultivation may be extended. Currently, the estimated growing stock is 2.4 percent of the regional total (113 million tonnes). The estimated sustainable yield is less at 1 percent of the regional total (1.3 million tonnes).

Class 77—Highland Cultivation Mosaic

Highland Cultivation Mosaic occurs on the plateau areas of Ethiopia between 2,000 and 3,500 meters above sea level and occupies approximately 17 percent of Ethiopia. Important areas are in Shewa, Gojam, Welo, Gonder, and Tigre, with smaller extents in Hārergē, Balē, Eritrea, Sidamo, Welega, Kefa, and Ilubabor. This class coincides with Von Breitenbach's "Mountain Savanna" (1963) and forms substantial areas of the drier sections of White's Class 19a, "Undifferentiated Afromontane" (1983). It lies in the area of "Woing Dega" (Von Breitenbach 1963), or temperate highlands, the moderate temperatures reflecting the effect of altitude.

Average precipitation is between 500 and 1,000 millimeters, increasing locally to 1,400 millimeters, and a 7-month dry season exists. Values of NDVI indicate high productivity between June and November, with a peak value of 0.6 (figure 10-3). The most important soil types on the lava plateaus are calcareous black soils with a tendency toward desiccation and leaching on flatter areas, and better red soils on steeper slopes (Von Breitenbach 1963).

A dominant characteristic of this class is the severe effect of deforestation spanning thousands of years in the most densely populated part of the country. Here the effects of cultivation, grazing, and fuel collection combine to make difficult the identification of primary and secondary vegetation. Little tree cover now remains over large areas, although Von Breitenbach (1963) and White (1983) believe that forest and woodland cover were formerly much more extensive. They identify remnants of the former cover from *Acacia xiphocarpa (abyssinica)*, *Apodytes dimidiata*, *Hagenia abyssinica*, *Juniperus procera*, *Nuxia* spp., *Olea africana*, and *Prunus africana* in mixed woodland or from single-dominants such as *Hagenia abyssinica*, *Juniperus procera*, or *Widdringtonia cupressoides*.

Figure 10-3. NDVI **Profiles, High Woody Biomass Mosaic (Classes 74 and 77)**

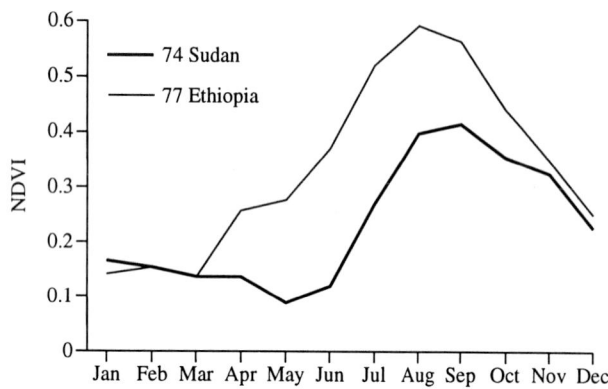

White (1983) believes that the single-dominant stands may depend on fire, either natural or anthropogenic. Von Breitenbach (1963) identifies "Grass Savanna" with Cyperaceae and Gramineae and "Scrub Savanna" as a more arid variant. A moister "Shrub and Tree Savanna" grades upward to an "Afro-Alpine Savanna" with, for example, *Lobelia rhynchopetalum.* In the same area, White (1983) describes areas of afromontane bamboo dominated by impressive stands of *Arundinaria alpina.*

Field investigations near Addis Ababa in 1988 revealed the complexity of landscapes of this class, with cultivated land dominating flatter areas and remnants of a more extensive tree cover evident on some hills. Nearby hills either were completely treeless or had only a partial tree cover. In a short distance, the landscape encompasses treeless steppe, dense cultivation, parkland savanna, small patches of farm trees, and plantations, particularly of *Eucalyptus* spp. (see Class 45). *Eucalyptus globus* was introduced in the late nineteenth century to provide fuel and building wood for urban areas, particularly Addis Ababa.

The presence of trees in the landscape, however, may give a false impression of the adequacy of fuelwood supplies. In 1988, near Addis Ababa and in the city itself, there were signs of a severe fuelwood problem. The indicators were extensive cutting of trees, particularly *Eucalyptus* spp.; the use of low-grade wood such as twigs for fuel; the carriage of fuelwood across long distances; the remarkably small individual parcels of wood and charcoal available for sale in some streets; the fact that men were involved in fuel collection as well as women and children; and that considerable time was spent in wood collection. Dung was sold as fuel, even in Addis Ababa, a clear indicator of fuel shortage. Even in the country 50 kilometers to the west of Addis Ababa, twigs and dung were being used as fuel.

Eucalyptus plantations on the edge of Addis Ababa showed evidence of severe cutting, and fuel was being imported from at least 50 kilometers away; at this distance, numerous coppiced plantations exist. The high price of a donkey load of small branches (5 birr) implied a severe fuel shortage: in comparison, the same load in Dirē Dawa on the edge of the desert would cost only 3 birr, confirming Kamweti's suggestion (1984) that fuelwood shortage is most severe in the highlands.

Historically, this has been a region of severe fuelwood shortage and there is every reason to expect that the shortage will become worse. Kamweti (1984) writes of "a worsening energy balance" in the area of this class. The estimated growing stock is still 7.5 percent of the regional total (350 million tonnes), however, and the estimated sustainable yield is 3 percent of the regional total (3.9 million tonnes).

Class 84—Montane Forest

This class covers 1 percent of the region. It occupies 3 percent of Ethiopia. Montane Forest is almost entirely restricted to southwestern Ethiopia, with the two largest areas divided by the Rift Valley. The largest area is to the east in the Mendebo Mountains, with an outlier on the Gogu and Badda mountains. It has extensive sections above 3,000 meters and occurs above 4,000 meters in three areas. To the west, the other more dissected area exists in the mountains of Ilubabor, Kefa, and southern Shewa, with an outlier rising to 3,300 meters at Wallel. In Sudan, small areas occur in the Didinga Hills and Dongotona Mountains of Eastern Equatoria.

Precipitation is between 1,000 and 1,500 millimeters in the west of Ethiopia, and much of the eastern area receives more than 1,400 millimeters. In the western zone, only 2 months are dry, but in the east this increases to 5 months. Values of NDVI are variable for this class (figure 10-4) with an October peak value of nearly 0.6.

In Von Breitenbach's classification (1963), the highest classes above 3,200 meters are mapped as "Montane Steppe," whereas most of the remainder is "Humid or Subhumid Mountain Woodland." According to White (1983), the individual trees are usually shorter than in the montane rain forests that exist at lower altitude. The majority of tree species are widespread and not distinctive to this zone. Included are *Apodytes dimidiata, Halleria lucida, Ilex mitis, Nuxia congesta, Ocotea bullata, Podocarpus falcatus, P. gracilior, Prunus africana,* and *Rapanea melanophloeos. Hagenia abyssinica* may form dominant stands, as may *Juniperus procera,* creating *Hagenia* thicket and *Juniperus* thicket.

Von Breitenbach (1963) refers to *Juniperus procera* forest 45 meters high, but in much of the forest, the upper story does not exceed 20 meters, although *Ekebergia* spp. may reach 35 meters. On wetter sites, *Arundinaria alpina* (bamboo) may form dense thickets. "Subalpine Forest," often festooned with lichens and

Figure 10-4. NDVI **Profile, Montane Forest (Class 84)**

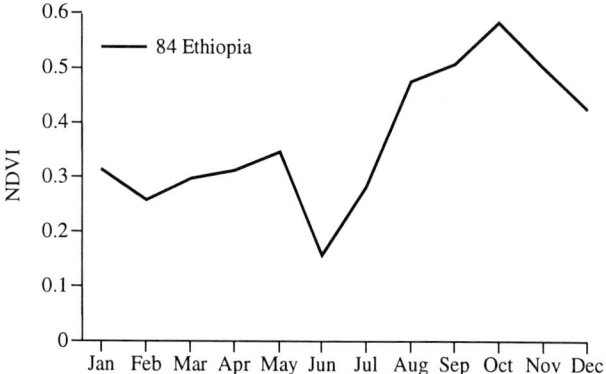

mosses, occurs in the highest areas (Beals 1968). "*Sideroxylon* Forest," with a dense canopy 10 to 20 meters high and "*Galiniera* Forest" 10 to 15 meters high may be precursors to, or degraded forms of, *Juniperus* forest (Von Breitenbach 1963). After fires, floristically mixed stands may be replaced by almost pure stands of *Juniperus procera, Hagenia abyssinica,* or *Widdringtonia cupressoides* (White 1983).

Much of the natural forest in this class has been removed during the past century with the extension of agriculture, and considerable areas are now used for coffee plantations. This is particularly true of the western plateau of Ethiopia, where virtually no montane forest remains. Elsewhere, crops are grown within the remnants of the forest. Timber extraction for commercial purposes has led to the removal of a wide range of the larger trees. Of those previously mentioned, *Apodytes dimidiata, Ekebergia* spp., *Juniperus procera,* and *Hagenia abyssinica* are used for sawn timber and plywood (FAO 1963).

Removal of timber commercially, or during land clearance for agriculture, has greatly reduced the fuelwood potential of this zone. This is one of the areas of Ethiopia where Kamweti (1984) concludes that, assuming suitable access, fuelwood supplies should be adequate to the end of the century, although that is not very long. At the present time the estimated growing stock is 7.6 percent of the regional total (356 million tonnes), and the estimated sustainable yield is 2.8 percent of the regional total (14.9 million tonnes).

Land Cover Class Tables

Tables 10-1 through 10-4, beginning on page 102, present summaries for each land cover class of the area, showing growing stock and sustainable yield for the Horn of Africa nations of Djibouti, Ethiopia, Somalia, and Sudan.

References

Every effort has been made to facilitate access to the documents listed here. Some documents, however, lack full bibliographic information because it was unavailable; also, some documents are of limited circulation.

Bally, P. K. 1968. "Somali Republic South." In I. Hedberg and O. Hedberg, eds., "Conservation of Vegetation in Africa South of the Sahara." *Acta Phytogeographica Suecica* 54:145–47.

Bari, E. A. 1968. "Sudan." In I. Hedberg and O. Hedberg, eds., "Conservation of Vegetation in Africa South of the Sahara." *Acta Phytogeographica Suecica* 54:59–64.

Beals, E. W. 1968. "Ethiopia." In I. Hedberg and O. Hedberg, eds., "Conservation of Vegetation in Africa South of the Sahara." *Acta Phytogeographica Suecica* 54:137–39.

Berry, L. 1983. *East Africa Country Profile—Sudan.* Programme for International Development. Worcester, Mass.: Clark University.

Berry, L., T. Taurus, and R. Ford. 1980. *East Africa Country Profiles.* Program for International Development. Worcester, Mass.: Clark University.

Bird, N. M., and G. Shepherd. 1988. *Charcoal in Somalia: A Woodfuel Inventory in the Bay Region of Somalia.* London: Overseas Development Administration.

Booth, G. A. 1984. *Assessment of Restocking of the Gum Belt for Desertification Control. Phase I.* Rome: FAO.

Douthwaite, R. J. 1987. "Lowland Forest Resources and Their Conservation in Southern Somalia." *Environmental Conservation* 14(1):29–35.

FAO. 1963. *Forest Resource Report for Ethiopia.* Rome: FAO.

Finlayson, W., G. S. Child, and J. J. Van Rensburg. 1972. *Forestry and Wildlife Development Survey Mission, 12 June 1972–10 July 1972. Somalia.* Rome: FAO.

Graham, A. 1969. "Man-Water Relations in East Central Sudan." In M. F. Thomas and G. W. Whittington, eds., *Environment and Land Use in Africa.* London: Methuen.

Grove, A. T. 1978. *Africa.* Oxford: Oxford University Press.

Hamilton, P., and J. Maizels. 1989. "Introducing Flow Charts to Forecast Famine." *Geographical Magazine* 61(7):38–40.

Hamrouni, A. E. 1984. *Rapport au gouvernement de la République de Djibouti sur le développement forestier et la lutte contre la désertification.* Rome: FAO.

Hemming, C. F. 1961. "The Ecology of the Coastal Area of Northern Eritrea." *Journal of Ecology* 49:55–78.

Hemming, C. F. 1966. "The Vegetation of the Northern Region of the Somali Republic." *Proceedings of the Linnaean Society of London* 177(2):173–250.

Hemming, C. F. 1968. "Northern Somalia." In I. Hedberg and O. Hedberg, eds., "Conservation of Vegetation in Africa South of the Sahara." *Acta Phytogeographica Suecica* 54:142–45.

Kamweti, D. M. 1984. *Fuelwood in Eastern Africa: Present Situation and Future Prospects.* Rome: FAO.

Kassas, M. 1970. "Desertification Versus Potential for Recovery in Circum-Saharan Territories." In H. E. Dregne, ed., *Arid Lands in Transition.* Washington: American Institute for the Advancement of Science.

Kebbede, G., and M. J. Jacob. 1988. "Drought, Famine and the Political Economy of Environmental Degradation in Ethiopia." *Geography* 73(1):65–70.

Lewis, L. A., and Leonard Berry. 1988. *African Environments and Resources.* Boston: Unwin Hyman.

Luchini, R. 1986. "Useful Timbers of Italian Somalia." Working paper number 6, Ministry of Livestock, Forestry, and Range, National Range Agency, Mogadishu.

Moghraby, A. I., O. M. M. Ali, and M. T. Seed. 1987. "Desertification in Western Sudan and Strategies for Rehabilitation." *Environmental Conservation* 14 (3):227–31.

Mooney, H. F. 1959. *Report on the Scope for Forestry in Somalia.* Addis Ababa: British Middle East Development Division.

Okafo, O. A. 1987. *Northern Rangelands Development Project* (NRDP). UTFN/SOM/022, Forestry component, Terminal Report. Rome: FAO.

Rochetti, G. 1961. *Problèmes de reboissement forestier et de conservation des sols dans les pays d'outre mer: Ethiopie.* Florence, Italy: Istituto Agronomico per l'Oltremare.

Smith, D. 1953. *Tree Growth in Sudan.* Publication 3. Khartoum: Sudan Department of Forestry.

Sudan Survey Department. 1954. *Vegetation of the Anglo-Egyptian Sudan.* Khartoum. (Map scale 1:4,000,000.)

Von Breitenbach, F. 1963. *The Indigenous Trees of Ethiopia.* Addis Ababa: Ethiopian Forestry Association.

White, F. 1983. "The Vegetation of Africa." *Natural Resources Research Series* 20. Paris: UNESCO/AETFAT/UNSO (United Nations Educational, Scientific and Cultural Organization/Association pour l'Etude Taxonomique de la Flore de l'Afrique Tropicale/United Nations Sudano-Sahelian Office).

Table 10-1. Land Cover Classes—Djibouti (Horn of Africa Region)

Land cover class	Area		Growing stock		Sustainable yield	
	km²	Percent	Thousand tonnes	Percent	Thousand tonnes per year	Percent
0	22,132	96.55	0.00	0.00	0.00	0.00
21	790	3.45	260.70	100.00	7.90	100.00
2	790	3.45	260.70	100.00	7.90	100.00
Total	22,922	100.00	260.70	100.00	7.90	100.00
(Percentage of region)	(0.53)		(0.01)		(0.01)	

Note: In the following tables, details may not add to totals because of rounding.
Source: Authors' calculations from data bases derived from land cover classification and table 4-1.

Table 10-2. Land Cover Classes—Ethiopia (Horn of Africa Region)

Land cover class	Area		Growing stock		Sustainable yield	
	km²	Percent	Thousand tonnes	Percent	Thousand tonnes per year	Percent
0	134,742	11.12	0.00	0.00	0.00	0.00
12	2,846	0.23	646.04	0.03	28.46	0.05
13	40,628	3.35	9,222.56	0.46	406.28	0.69
1	43,474	3.58	9,868.60	0.49	434.74	0.74
21	112,241	9.26	37,039.53	1.85	1,122.41	1.91
22	28,034	2.31	9,251.22	0.46	280.34	0.48
23	632	0.05	208.56	0.01	6.32	0.01
2	140,907	11.62	46,499.31	2.33	1,409.07	2.40
41	283,342	23.38	393,845.38	19.71	5,950.18	10.13
43	5,480	0.45	9,272.16	0.46	115.08	0.20
44	31,617	2.61	4,458.00	0.22	1,106.60	1.88
45	21,552	1.78	39,892.75	2.00	452.59	0.77
4	341,991	28.22	447,468.29	22.39	7,624.45	12.98
52	60,178	4.97	129,382.70	6.47	3,791.21	6.46
5	60,178	4.97	129,382.70	6.47	3,791.21	6.46
62	48,163	3.97	128,595.21	6.43	3,708.55	6.32
63	47,478	3.92	126,766.26	6.34	2,089.03	3.56
64	14,122	1.17	37,705.74	1.89	649.61	1.11
65	22,079	1.82	58,950.93	2.95	1,059.79	1.80
6	131,842	10.88	352,018.14	17.61	7,506.98	12.78
73	63,445	5.24	106,841.38	5.35	1,205.45	2.05
74	35,095	2.90	59,099.98	2.96	666.80	1.14
77	207,936	17.16	350,164.22	17.52	3,950.78	0.00
7	306,476	25.29	516,105.58	25.82	5,823.03	9.92
84	35,359	2.92	350,054.10	17.52	14,037.52	23.91
85	6,271	0.52	80,130.84	4.01	9,281.08	15.81
86	2,002	0.17	25,581.56	1.28	2,962.96	5.05
87	3,952	0.33	41,452.53	2.07	5,848.96	9.96
8	47,584	3.94	497,219.03	24.88	32,130.52	54.72
Lakes	4,690	0.39	0.00	0.00	0.00	0.00
Total	1,211,884	100.00	1,998,561.64	100.00	58,720.03	100.00
(Percentage of region)	(27.88)		(42.78)		(42.84)	

Source: Authors' calculations from data bases derived from land cover classification and table 4-1.

Table 10-3. Land Cover Classes—Somalia (Horn of Africa Region)

Land cover class	Area		Growing stock		Sustainable yield	
	km²	*Percent*	*Thousand tonnes*	*Percent*	*Thousand tonnes per year*	*Percent*
0	118,248	18.65	0.00	0.00	0.00	0.00
21	212,257	33.48	70,044.81	13.30	2,122.57	23.31
22	1,475	0.23	486.75	0.09	14.75	0.16
25	422	0.07	139.26	0.03	4.22	0.05
2	214,154	33.78	70,670.82	13.42	2,141.54	23.52
33	685	0.11	685.00	0.13	34.25	0.38
3	685	0.11	685.00	0.13	34.25	0.38
41	215,049	33.92	298,918.11	56.77	4,516.03	49.60
43	64,657	10.20	109,399.64	20.78	1,357.80	14.91
45	5,796	0.91	10,728.40	2.04	121.72	1.34
4	285,502	45.03	419,046.15	79.59	5,995.55	65.85
51	2,477	0.39	5,523.71	1.05	156.05	1.71
52	7,641	1.21	16,428.15	3.12	481.38	5.29
5	10,118	1.60	21,951.86	4.17	637.43	7.00
65	4,110	0.65	10,973.70	2.08	197.28	2.17
6	4,110	0.65	10,973.70	2.08	197.28	2.17
73	1,054	0.17	1,774.94	0.34	20.03	0.22
7	1,054	0.17	1,774.94	0.34	20.03	0.22
83	158	0.02	1,402.72	0.27	78.68	0.86
8	158	0.02	1,402.72	0.27	78.68	0.86
Total	634,029	100.00	526,505.19	100.00	9,104.76	100.00
(Percentage of region)	(14.58)		(11.27)		(6.64)	

Source: Authors' calculations from data bases derived from land cover classification and table 4-1.

Table 10-4. Land Cover Classes—Sudan (Horn of Africa Region)

Land cover class		Area km²	Area Percent	Growing stock Thousand tonnes	Growing stock Percent	Sustainable yield Thousand tonnes per year	Sustainable yield Percent
0		840,068	33.89	0.00	0.00	0.00	0.00
	12	90,478	3.65	20,538.51	0.96	904.78	1.31
1		90,478	3.65	20,538.51	0.96	904.78	1.31
	21	339,463	13.70	112,022.79	5.22	3,394.63	4.90
	22	66,185	2.67	21,841.05	1.02	661.85	0.96
	23	316	0.01	104.28	0.00	3.16	0.00
2		405,964	16.38	133,968.12	6.24	4,059.64	5.86
	41	26,400	1.07	36,696.00	1.71	554.40	0.80
	43	211	0.01	357.01	0.02	4.43	0.01
	44	389,049	15.70	54,855.91	2.56	13,616.72	19.67
	45	9,854	0.40	18,239.75	0.85	206.93	0.30
4		425,514	17.18	110,148.67	5.14	14,382.48	20.77
	52	36,360	1.47	78,174.00	3.64	2,290.68	3.31
5		36,360	1.47	78,174.00	3.64	2,290.68	3.31
	62	363,387	14.66	970,243.29	45.20	27,980.80	40.42
	63	49,586	2.00	132,394.62	6.17	2,181.78	3.15
	64	57,069	2.30	152,374.23	7.10	2,625.17	3.79
	65	149,022	6.01	397,888.74	18.53	7,153.06	10.33
6		619,064	24.97	1,652,900.88	77.00	39,940.81	57.69
	73	23,924	0.97	40,288.02	1.88	454.56	0.66
	74	32,144	1.30	54,130.50	2.52	610.74	0.88
7		56,068	2.27	94,418.52	4.40	1,065.30	1.54
	84	685	0.03	6,781.50	0.32	271.94	0.39
	85	1,370	0.06	17,505.86	0.82	2,027.60	2.93
	86	843	0.03	10,771.85	0.50	1,247.64	1.80
	87	2,055	0.08	21,554.90	1.00	3,041.40	4.39
8		4,953	0.20	56,614.11	2.64	6,588.58	9.52
Total		2,478,469	100.00	2,146,762.80	100.00	232.27	100.00
(Percentage of region)		(57.01)		(45.95)		(50.51)	

Source: Authors' calculations from data bases derived from land cover classification and table 4-1.

11

Central Africa

Terry D. Douglas

This chapter presents a detailed description of the most important land cover classes in this region. Helpful figures in other chapters include figure 3-1 (cloud cover); figures 3-2, 3-3, and 3-4 (NDVI summary land cover profiles); figure 3-5 (regional summary map of land cover classes); figures 7-1 and 7-2 (continental maps of growing stock and sustainable yield); and the "Regional Land Cover Class Map of Central Africa" at the end of this volume.

Helpful tables in other chapters include table 3-2 (land cover classes); table 4-1 (data and sources for growing stock and sustainable yield); and table 6-4 (central Africa estimated woody biomass by summary class).

Class 11—*Veld* Grassland

This class is of only minor importance in central Africa, occurring only in Zaire where it covers 0.6 percent of the land area (14,017 square kilometers). The class takes its name from the extensive (but floristically different) grasslands of South Africa. Many Belgian writers have described the Zaire examples as "steppes." These are distributed in areas along the Angola border, most notably south of Kananga, on interfluves in the Kasai Basin, and in the upper Kwango Basin. These locations lie between 6° S and 9° S.

Debate continues about the origin of these grasslands. According to Mullenders (1954) and Devred (1958), they are secondary grasslands, replacing woodland degraded by fires that maintain the grassland. The dry season in these areas lasts 90 to 120 days (June to September) and annual rainfall ranges from 1,600 to 1,800 millimeters. In Kwango, this class occurs on the extensive Kalahari Sand-covered plateau where edaphic factors are very influential. The principal grasses here are *Aristida vanderystii* and *Loudetia demensii*.

In places the grasslands are very lightly wooded, with shrubby and stunted species such as *Burkeu africana* and *Hymenocardia acida*. In Kaniama, where the grasslands occur on the Plateau of Kasai 600 to 900 meters above sea level, the principal grasses are *Andropogon schirensis* and *Hyparrhenia confinis*. The woody flora is Zambezian in affinity and very sparse, although *Acacia* spp. have been reported. Woody biomass is virtually nonexistent in this class and there is pressure on surrounding land cover classes (*miombo* woodland, Classes 66 and 67) for fuelwood.

Class 25—Edaphic Wooded Grassland

This wooded grassland class represents 1 percent of the land area of central Africa, occurring largely in Congo (22,922 square kilometers) with a contiguous area in Gabon (6,323 square kilometers) and a few isolated occurrences in Zaire (1,686 square kilometers). It is just 0.04 percent of the region's growing stock. The majority of this class occurs in one specific area between 1° S and 3° S in the Plateaux Province of Congo and extending into southeastern Gabon to the east of Franceville. Other small occurrences are immediately to the north of Brazzaville.

Bounding these areas is Ombrophilous Humid Tropical Forest (Class 87) to the north, Seasonal *Miombo* Woodland (Class 66) to the south and west, and forest mosaics (Classes 72 and 73) to the east. The class is almost entirely restricted to the Bateké Plateau, which is composed of soft sandstones. The plateau surface is gently undulating at about 600 to 800 meters above sea level and is dissected by tributaries of the Zaire River, notably the Léfini.

A remarkable feature of this land cover class is that it exists in the wettest part of Congo, where annual

Figure 11-1. NDVI **Profile, Edaphic Wooded Grassland (Class 25)**

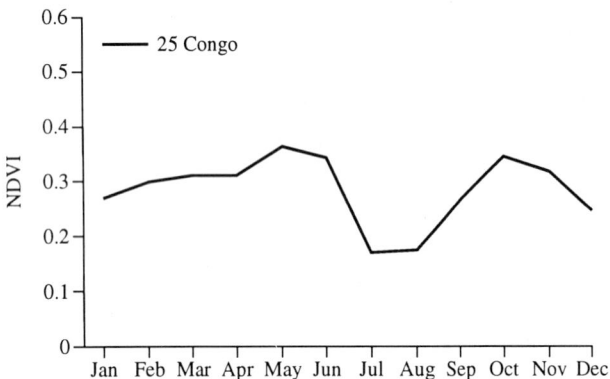

precipitation is between 1,800 and 2,200 millimeters, yet NDVI values are remarkably small for a site so near the equator within the Guineo-Congolian zone. White's map (1983) does not distinguish this area from the surrounding higher biomass groups, although it is particularly distinctive.

For a station near Djambala in the center of this class, NDVI values are consistently below 0.38, falling to 0.17 in the driest month (July) and to 0.26 in December and January (figure 11-1). Although the prime reason for the low productivity of this class is the droughty nature of the plateau soils (hence the name Edaphic Wooded Grassland), regular burning has helped to maintain the grassland area.

The mix of grassland and trees is varied. On the plateau, large areas are devoid of trees or have only a thin scattering, and have been described as savannas. Elsewhere, on steeper valley slopes in the Léfini and Mpama catchments and along the watercourses, dense woodland and gallery forest exist. In addition, small areas of forest occur on the plateau, usually in patches of less than 1 square kilometer. The grasses grow to heights exceeding 2 meters and include *Andropogon* spp., *Hyparrhenia* spp., and *Loudetia demensii*. Many of the tree and shrub species are Sudanian in nature; of particular importance are *Acacia* spp., *Annona senegalensis, Hymenocardia acida*, and *Khaya senegalensis*. In the islands of relict forest there exist *Gilbertiodendron dewevrei, Musanga cecropioides*, and other representatives of the Guineo-Congolian Domain.

This class coincides with scant population density, particularly in the Plateaux Province of Congo. The growing stock estimate for the region is 1 million tonnes with a sustainable yield of 0.3 million tonnes. Near towns such as Djambala in Congo, some pressure for woodfuel clearly exists on the stock, although this is localized.

Class 62—Dry Sudanian Woodland

This class represents 1.6 percent of the land cover of central Africa, but only 0.3 percent of the estimated growing stock. It occurs at the very north of Cameroon and Central African Republic (CAR) around 10° N. In Cameroon, it exists between the slopes of the Mandara Mountains and the lowlands south of Lake Chad. In CAR, it occupies the lowland of the northern tip around Birao. It coincides with White's "Sudanian Undifferentiated Woodland" class; it is regarded as being within the Sahelian domain (Vennetier and Laclavère 1984). Within Cameroon, it covers 35,000 square kilometers (7.7 percent of the land cover) and 28,000 square kilometers (4.5 percent) in CAR.

The dry season is at least as long as the wet, and in northern CAR it lasts from November to May. Rainfall is less than 1,000 millimeters a year, even in the elevated areas of northern Cameroon, and is less than 700 millimeters in northern CAR. The rainfall is sufficient for cultivation, but is unable to support a dense vegetation cover. For northern CAR, NDVI values are less than 0.15 from January to June, increasing to a little more than 0.3 in August and September (figure 11-2). The phenological curves show clearly the basis on which the various Sudanian woodlands have been differentiated.

This class is encroaching southward as the Sahel undergoes desertification in response to cultivation, grazing, and climate change. In CAR, much of this zone has been designated a "Zone d'alarme" by the authorities in recognition of the desertification threat.

Characteristic tree species include *Acacia seyal, Combretum aculeatum*, and *Ziziphus abyssinica*. The thorn woodland is sparse and in places discontinuous, much of it being replaced by secondary thicket. Cultivation in this class is predominantly of sorghum and groundnuts. With increasing population and shifting agriculture, the fuelwood resource is threatened. The esti-

Figure 11-2. NDVI **Profiles, Sudanian Woodland (Classes 62, 64, and 65)**

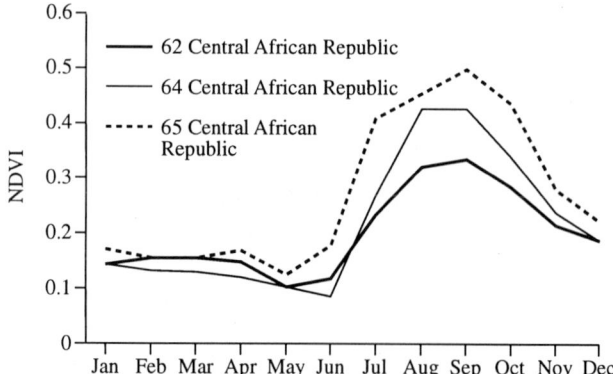

mated growing stock in CAR is 75 million tonnes and in Cameroon 93 million tonnes, with sustainable yields of 2.2 million tonnes and 2.7 million tonnes, respectively.

Class 64—Sudanian Woodland

This class covers just 1 percent of central Africa but accounts for only 0.5 percent of the growing stock. It is restricted to the northernmost fringes of the region. In Cameroon, it occupies 21,974 square kilometers (4.9 percent of the area). In CAR, it occupies 27,000 square kilometers (4.5 percent of the area). It occurs as a belt between the Dry Sudanian Woodland (Class 62) to the north and the Moist Sudanian Woodland (Class 65) to the south.

It falls within the class described by White (1983) as "Sudanian Woodland with Abundant *Isoberlinia.*" In Cameroon, it occurs up to 1,800 meters altitude in the Alantika Mountains and up to 1,400 meters in the Mandara Mountains. It stretches as a belt into the south of Chad and thence into northern CAR and southern Sudan. In CAR, it exists in the Sudan-Sahelian climatic zone to the north of Ndélé, with annual rainfall of 1,000 to 1,300 millimeters. Here the topography is largely below 500 meters and is developed on the Quaternary deposits of the basin, which drains northward into Lake Chad. The dry season extends from November to March and thus seasonality is quite pronounced. For this class in eastern CAR, NDVI values are similar to Dry Sudanian Woodland (Class 62) during the dry season, but increase to more than 0.4 in August and September (figure 11-2).

This class is intermediate between "*Isoberlinia* Woodland" and the "Dry Sudanian Woodland" to the north. It is deciduous and the woody species in CAR include *Acacia* spp., *Combretum* spp., *Commiphora* spp., *Isoberlinia doka*, *Khaya senegalensis*, and *Terminalia laxiflora*. At a local scale, the vegetation may be linked to catenary patterns (Barber, Buchanan, and Galbreath 1980), which are dependent on the distribution of laterite, floodplains, and drainage. Cultivation has substantially modified much of this class, with tree degradation occurring in the more populated areas. The French term for such areas is "Mosaïque de Savanne Arborée, de Savanne Arbustive et de Culture." In the more elevated parts of the class in northwestern Cameroon, *Isoberlinia doka* and other typical trees of the Sudanian sector often have been replaced by domesticated trees, which include *Acacia albida*, *Celtis integrifolia*, *Khaya senegalensis*, and *Parkia biglobosa* (Laclavère 1980).

The fuelwood potential of this class is partly threatened by the extensive cultivation within it, notably sorghum with some maize, cassava, and groundnuts.

Estimates of growing stock are 133 million tonnes for the region and a sustainable yield of 2.3 million tonnes.

Class 65—Moist Sudanian Woodland

This is the most extensive woodland class in central Africa north of the equator and represents 9 percent of the land area of the region. The growing stock is estimated to account for just 4.58 percent. The class occurs largely within White's "Sudanian *Isoberlinia* and Related Woodlands" class, but overlaps the northern edge of the Guineo-Congolian mosaic. It occupies a broad latitudinal belt from 2° N to 9° N. It is by far the largest land cover class in CAR, covering 263,000 square kilometers (42.1 percent) and is important in Cameroon (65,000 square kilometers, 14.4 percent) and northern Zaire (43,157 square kilometers, 1.9 percent). In Cameroon, it impinges on the northern edge of the Mesophilous Humid Tropical Forest (Class 85). In CAR, it occupies most of the central area on plateau surfaces between 500 and 900 meters, where it exists alongside Guinean Woodland (Class 74). In northern Zaire, it occurs in Oubangui and Haut Uele in the extreme northeast near the Sudan border.

This broad belt of woodland is characterized by annual rainfalls of 1,400 to 1,600 millimeters and dry seasons usually lasting 3 to 4 months from November to at least March. The values for NDVI increase to 0.5 in the wettest month (figure 11-2) and show smaller values during the dry season.

As is the case with many of the Sudanian classes in central Africa, human activity has played an important role in altering the natural Moist Sudanian Woodland, with pressure on tree cover and considerable evidence of burning and the creation of secondary savanna. The natural woodland often is dominated by *Isoberlinia* spp., which form a discontinuous woodland cover with a canopy 10 to 15 meters high. Other tree species include *Burkea africana*, *Daniellia oliveri*, *Erythrophleum africanum*, and *Monotes kerstingii*.

Jongen and others (1960) studied an area south of the River Oubangui in northwestern Zaire (Bangala). They identified several communities at a small scale, with riparian forests containing *Guibourtia demeusei* and *Uapaca heudelotii*. They observed more open woodland on the sandy interfluves and yet other variants on the laterite crusts. In this region, *Isoberlinia* spp. are rare, and many of the tree species, such as *Gilbertiodendron dewevrei*, are derived from the humid tropical forests to the south.

In many areas, for example in western CAR, this class contains a quite dense rural population and, although woody biomass growing stock is estimated to be quite substantial, in some areas of greater population its sustainability may be threatened. In eastern CAR, along the border with Sudan, the population density is much lower and fuelwood depletion is not currently a problem. Growing stock and sustainable yield are estimated as follows:

Country	Growing stock (million tonnes)	Sustainable yield (million tonnes)
CAR	702	12
Cameroon	174	3
Zaire (northern)	115	2

Class 66—Seasonal *Miombo* Woodland

Seasonal *Miombo* Woodland is the most extensive land cover class recognized in this study in central Africa. It covers 21 percent of the region and accounts for 12.4 percent of the growing stock. This class exists in a broad belt stretching from Gabon (19,000 square kilometers) through southern Congo (62,000 square kilometers) and across Zaire (768, square kilometers) from Bas Zaire through Kasai and Shaba. The class has been widely studied in Zaire (Malaisse 1978) and has been identified by Millington and Townshend (1989) in the SADC region from 1984 AVHRR NDVI imagery.

Because this land cover class spans such a wide area in central, southern, and East Africa, it does not coincide with a particular climatic type, although its marked seasonality is a diagnostic feature. Rainfall commonly is between 900 and 1,300 millimeters a year, but this disguises considerable year-to-year variability, which in the Lubumbashi area of southern Zaire may be as much as 700 to 1,500 millimeters (Malaisse 1978). *Miombo* woodlands have been divided into wetter and drier types. White (1983) separated these at the 1,000-millimeter mean annual rainfall isohyet, whereas Chidumayo (1987) used 1,100 millimeters.

The values of NDVI vary according to latitudinal position and other climatic factors. Southern Gabon woodlands exhibit the least seasonality in the region, with values between 0.27 and 0.5. In southern Zaire, the dry season is longer (May to September), with NDVI values reducing to 0.22, but increasing to as much as 0.57 at the wet season of November to March (figure 11-3).

Seasonal *Miombo* Woodland exhibits a structure of spaced trees, which sometimes nearly interlock to form umbrellalike canopies. Where agricultural clearance and burning are extensive, the woodland is often more open. The trees are deciduous in response to the marked dry-season lull in productivity. The undergrowth is rarely as well developed as in savannas. The Seasonal *Miombo* Woodlands are mainly restricted to the Zambezian phytogeographical region, although

Figure 11-3. NDVI **Profiles,** *Miombo* **Woodland (Classes 66 and 67)**

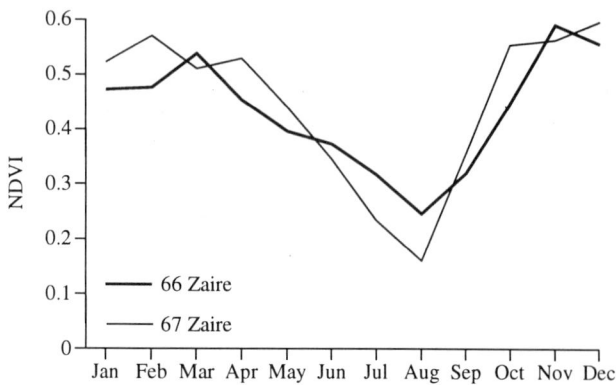

the northern edge of this class is within the Guineo-Congolian region. Hence, trees such as *Marquesia macroura* are dominant to the north and *Brachystegia* spp., *Isoberlinia* spp., and *Julbernardia* are dominant in southern Zaire. Where an understory is present, *Protea* spp. and *Uapaca* spp. exist. Malaisse (1984, 1985) has discussed the structure of this woodland in the Shaba Province of Zaire near Lubumbashi. He describes a transect at Luisnishi, 28 kilometers northeast of Lubumbashi, where the tree layer attains 9 to 19 meters in height with a few dominant species reaching 22 meters and, exceptionally, 29 meters. An inventory of tree species with diameters greater than 10 centimeters showed *Brachystegia spiciformis* to be the most abundant species, with 150 individuals per hectare, followed by *Aidia micrantha* (63), *Syzygium guineense* (52), *Parinari excelsa* (49), *Brachystegia taxifolia* (43), and twelve other species with densities of more than six per hectare.

The Seasonal *Miombo* Woodlands are a very important source of fuelwood. Malaisse and Binzangi (1985) have considered the depletion of trees by fuelwood gathering in Upper Shaba. For principal centers such as Lubumbashi, surrounding woodland has been clear-cut throughout a circle of radius 30 kilometers, and this zone is estimated to be expanding at about 1 kilometer a year. Projections indicate complete deforestation of more than 10,000 square kilometers surrounding Lubumbashi by the year 2050, although just 20 percent of this area applied to forestry would be sufficient to satisfy the woodfuel needs of the population. Growing stock in this class in Zaire is 4,270 million tonnes, with a sustainable yield of 68 million tonnes.

Class 67—Wet *Miombo* Woodland

Wet *Miombo* Woodlands form an intermittent band largely to the north of the Seasonal *Miombo* Woodland (Class 66). They occur almost entirely in Zaire where they cover 268,000 square kilometers and stretch from the Atlantic coast in Bas Zaire to the Rift Valley. This class covers 7 percent of central Africa and accounts for an estimated 4.7 percent of the growing stock. It occurs in Kwango, on the Angolan border, and as several patches throughout Kasai and Shaba on the plateau at elevations of about 1,000 meters. It also is prominent on the hills near Kalemie to the west of Lake Tanganyika, where it extends over 1,500 meters above sea level. In southern Zaire, near the border with Zambia, Wet *Miombo* and Seasonal *Miombo* Woodlands are intermixed on the higher plateau area to the south of Lubumbashi.

This class often occurs on light sandy soils of the Kasai and Shaba plateaus and on the well-drained soils of the hills near Lake Tanganyika. Such edaphic factors have created an open woodland with extensive grassland cover. Human interference has meant that woody species are under pressure, and a well-developed shrub layer is rare within these woodlands.

Wet *Miombo* Woodland in Zaire generally exists in wetter areas than the Seasonal *Miombo* Woodland (Class 66) and in areas where the effect of seasonality is less. Rainfall is generally 1,000 to 1,700 millimeters a year and the dry season is usually shorter than 3 months, although it is longer in Bas Zaire between Cabinda and Angola. As a result, annual productivity is greater than for Seasonal *Miombo* Woodland, with NDVI values falling below 0.35 for only 2 months (usually July and August). In the wetter parts of central Zaire, NDVI values reach 0.58 during the wet season. The phenology shows marked seasonality, with a big reduction in productivity during the dry season. The values of NDVI for a site in eastern Zaire show wet season values well above 0.5 from October to April, but a sharp reduction in NDVI to 0.15 in August (figure 11-3).

The species occurring at the northern limit of this class often include many from the humid tropical forest, but *Brachystegia* spp., *Isoberlinia* spp., *Julbernardia* spp., and *Marquesia macroura* are dominant. Also common, however, are *Acacia* spp., *Annona* spp., *Piliostigma thonningii*, and *Terminalia sericea*. Quite dense thickets exist in the understory together with a mixed herbaceous layer.

This class accounts for an estimated 3,191 million tonnes of growing stock in Zaire alone. The class is associated with quite high human population densities in some areas, especially near Kananga, Mbuji-Mayi, and Kalemie. Maize, cassava, groundnuts, and some cotton are the main agricultural crops.

Class 72—Cultivation and Forest/Woodland Mosaic

This class occurs south of the equator, generally at the southerly edge of the Ombrophilous Humid Tropical Forest (Class 87) and toward the northern limit of the

miombo woodlands (Classes 66 and 67). Within central Africa it represents 5 percent of the land area and accounts for an estimated 6.7 percent of the growing stock. It is particularly important in Congo (15.2 percent, 51,000 square kilometers) and Zaire (5.3 percent, 120,000 square kilometers); but it also occurs in Gabon. The greatest extent of this land cover class is in central Congo, to the north of Brazzaville and in the neighboring parts of Bandundu Province in Zaire. The class also occurs in scattered pockets from central Gabon eastward throughout central Zaire to Rwanda and Burundi. This class exists within the area mapped by White (1983) as "Guineo-Congolian Mosaic of Lowland Rain Forest and Secondary Grassland."

The climate of this belt is characterized by a dry season which extends for 90 to 120 days between June and December (Laclavère 1978) and has a rainfall between 1,200 and 1,800 millimeters a year. The values of NDVI are consistently about 0.35 in the growing season, diminishing to 0.23 in the dry season.

The high woody biomass of this class results from being a forest remnant, secondary forest, or well-wooded agricultural land with large patches of cultivation. The main crops include manioc, maize, groundnuts, bananas, and vegetables. The trees are varied, with some species occurring in the forest belts to the north—*Celtis zenkeri*, *Chlorophora excelsa*, *Terminalia superba*, and *Triplochiton scleroxylon*, and some that have been introduced, such as the oil palm (*Elaeis guineensis*). The presence of many light-demanding species is attributed to the open nature of the forest in this mosaic, which is probably a function of disturbance and regrowth.

Woody biomass productivity and sustainable yield are quite significant and fuelwood shortages are likely to occur only in areas of concentrated cultivation. The regional growing stock estimated for this class is 320 million tonnes, and the sustainable yield is 3.6 million tonnes.

Class 73—Cultivation and Forest Regrowth Mosaic

This land cover class occurs in two extensive areas and as a series of smaller patches. The area of greatest extent exists in Gabon, where it extends from Río Muni (Equatorial Guinea) in the north, occupying a sector from the coast near Libreville and through much of the center of the country in the basin of the Ogooué River. In this area, commercial timber has been exploited (Walker and Sillars 1967), and much of the area is now cultivated.

It is the second most important land cover class in Gabon, covering 21 percent of the land area (56,000 square kilometers). In Cameroon, 47,000 square kilometers are covered with this mosaic, some 9.4 percent of the land area. It occurs in the coastal lowlands to the south of Douala and on the border with Nigeria to the north of Mount Cameroon, where it attains altitudes to 1,500 meters. Elsewhere it exists in patches of varying size in eastern and southeastern Cameroon on the plateau at 500 to 700 meters and in patches throughout the Guinean-Sudanian transition in CAR and northernmost Zaire. In CAR it covers 34,000 square kilometers and in Zaire 32,000 square kilometers.

Within central Africa as a whole, it covers 4 percent of the land area and accounts for 2 percent of the growing stock. As a class, it does not correspond to any of White's mapping units (White 1983). It occurs in conjunction with the forest classes and other high woody biomass mosaics (most notably Guinean Woodland, Class 74, in CAR) and at the southern margin of the Sudanian Woodland. The values of NDVI for this class in Gabon near the equator show a range of moderate values attaining 0.5 (figure 11-4).

The key to understanding this extensive mosaic is that in most locations it has been greatly affected by human interference. In West Africa, this land cover class is sometimes known as "Farm Bush." Elsewhere it corresponds with the "secteur Préforestier" (see Chapter 9, Classes 75 and 76). It is largely evergreen in the wetter, western parts of central Africa and semi-deciduous at the drier end of the range in eastern CAR and northeastern Zaire. Extremely high rainfalls of 2,000 to 4,000 millimeters a year are associated with this class on the Cameroon-Nigeria border, whereas in the east, annual rainfall is only 1,300 millimeters.

In a class that covers many ecological zones, the tree species are of course quite varied. In Cameroon, species from the "Atlantic Evergreen Forest" dominate, notably Caesalpinaceae, whereas in Gabon and northern Zaire, species from the Mesophilous Humid Tropical Forest (Class 85) occur. The remaining forest in this class is inevitably secondary, with light-demanding tree species such as *Chlorophora excelsa*, *Khaya anthotheca*, *Musanga cecropioides*, and *Terminalia superba*.

Figure 11-4. NDVI Profile, Cultivation and Forest Regrowth Mosaic (Class 73)

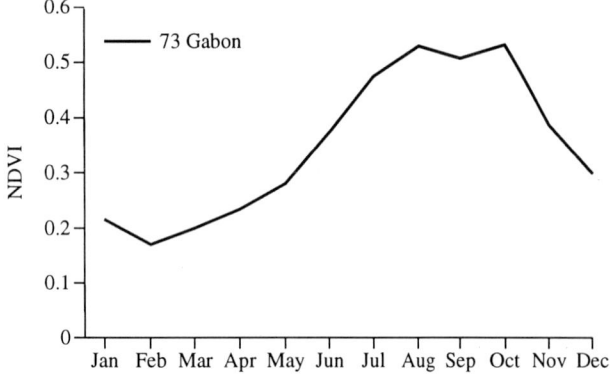

Cultivation throughout this class includes large areas of manioc. Other crops often are grown in areas of secondary forest.

Fuelwood is relatively plentiful, as indicated by the large growing stock of 285 million tonnes for the region. Sustainable yield is estimated at 3.1 million tonnes.

Class 74—Guinean Woodland

This class occurs in a belt to the north of the Mesophilous Humid Tropical Forests (Class 85) where it shares a broad zone between 4° N and 8° N with Moist Sudanian Woodland (Class 65). This zone runs from the west of Cameroon, where the Guinean Woodlands are rather fragmented, through central CAR. It continues to its southern extremity in northern Zaire to the south of the Oubangui River and in the vicinity of the Parc National de la Garamba on the border with Sudan. The class constitutes 6 percent of the land area of central Africa and is 1.6 percent of the growing stock. This includes 26.3 percent of CAR (165,000 square kilometers) and 13 percent of Cameroon (60,000 square kilometers).

With the exception of occurrences in Cameroon on Massif de l'Adoumaoua above 1,000 meters, the Guinean Woodlands generally exist at altitudes between 400 and 800 meters. The rainfall of this area exceeds 2,000 millimeters a year in central Cameroon but is typified by annual ranges of 1,300 to 1,600 millimeters in CAR. The wettest months are July to October and the driest December to March. This is reflected in the vegetation productivity with NDVI values in central CAR ranging from 0.16 in February to 0.52 in August.

Ecologically, this class occupies a transitional zone between the Guinean-Congolian floristic region to the south and the Sudanian floristic region to the north (White 1983). As a result of the greater rainfall, woody vegetation is quite dense in places, with trees and shrubs separated by grasslands where burning for agriculture often has created local patches of herbaceous wooded savanna (Laclavère 1980). Gallery forests often exist along water courses, being more frequent at the southern edge of the zone. The main tree and shrub species include *Burkea africana*, *Daniellia oliveri*, *Hymenocardia acida*, *Lophira lanceolata*, and *Uapaca heudelotii*. Many of these species have high calorific values and are important sources of fuelwood and high-quality charcoal.

Growing stock for this class is estimated at 440 million tonnes and sustainable yield at 4.8 million tonnes. The combination of closed canopy woodland and high sustainable yield with low to moderate population density indicates few woody biomass supply problems. Only when supplies from the neighboring Sudanian Woodlands become scarce is this class likely to come under pressure.

Class 82—Evergreen Forest

This class occupies less than 2 percent of central Africa and represents an estimated 1.9 percent of the growing stock. It occurs almost exclusively in eastern Zaire, where it represents a transition between the lowland humid tropical forests of the Zaire Basin and the montane classes with less woody biomass on the western wall of the Rift Valley. Leonard (1965) identifies this area as representing the eastern extremity of the Guineo-Congolian phytogeographic zone.

The class is represented by a scattering of small areas among woodlands and areas of secondary forest and cultivation; no one area is particularly extensive. This pattern probably results from the widespread clearance this class has undergone because it occurs on accessible terrain between 1,100 and 1,800 meters above sea level. The greatest concentrations exist on the Dorsale du Kivu and have been described in some detail by Pecrot and Leonard (1960) and mapped by Devred (1960).

The climate of this area is moist with the high humidities typical of cloudy montane regions. Rainfall is greatest on the upper slopes and temperatures highest toward the lowland margin, so a complex mix of forest types that changes rapidly with altitude is encountered; this pattern is largely controlled by climate and altitude. The phenology shows a wet season of 6 months from November to April, with NDVI values reaching 0.58 and a drier period from June to October with NDVI values below 0.3 (figure 11-5).

The canopies of these forests are rarely as high as those in the Zaire Basin, with the upper stratum attaining 20 to 25 meters or less at higher altitudes. The lower slopes are characterized by contact with the lowland rain forest and species such as *Cynometra alexandri, Julbernardia seretii, Lebrunia bushaie, Pentadesma lebrunii,* and *Symphonia globulifera* are commonly encountered. Higher up, *Aningeria adolfi–friedericii, Ficalhoa laurifolia, Ocotea usambarensis,* and *Podocarpus* spp.

Figure 11-5. NDVI **Profiles, Forests (Classes 82, 85, and 87)**

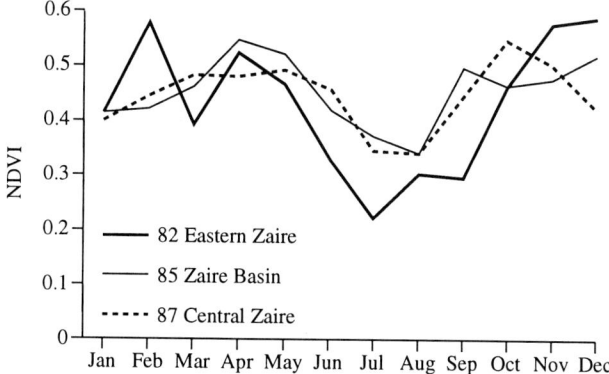

occur. On the higher slopes, the forest often exists only on the steepest rocky slopes where it has escaped clearance.

In eastern Zaire, this zone is moderately well populated and much of the forest has been cleared. Growing stock is estimated for Zaire at 434 million tonnes, with a sustainable yield of 36 million tonnes.

Most of the areas occupied by this land cover class are rural, with few large urban centers. Wood is not as a rule exported from here and the only threat is further clearance for agriculture.

Class 85—Mesophilous Humid Tropical Forest

Mesophilous Humid Tropical Forest is marginal to the narrow belt of the Ombrophilous Humid Tropical Forest (Class 87). It occupies 10 percent of the land cover of central Africa and is an estimated 18 percent of the growing stock. Its distribution in central Africa is best described with reference to those areas north and south of the equator.

To the south of the equator, a coastal belt stretches from Cabinda in an unbroken swathe across Congo and reaches its broadest development in southern Gabon, where it stretches from the coast for 200 kilometers inland. A few other small patches exist in Zaire to the south of the main area of Ombrophilous Humid Tropical Forest (Class 87).

To the north of the equator, this class is much more extensive. Here it is an eastward continuation of the West African peripheral domain of the Guineo-Congolian phytogeographic region (Leonard 1965). It extends from the Nigerian border, where the pattern is broken by the uplands of the Bamileke Plateau, throughout central Cameroon into the north of Río Muni, Gabon, and Congo. It continues via the southern tip of CAR to northern Zaire, where it occurs as a large area in Haut Zaire along the tributaries of the Oubangui.

Zaire has by far the largest area within this land cover class of any country in Africa—more than 200,000 square kilometers (8.9 percent), whereas Cameroon has 19.6 percent of its area within this class, about 88,000 square kilometers. Gabon has 36,000 square kilometers; Congo 26,000 square kilometers; CAR 40,000 square kilometers, and Río Muni 2,500 square kilometers.

Phenologically, this class demonstrates greater seasonality than Ombrophilous Humid Tropical Forest (Class 87). Because some of the area of this class corresponds to White's "Drier Guineo-Congolian Rain Forest," it is not surprising that NDVI values are lower. For a Southern Hemisphere site in southern Gabon, NDVI values increase from 0.32 in January to 0.46 in June, reducing dramatically in the dry season to about 0.12 between July and September, and increasing to 0.35 again by November.

In Haut Zaire, along the border with CAR in the Oubangui Basin, these forests occupy land above 500 meters altitude. Here the NDVI values increase to greater than 0.48 from September through to May, and reduce to 0.36 in July and August (figure 11-5). In central Cameroon, to the east of Yaoundé, on the south Cameroon plateau above 500 meters, a similar NDVI pattern exists to that in Haut Zaire, but with slightly increased seasonality, indicating 2 months with low productivity—June (0.35) and December (0.27).

These Mesophilous Humid Tropical Forests, being farther from the equator than either the Humid Tropical Swamp Forests (Class 86) or the Ombrophilous Humid Tropical Forests (Class 87), are generally characterized by a more definite dry season, often with a distinct period of 2 dry months. Rainfall for this land cover class ranges from 1,200 to 1,600 millimeters a year, although some parts of western Cameroon are wetter. Most tree species are briefly deciduous, although few trees are leafless at the same time; hence the widely applied term "semideciduous."

The broad extent of this forest type demonstrates a wide range of species. Some areas have dominant species, whereas others are characterized by a considerable mixture. In the Mayumba Forest of western Congo and southern Gabon, the most important trees are *Chlorophora excelsa, Klainedoxa gabonensis, Petersianthus macrocarpus, Sarcocephalus diderrichii,* and *Terminalia superba.* To the north of the equator, these forests often are floristically better characterized than in Gabon-Congo, with the occurrence of species such as *Celtis zenkeri, Chlorophora excelsa, Musanga cecropioides,* and *Piptadeniastrum africanum.* Most of these more common species are widely distributed, from Cameroon through southernmost CAR, in northern Congo, and in the north of the Zaire Basin toward the border with Sudan.

These forests often are inaccessible and clearly represent a significant woody biomass resource. Agriculture is practiced in many areas of this land cover class, but forest clearance has been local rather than extensive. Growing stock for the region is an estimated 5,033 million tonnes, with a sustainable yield of 583 million tonnes.

Class 86—Humid Tropical Swamp Forest

This class is restricted to a narrow belt near the equator, running eastward from the coast of Río Muni through northern Gabon, southernmost Cameroon, Congo, and into Zaire in the province of Equateur. Here it extends to the north of Mbandaka between the Oubangui and Zaire rivers. Other isolated areas exist within the Zaire Basin. It is an important class, representing 6 percent of the area of central Africa and 10.1 percent of the growing stock.

It is principally surrounded by Ombrophilous Humid Tropical Forest (Class 87) and includes much of the area mapped by White as "Swamp Forest." Many authors (for example, Lebrun and Gilbert 1954) have subdivided swamp forest to distinguish between riparian or floodplain forest and forest that occupies drained sites (often with waterlogged, clay soils). Many swamp forests are periodically inundated with water, especially during the wet season.

Climatically, this class occurs in conditions very similar to those of Ombrophilous Humid Tropical Forest (Class 87), with annual rainfall exceeding 1,500 millimeters. The NDVI values show little seasonality, as would be expected of a class occupying an area so close to the equator.

Phenologically, these forests are separated from the Ombrophilous Humid Tropical Forests (Class 87) of the drier soils by NDVI levels that are slightly smaller overall. The depression of NDVI values may be attributed in part to the presence of standing surface water, which may appear through gaps in the evergreen tree canopy. Its low reflection coefficient will depress NDVI values in mixed pixels of standing water and tree canopy.

Wetter sites are host to species such as *Guibourtia demeusei, Mitragyna* spp., *Raphia* spp., *Symphonia globulifera,* and *Uapaca guineensis.* Badly drained sites support similar species, as well as *Phoenix reclinata* and *Xylopia rubescens.*

Accessibility is very limited and population density is generally very low in this class. Growing stock is estimated at 2,830 million tonnes for the region, with a sustainable yield of 327 million tonnes.

Class 87—Ombrophilous Humid Tropical Forest

This class represents areas with consistently greater NDVI values and corresponds to White's "Wetter Guineo-Congolian Rain Forest" and "Les Forêts Ombrophiles Sempervirentes Equitoriales" of Lebrun and Gilbert (1954). It occupies a belt centered on the equator and largely within 3° or 4° of it, although a large outlier exists in central CAR at 8° N. Its broadest extent is in the center of the Zaire Basin on terrain between 300 and 500 meters above sea level. It is a particularly important class for woody biomass, covering 20 percent of central Africa and providing an estimated 30.9 percent of the growing stock.

It forms a significant element in the vegetation of all central African countries, with more than half a million square kilometers in Zaire (24 percent of the area); 95,000 square kilometers in Congo (28 percent); 55,000 square kilometers in Gabon (21 percent); 63,000 square kilometers in Cameroon (14 percent); 52,000 square kilometers in CAR (8 percent) and 5,000 square kilometers in Río Muni (22 percent).

To the east, this class merges with Evergreen Forest (Class 82), to the north and west with the Mesophilous Humid Tropical Forest (Class 85), and to the south with the two *miombo* woodlands (Classes 66 and 67). Ombrophilous Humid Tropical Forest often occurs on better-drained soils of the Zaire Basin. Within it, on riverine sites, patches of Humid Tropical Swamp Forest (Class 86) exist.

The phenology of Ombrophilous Humid Tropical Forest reflects the generally even distribution of rainfall of the equatorial belt and the lack of environmental stress on plant growth. The values of NDVI are consistently above 0.35 and usually are about 0.5 in the wettest months. The least variation is displayed by sites in the vicinity of Kisangani (Zaire), whereas with increasing distance from the equator, a measure of seasonality occurs, with smaller values notable in July and January. This class as a whole, however, is differentiated from surrounding land cover classes on two criteria, namely the lack of appreciable seasonality and consistently large NDVI values (figure 11-5). Rainfall within the region covered by Ombrophilous Humid Tropical Forest is evenly distributed throughout the year, with a mean annual total in the range 1,500 to 1,900 millimeters, although it may be locally greater.

The ecology of this type of rain forest is dominated by tall, closed-canopy tree species which typically attain 35 to 45 meters in height. The fairly evenly distributed rainfall with only a short dry season creates a semi-evergreen forest with a mixture of evergreens and some species that briefly shed their leaves.

Few detailed descriptions of this forest have been published from the area of its largest extent in central Zaire. Here, as in much of the Zaire Basin, *Brachystegia laurentii* and *Gilbertiodendron dewevrei* often are dominant in the upper stratum of the forest.

Fahem (1978) reports that *Gilbertiodendron* spp. in northeastern Zaire (Ituri) sometimes represent 80 percent of the canopy species. *Cynometra alexandri, Julbernardia seretii, Oxystigma ozyphyllum,* and *Scorodophloeus zenkeri* also are important, but gradual changes occur in composition throughout the large area covered by this class. For instance, in the area of this land cover class occurring in southwestern Cameroon, Cesalpinaceae are dominant. A detailed study of species in southern Cameroon is provided by Vivien and Faure (1985). The prominent northerly outlier of this class in central CAR is less dense than that of the Zaire Basin and includes species such as *Erythrina tomentosa* and *Ochthocosmus africanus*.

Clearly, this belt of Ombrophilous Humid Tropical Forest currently is one of the world's greatest resources of woody biomass. Growing stock is estimated to be 8,644 million tonnes, with sustainable yields of 1,218 million tonnes. Much of this forest remains generally intact with very low population densities.

Land Cover Class Tables

Tables 11-1 through 11-6, beginning on page 115, present summaries for each land cover class of the area, showing growing stock and sustainable yield for the central African nations of Cameroon, Central African Republic, Congo, mainland Equatorial Guinea, Gabon, and Zaire.

References

Every effort has been made to facilitate access to the documents listed here. Some documents, however, lack full bibliographic information because it was unavailable; also, some documents are of limited circulation.

Barber, K. B., S. A. Buchanan, and P. F. Galbreath. 1980. *An Ecological Survey of the St. Floris National Park, Central African Republic.* Washington, D.C.: International Park Affairs Division, National Park Service, U.S. Department of the Interior.

Chidumayo, E. N. 1987. "Species Structure in Zambian *Miombo* Woodland." *Journal of Tropical Ecology* 3(2): 109–18.

Devred, R. 1958. "La végétation forestière du Congo et du Ruanda-Urundi." *Bulletin de la Société Forestière Belge* 65(6):409–68.

Devred, R. 1960. "La cartographie de la végétation au Congo Belge." *Bulletin Agricole de Congo Belge* 51(3): 529–41.

Fahem, A. K. 1978. "Végétation." In G. Laclavère, *Atlas de la République du Zaire.* Paris: Les Editions Jeune Afrique.

Jongen, P., M. Van Oosten, C. Evrard, and J. M. Berce. 1960. *Notice explicative de la carte des sols et de la végétation.* No. 11, Ubangi [region]. Brussels: INEAC (L'Institut National pour l'Etude Agronomique du Congo Belge).

Laclavère, G. 1978. *Atlas de la République du Zaire.* Paris: Les Editions Jeune Afrique.

Laclavère, G. 1980. *Atlas of the United Republic of Cameroon.* Paris: Les Editions Jeune Afrique.

Lebrun, J., and G. Gilbert. 1954. "Une classification écologique des forêts du Congo." *Série Scientifique* No. 63. Brussels: INEAC (L'Institut National pour l'Etude Agronomique du Congo Belge).

Leonard, J. 1965. "Contribution à la subdivision phytogéographique de la région Guineo-Congolaise d'après la répartition géographique d'euphorbiacées d'Afrique tropicale." *Webbia* 19(2): 627–49.

Malaisse, F. 1978. "The *Miombo* Ecosystem." In UNESCO (United Nations Educational, Scientific and Cultural Organization), *Tropical Forest Ecosystems.* Paris: UNESCO.

Malaisse, F. 1984. "Structure d'une forêt dense sèche Zambezienne des environs de Lubumbassi." *Bulle-*

tin de la Société Royale de Botanique de Belgique 117(2): 428–58.

Malaisse, F. 1985. "Comparison of the woody structure in a regressive Zambezian succession." *Bulletin de la Société Royale de Botanique de Belgique* 118(2): 244–65.

Malaisse, F., and K. Binzangi. 1985. "Wood as a Source of Fuel in Upper Shaba." *Commonwealth Forestry Review* 64(3):227–39.

Millington, Andrew, and J. Townshend. 1989. *Biomass Assessment*. London: Earthscan.

Mullenders, W. 1954. "La végétation de Kaniama." *Série Scientifique* No. 61. Brussels: INEAC (L'Institut National pour l'Etude Agronomique du Congo Belge).

Pecrot, A., and A. Leonard. 1960. *Notice explicative de la carte des sols et de la végétation*. No. 16, Dorsale du Kivu [region]. Brussels: INEAC (L'Institut National pour L'Etude Agronomique du Congo Belge).

Vennetier, P., and G. Laclavère. 1977. *Atlas de la République Populaire du Congo*. Paris: Les Editions Jeune Afrique.

Vivien, J., and J. J. Faure. 1985. *Arbres des forêts denses d'Afrique Centrale*. Paris: Agence de Coopération Culturelle et Technique.

Walker, A., and R. Sillars. 1967. *Les plantes utiles du Gabon*. Paris: Editions Paul Lechevalier.

White, F. 1983. "The Vegetation of Africa." *Natural Resources Research Series* 20. Paris: UNESCO/AETFAT/UNSO (United Nations Educational, Scientific and Cultural Organization/Association pour l'Etude Taxonomique de la Flore de l'Afrique Tropicale/United Nations Sudano-Sahelian Office).

Table 11-1. Land Cover Classes—Cameroon (Central Africa Region)

		Area		Growing stock		Sustainable yield	
Land cover class		km^2	Percent	Thousand tonnes	Percent	Thousand tonnes per year	Percent
0		105	0.02	0.00	0.00	0.00	0.00
	12	2,687	0.59	609.95	0.02	26.87	0.01
1		2,687	0.59	609.95	0.02	26.87	0.01
	21	105	0.02	34.65	0.00	1.05	0.00
	22	2,635	0.58	869.55	0.03	26.35	0.01
	23	790	0.17	260.70	0.01	7.90	0.00
2		3,530	0.77	1,164.90	0.04	35.30	0.01
	41	158	0.03	219.62	0.01	3.32	0.00
	43	632	0.14	1,069.34	0.03	13.27	0.00
	44	10,961	2.43	1,545.50	0.05	383.64	0.12
4		11,751	2.60	2,834.46	0.09	400.23	0.12
	52	685	0.15	1,472.75	0.05	43.16	0.01
5		685	0.15	1,472.75	0.05	43.16	0.01
	62	34,779	7.70	92,859.93	3.00	2,677.98	0.82
	63	685	0.15	1,828.95	0.06	30.14	0.01
	64	21,974	4.86	58,670.58	1.89	1,010.80	0.31
	65	65,289	14.45	174,321.63	5.62	3,133.87	0.96
6		122,727	27.16	327,681.09	10.57	6,852.89	2.10
	73	42,525	9.41	71,612.10	2.31	807.98	0.25
	74	59,915	13.26	100,896.86	3.26	1,138.38	0.35
7		102,440	22.67	172,508.96	5.57	1,946.36	0.60
	81	8,906	1.97	219,800.08	7.09	26,263.79	8.03
	84	53	0.01	524.70	0.02	21.04	0.01
	85	88,370	19.55	1,129,191.86	36.44	130,787.60	39.99
	86	45,792	10.13	585,130.18	18.88	67,772.16	20.72
	87	62,750	13.88	658,184.75	21.24	92,870.00	28.40
8		205,871	45.54	2,592,831.57	83.66	317,714.59	97.15
Lakes		2,161	0.48	0.00	0.00	0.00	0.00
Total		451,957	100.00	3,099,103.68	100.00	327,019.30	100.00
(Percentage of region)		(11.42)		(11.07)		(13.75)	

Note: In the following tables, details may not add to totals because of rounding.
Source: Authors' calculations from data bases derived from land cover classification and table 4-1.

Table 11-2. Land Cover Classes—Central African Republic (Central Africa Region)

Land cover class		Area km²	Percent	Growing stock Thousand tonnes	Percent	Sustainable yield Thousand tonnes per year	Percent
	12	105	0.02	23.84	0.00	1.05	0.00
1		105	0.02	23.84	0.00	1.05	0.00
	23	105	0.02	34.65	0.00	1.05	0.00
2		105	0.02	34.65	0.00	1.05	0.00
	44	105	0.02	14.80	0.00	3.68	0.00
	45	685	0.11	1,267.93	0.05	14.39	0.01
4		790	0.13	1,282.74	0.05	18.06	0.01
	52	1,212	0.19	2,605.80	0.11	76.36	0.04
5		1,212	0.19	2,605.80	0.11	76.36	0.04
	62	28,297	4.53	75,552.99	3.17	2,178.87	1.27
	63	2,108	0.34	5,628.36	0.24	92.75	0.05
	64	27,928	4.47	74,567.76	3.13	1,284.69	0.75
	65	263,107	42.15	702,495.69	29.51	12,629.14	7.38
6		321,440	51.49	858,244.80	36.05	16,185.45	9.46
	73	33,936	5.44	57,148.22	2.40	644.78	0.38
	74	164,673	26.38	277,309.33	11.65	3,128.79	1.83
7		198,609	31.81	334,457.55	14.05	3,773.57	2.21
	85	40,048	6.42	511,733.34	21.50	59,271.04	34.65
	86	9,591	1.54	122,553.80	5.15	14,194.68	8.30
	87	52,379	8.39	549,403.33	23.08	77,520.92	45.32
8		102,018	16.34	1,183,690.47	49.73	150,986.64	88.27
Total		624,279	100.00	2,380,339.85	100.00	171,042.17	100.00
(Percentage of region)		(15.77)		(8.50)		(7.19)	

Source: Authors' calculations from data bases derived from land cover classification and table 4-1.

Table 11-3. Land Cover Classes—Congo (Central Africa Region)

Land cover class		Area km²	Percent	Growing stock Thousand tonnes	Percent	Sustainable yield Thousand tonnes per year	Percent
	24	474	0.14	29.39	0.00	4.74	0.00
	25	22,922	6.76	7,564.26	0.29	229.22	0.08
2		23,396	6.90	7,593.65	0.29	233.96	0.09
	33	1,423	0.42	1,423.00	0.05	71.15	0.03
3		1,423	0.42	1,423.00	0.05	71.15	0.03
	42	3,952	1.17	7,315.15	0.28	82.99	0.03
	43	2,529	0.75	4,279.07	0.16	53.11	0.02
4		6,481	1.91	11,594.22	0.45	136.10	0.05
	61	1,107	0.33	3,254.58	0.13	48.71	0.02
	65	2,002	0.59	5,345.34	0.21	96.10	0.04
	66	62,180	18.35	345,907.34	13.30	5,534.02	2.03
	67	4,005	1.18	47,659.50	1.83	532.67	0.20
6		69,294	20.44	402,166.76	15.47	6,211.50	2.28
	71	1,423	0.42	730.00	0.03	27.04	0.01
	72	51,589	15.22	86,875.88	3.34	980.19	0.36
	73	2,846	0.84	4,792.66	0.18	54.07	0.02
	74	685	0.20	1,153.54	0.04	13.01	0.00
7		56,543	16.68	93,552.08	3.59	1,074.31	0.39
	81	632	0.19	15,597.76	0.60	1,863.77	0.69
	82	2,687	0.79	16,122.00	0.62	1,338.13	0.49
	83	3,109	0.92	27,601.70	1.06	1,548.28	0.57
	85	26,348	7.77	336,674.74	12.95	38,995.76	14.34
	86	54,487	16.08	696,234.89	26.78	80,640.76	29.64
	87	94,535	27.89	991,577.62	38.14	139,911.80	51.43
8		181,798	53.64	2,083,808.71	80.15	264,297.78	97.16
Total		338,935	100.00	2,600,138.41	100.00	272,024.79	100.00
(Percentage of region)		(8.56)		(9.29)		(11.44)	

Source: Authors' calculations from data bases derived from land cover classification and table 4-1.

Table 11-4. Land Cover Classes—Equatorial Guinea (Mainland; Central Africa Region)

Land cover class		Area — km^2	Area — Percent	Growing stock — Thousand tonnes	Growing stock — Percent	Sustainable yield — Thousand tonnes per year	Sustainable yield — Percent
	22	211	0.94	69.63	0.03	2.11	0.01
2		211	0.94	69.63	0.03	2.11	0.01
	41	105	0.47	145.95	0.05	2.21	0.01
4		105	0.47	145.95	0.05	2.21	0.01
	52	738	3.27	1,586.70	0.59	46.49	0.15
5		738	3.27	1,586.70	0.59	46.49	0.15
	64	211	0.94	563.37	0.21	9.71	0.03
	65	422	1.87	1,126.74	0.42	20.26	0.06
6		633	2.81	1,690.11	0.63	29.96	0.09
	73	1,528	6.77	2,573.15	0.96	29.03	0.09
	74	53	0.23	89.25	0.03	1.01	0.00
7		1,581	7.01	2,662.40	0.99	30.04	0.09
	81	2,424	10.75	59,824.32	22.23	7,148.38	22.47
	84	369	1.64	3,653.10	1.36	146.49	0.46
	85	2,529	11.21	32,315.56	12.01	3,742.92	11.76
	86	9,064	40.19	115,819.79	43.03	13,414.72	42.16
	87	4,901	21.73	51,406.59	19.10	7,253.48	22.80
8		19,287	85.51	263,019.36	97.73	31,705.99	99.65
Total		22,555	100.00	269,174.16	100.00	31,816.80	100.00
(Percentage of region)		(0.57)		(0.96)		(1.34)	

Source: Authors' calculations from data bases derived from land cover classification and table 4-1.

Table 11-5. Land Cover Classes—Gabon (Central Africa Region)

Land cover class		Area — km^2	Area — Percent	Growing stock — Thousand tonnes	Growing stock — Percent	Sustainable yield — Thousand tonnes per year	Sustainable yield — Percent
	11	422	0.16	95.79	0.00	4.22	0.00
1		422	0.16	95.79	0.00	4.22	0.00
	24	1,054	0.40	65.35	0.00	10.54	0.00
	25	6,323	2.42	2,086.59	0.10	63.23	0.03
2		7,377	2.82	2,151.94	0.11	73.77	0.03
	33	4,005	1.53	4,005.00	0.20	200.25	0.09
3		4,005	1.53	4,005.00	0.20	200.25	0.09
	42	738	0.28	1,366.04	0.07	15.50	0.01
	43	3,689	1.41	6,241.79	0.31	77.47	0.04
4		4,427	1.69	7,607.83	0.38	92.97	0.04
	51	263	0.10	586.49	0.03	16.57	0.01
	52	1,212	0.46	2,605.80	0.13	76.36	0.04
5		1,475	0.56	3,192.29	0.16	92.92	0.04
	61	474	0.18	1,393.56	0.07	20.86	0.01
	65	422	0.16	1,126.74	0.06	20.26	0.01
	66	19,339	7.40	107,582.86	5.40	1,721.17	0.80
	67	1,528	0.58	18,183.20	0.91	203.22	0.09
6		21,763	8.32	128,286.36	6.44	1,965.51	0.91
	71	1,423	0.54	730.00	0.04	27.04	0.01
	72	18,812	7.19	31,679.41	1.59	357.43	0.17
	73	56,279	21.52	94,773.84	4.75	1,069.30	0.50
	74	580	0.22	976.72	0.05	11.02	0.01
7		77,094	29.47	128,159.97	6.43	1,464.79	0.68
	81	4,743	1.81	117,057.24	5.87	13,987.11	6.50
	82	8,273	3.16	49,638.00	2.49	4,119.95	1.92
	83	1,107	0.42	9,827.95	0.49	551.29	0.26
	84	949	0.36	9,395.10	0.47	376.75	0.18
	85	36,307	13.89	463,930.85	23.27	53,734.36	24.98
	86	36,836	14.85	496,246.41	24.90	57,477.28	26.72
	87	54,693	20.92	573,674.88	28.78	80,945.64	37.63
8		144,908	55.41	1,719,770.42	86.28	211,192.38	98.19
Total		261,471	100.00	1,993,269.58	100.00	215,086.80	100.00
(Percentage of region)		(6.61)		(7.12)		(9.04)	

Source: Authors' calculations from data bases derived from land cover classification and table 4-1.

Table 11-6. Land Cover Classes—Zaire (Central Africa Region)

Land cover class	Area km²	Area Percent	Growing stock Thousand tonnes	Growing stock Percent	Sustainable yield Thousand tonnes per year	Sustainable yield Percent
11	14,017	0.62	3,181.86	0.02	140.17	0.01
14	105	0.00	23.84	0.00	1.05	0.00
1	14,122	0.63	3,205.69	0.02	141.22	0.01
23	580	0.03	191.40	0.00	5.80	0.00
24	4,690	0.21	290.78	0.00	46.90	0.00
25	1,686	0.07	556.38	0.00	16.86	0.00
2	6,956	0.31	1,038.56	0.01	69.56	0.01
33	2,002	0.09	2,002.00	0.01	100.10	0.01
3	2,002	0.09	2,002.00	0.01	100.10	0.01
42	18,233	0.81	33,749.28	0.19	382.89	0.03
43	6,060	0.27	10,253.52	0.06	127.26	0.01
45	1,950	0.09	3,609.45	0.02	40.95	0.00
4	26,243	1.17	47,612.25	0.27	551.10	0.04
51	211	0.01	470.53	0.00	13.29	0.00
52	790	0.03	1,698.50	0.01	49.77	0.00
5	1,001	0.04	2,169.03	0.01	63.06	0.00
61	5,322	0.24	15,646.68	0.09	234.17	0.02
65	43,157	1.91	115,229.19	0.65	2,071.54	0.15
66	767,717	33.98	4,270,809.67	24.18	68,326.81	5.02
67	268,219	11.87	3,191,806.10	18.07	35,673.13	2.62
6	1,084,415	48.00	7,593,491.64	43.00	106,305.64	7.81
71	21,025	0.93	10,785.83	0.06	399.47	0.03
72	119,934	5.31	201,968.86	1.14	2,278.75	0.17
73	32,250	1.43	54,309.00	0.31	612.75	0.05
74	35,622	1.58	59,987.45	0.34	676.82	0.05
7	208,831	9.25	327,051.14	1.85	3,967.79	0.29
82	72,403	3.20	434,418.00	2.46	36,056.69	2.65
83	1,950	0.09	17,312.10	0.10	971.10	0.07
84	3,794	0.17	37,560.60	0.21	1,506.22	0.11
85	200,295	8.87	2,559,369.51	14.49	296,436.60	21.77
86	63,709	2.82	814,073.60	4.61	94,289.32	6.92
87	554,851	24.56	5,819,832.14	32.96	821,179.48	60.31
8	897,002	39.71	9,682,565.95	54.83	1,250,439.41	91.83
Lakes	18,707	0.83	0.00	0.00	0.00	0.00
Total	2,259,279	100.00	17,659,136.26	100.00	1,361,637.89	100.00
(Percentage of region)	(57.07)		(63.07)		(57.24)	

Source: Authors' calculations from data bases derived from land cover classification and table 4-1.

12

East Africa

Phil O'Keefe, Ian Ryle, and John Kirkby

This chapter presents a detailed description of the most important land cover classes in this region. Helpful figures in other chapters include figure 3-1 (cloud cover); figures 3-2, 3-3, and 3-4 (NDVI summary land cover profiles); figure 3-5 (regional summary map of land cover classes); figures 7-1 and 7-2 (continental maps of growing stock and sustainable yield); and the "Regional Land Cover Class Map of East Africa" at the end of this volume.

Helpful tables in other chapters include table 3-2 (land cover classes); table 4-1 (data and sources for growing stock and sustainable yield); and table 6-5 (East Africa estimated woody biomass by summary class).

Class 0—Desert

The only significant areas of this land cover class in the East African region occur in northern and eastern Kenya, especially in Marsabit District and on the western side of Lake Turkana, with smaller areas running parallel to the Kenyan coast. Desert covers 2.4 percent of Kenya, an area of 14,439 square kilometers.

According to Morgan (1973), only the Chalbi Desert of Marsabit District is classed as true desert. It is a rock desert of a type unique in East Africa and forms part of a closed drainage basin. The edge of the desert is marked by numerous springs, supplied by subsurface water from the surrounding mountains. The desert is liable to occasional flooding, and subsequent evaporation of the floodwater results in an accumulation of salt, which inhibits plant growth. Locally, however, outlets of significant tributary streams formed by seasonal floods may support a variety of annual herbs and grasses, such as the halophytic grass *Drakebrockmania somalensis*. Rainfall is less than 100 millimeters a

year but is unreliable. A slight tendency exists for rainfall maxima in April and November.

Much of the area classified as Desert in this study has a surface largely of bare rock. Other areas in this land cover class probably represent White's "semidesert Annual Grassland" (1983) which, along with true desert, occupies the driest areas of Kenya, nearly always below 1,000 meters. These areas are dominated by the grasses *Aristida adscensionis* and *A. mutabilis*, but these may disappear completely during drought periods lasting several years.

Even here, however, woody plants are rarely absent, and they provide 2 to 20 percent of the vegetation cover. Dominants may be either shrubs, such as *Duosperma eremophilum* or bushes and small trees, notably *Acacia horrida, A. reficiens, A. seyal, A. tortilis, A. senegal,* and *Commiphora* spp. It is worth remembering that the transition from bushland to desert through semidesert is very gradual, and many so-called desert areas may differ only in vegetation density from the land cover class that surrounds them. This may be a result of local and climatic factors at the time.

Some vegetation is present in most areas classed as desert, and this presents opportunities for migrant pastoralists, who use grasses and leaves as fodder. The vegetation is resident, adapted to survive irregular periods of severe drought and to respond quickly to any water that becomes available. Locally, overuse may lead to at least the early stage of desertification.

Class 21—Semidesert Wooded Grassland

Semidesert Wooded Grassland is restricted to the arid areas of north Kenya, where the annual rainfall is less than 250 millimeters, with maxima in April and November. The greatest area is around Lake Turkana in Kenya, extending southward and eastward toward the border with Somalia. This class accounts for 17.7 percent of Kenya, an area of 104,600 square kilometers. Growing stock for the region is 34 million tonnes and sustainable yield is 1 million tonnes a year.

The transition is gradual from Dry *Acacia-Commiphora* Bushland and Thicket (Class 41) to Semidesert Wooded Grassland (Class 21). The latter consists mainly of widely spaced dwarf shrubs and bushes with a grass cover that is barely discernible except in the wet season. The important woody species are *Acacia mellifera*, *A. reficiens*, *A. senegal*, and *A. tortilis* ss. *spirocarpa*, along with *Commiphora* spp. Other genera include *Balanites*, *Boscia*, *Euphorbia*, *Jatropha*, and *Sansevieria*. Various succulents also grow in these areas, and the grasses include *Aristida* spp. and *Chrysopogon aucheri* var. *quinquepluris* (Trapnell and Langdale-Brown 1972).

Crown cover is usually less than 10 percent (White 1983) and continuous areas of taller trees are almost entirely restricted to rocky outcrops and sites where water is available, for example near streams. In drier areas, with about 100 millimeters of rainfall, woody plant cover is less than 3 percent, and individual plants usually are less than 3 meters high (White 1983). Grasses such as *Cenchrus biflorus* and *Panicum turgidum* dominate the drier areas.

Much of the area of Class 21 has long been used for grazing and browsing. This has strongly influenced the grasses, herbs, and trees, because the more palatable species such as *Acacia senegal* are suppressed. Fluctuations in rainfall and occasional droughts can be tolerated by the resistant vegetation, but overuse of the biomass by pastoralists has caused general degradation of the environment, with severe social consequences.

Class 24—Transitional Wooded Grassland

Transitional Wooded Grassland occurs at heights of approximately 1,000 meters, mainly in north Tanzania and south Kenya, but occurs also in scattered areas of central and southern Tanzania. In Kenya, it grows extensively on the southern and eastern flanks of the highlands in the south, and forms a distinct band along the coastal hinterland separating the Moist *Acacia-Commiphora* Bushland and Thicket (Class 43) toward the coast from the Dry *Acacia-Commiphora* Bushland and Thicket (Class 41) of the interior.

Rainfall in this area is about 250 millimeters, with a November-December maximum. This class accounts for 1.1 percent (6,376 square kilometers) of Kenya. The Transitional Wooded Grassland of Tanzania is more scattered than in Kenya, although it does cover a larger area—23,607 square kilometers, or 2.5 percent of the country. The largest blocks of this vegetation are in the northern provinces of Arusha and Tanga, but significant outliers are scattered throughout the country, notably in Dodoma, Morogoro, and Tabora districts. The dry season here lasts about 6 months.

Transitional Wooded Grassland also occurs in Burundi, especially in the northwest, and covers 2.6 percent (658 square kilometers) of the country. Growing stock for the region is about 2 million tonnes and sustainable yield is 300,000 tonnes a year.

In Tanzania, this class is probably similar to the "*Acacia-Commiphora* Deciduous Wooded Grassland" described by White (1983). Canopy cover is less than 40 percent, with the woody vegetation dominated by *Acacia* spp. or *Commiphora* spp. thorn trees that form a single, open stratum. Trees are usually between 4 and 7 meters in height, although they may attain between 9 and 20 meters in a few species. Very few bushes and shrubs exist, although scattered bushes or small groups of *Grewia fallax* and *Cordia ovalis* occasionally form a very open understory.

The grass cover is fairly well developed and ranges from 0.5 to 1.5 meters in height. Principal species include *Digitaria macroblephara*, *Eustachys paspaloides*, and *Themeda triandra* on well-drained soils and *Pennisetum mezianum* on poorly drained soils. The small amount of Transitional Wooded Grassland in northwestern Burundi lies in an area of "Wooded Savanna" dominated by *Acacia* spp. (M'Hirit 1986) and will be described more fully under Open Woodland (Class 61) in this chapter.

The other areas probably correspond to the most important East African savanna type, the "small Tree Savanna." This is dominated by various broad-leaved *Combretum* species, most commonly *C. binderanum*, *C. ghasalense*, *C. molle*, and *C. zeyheri*. In wetter areas, the "small Tree Savanna" also includes large-leaved *Terminalia* spp., especially *T. glaucescens* and *T. mollis*. These are replaced in drier areas by smaller-leaved species—*T. brownii* in Kenya and *T. sericea* in Tanzania. The ground cover between the trees is mainly tall grassland, dominated by *Hyparrhenia* spp. It is possible that the savanna flanking the Kenya Highlands includes certain other *Acacia* spp. observed in southern Kenya (Trapnell and Langdale-Brown 1972); species involved include *Acacia gerrardii*, *A. nilotica*, *A. senegal*, and *A. seyal* ss. *subalata*, in a grass layer dominated by *Themeda* spp.

Land use in Transitional Wooded Grassland is more variable than in the groups previously described. Pastoralism dominates, although extensive areas are unused. Extensive smallholder farming is carried on in more favorable areas (Berry, Taurus, and Ford 1980). The pressure of use in this class is fairly light.

Class 33—Bushy Shrubland

Most Bushy Shrubland in the East African region occurs at elevations between 500 meters and 1,000 meters along the equator in Kenya, where it grows extensively on the lower eastern and southern slopes of the Central Highlands, and to a lesser extent at lower elevations in the lower Tana Valley. Elsewhere, significant areas exist on the eastern slopes of Mount Kilimanjaro and in the highlands on the Tanzanian border, east of Lake Victoria. In all, this class accounts for 3.7 percent of Kenya, an area of 21,552 square kilometers.

Elsewhere in the region, Bushy Shrubland is confined to Tanzania, where its distribution is more scattered. Patches tend to occur within and around the areas of Moist *Acacia-Commiphora* Bushland and Thicket in central Tanzania, notably in Arusha, Kilimanjaro, Tabora, and Tanga districts, with outliers in Mbeya District on the eastern shore of Lake Rukwa. The area of Bushy Shrubland in Tanzania is 10,065 square kilometers, 1.1 percent of the country.

The large block of Bushy Shrubland in eastern Kenya lies at the southern fringe of the Semidesert Wooded Grassland (Class 21), which dominates the arid interior of the country. It is likely that this vegetation is intermediate between Semidesert Wooded Grassland (Class 21) and Dry *Acacia-Commiphora* Bushland and Thicket (Class 41) in areas receiving annual rainfall between 200 and slightly less than 250 millimeters. Reaching only 2 to 3 meters in height, the vegetation consists of small bushes and stunted trees, dominated by *Acacia reficiens* ss. *misera*, forming a sparse canopy over a ground layer of small shrubs. The grass cover is ephemeral, with annual species appearing only briefly after rain.

The largest single area of this vegetation is in the Aberdare Mountains of south Kenya, mostly concentrated around the lower eastern slopes but attaining higher altitudes. The recording of Bushy Shrubland at such altitudes may result from amalgamation of a wide variety of montane vegetation types, ranging from heathland to forest, an artifact of the coarse spatial resolution of the AVHRR imagery used in the study.

The Aberdare Mountains host a well-developed montane vegetation, with the highest altitudes supporting Afro-alpine communities characterized by giant *Senecio* and *Lobelia* species. Farther down is the "Shrub and Moorland Zone," dominated by *Erica* spp. Next comes the uppermost layer of the montane rainforest belt, the "*Hagenia-Hypericum* Zone," consisting of vegetation stands between 9 and 15 meters in height, dominated by *Hagenia abyssinica* and *Hypericum revolutum*. The middle of the forest zone is characterized by giant bamboo, *Arundinaria alpina*, which is more extensive on the Aberdare Mountains than any-

where else in East Africa. The "Montane Moist Forest" itself is dominated by *Ocotea usambarensis* and *Podocarpus milanjianus,* among other species. All of these zones fall within the boundaries of the Aberdare National Park.

On the lower slopes of the Aberdare Mountains, as well as the area covered by this class in the highlands of southwestern Kenya, the vegetation is probably of evergreen and semi-evergreen bushland, forming a transition between the montane vegetation above and *Acacia-Commiphora* bushland below. This transitional vegetation has been observed on a number of drier mountain slopes in East Africa, and is described more fully in the section on East African Low Woody Biomass Mosaic (Class 52). A similar vegetation occurs on the lower slopes of Mount Kilimanjaro, although the moister conditions here give rise to upland rain forest-type vegetation at slightly higher altitudes (see Coastal and Gallery Forest, Class 83).

The remaining areas of Bushy Shrubland are rather small, and scattered throughout the Moist *Acacia-Commiphora* Bushland and Thicket (Class 43) of Tanzania. The vegetation in these areas is probably of the scrub forest type, more open than the surrounding bushland and thicket. A dense understory of 3 to 5 meters height exists, dominated by *Commiphora* spp., among others. The main emergents are *Adansonia digitata* and *Euphorbia* spp.

Land use in Class 33 is limited by the sparse and seasonal rainfall. Grazing is of poor quality and the carrying capacity low. Given the small population of the areas of Class 33, little degradation of the biomass occurs. Regional growing stock is 34 million tonnes and sustainable yield is 1.7 million tonnes a year.

Class 41—Dry *Acacia-Commiphora* Bushland and Thicket

This class is most characteristic of Kenya, of which it covers 36.4 percent (214,312 square kilometers). Dominating the eastern third of Kenya, it also extends west along the north of the Central Highlands as far as Uganda, where it covers 8,115 square kilometers in the northeast of that country. In Tanzania, it is confined to the north, occurring mainly in Arusha and Shinyanga districts. Considerable areas also exist as inclusions within the Moist *Acacia-Commiphora* Bushland and Thicket (Class 43) of Tabora and Singida districts and in the border districts of Tanga and Kilimanjaro. Dry *Acacia-Commiphora* Bushland and Thicket covers 36,044 square kilometers in Tanzania, 3.9 percent of the country.

Values of NDVI are quite small and reflect the semi-annual rainy season (figure 12-1). Class 41 occurs at elevations between 200 and 1,000 meters.

Figure 12-1. NDVI **Profiles, Bushland and Thicket (Classes 41 and 43)**

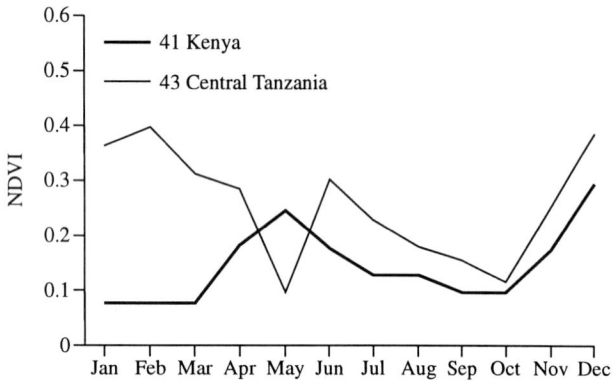

This area coincides with Moore's "semidesert" categories (1971), being characterized by deciduous bushland and thicket of markedly seasonal growth and extensive grassy plains with no woody vegetation. The bushland is dense, varying between 3 and 5 meters in height, with very occasional emergent trees to 10 meters. Tree species are dominated by *Acacia* spp. and *Commiphora* spp., or saltbush (*Suaeda monoica*), in vegetation that has been described as "*Acacia-Commiphora* thorn savanna" (Lucas 1968).

However, studies of this vegetation class on the Serengeti Plain of northern Tanzania indicate a much greater dominance of grassland communities than such a name would suggest, especially on soils formed on young volcanic ash, hardpan soils, and black cracking soils (*mbuga*). The grasses provide only 15 to 45 percent ground cover and trees are rare, restricted to *Acacia mellifera* bushes on young volcanic ash. These secondary grassland communities are controlled by burning and, to a lesser extent, by browsing and grazing. When the interval between burning increases beyond 5 years, *Acacia* spp. are able to regenerate, becoming fire resistant 3 to 4 years later.

The areas of "*Acacia-Commiphora* thorn savanna" (Lucas 1968) are dominated by *Acacia* spp., especially *A. mellifera*, *A. seyal*, and *A. tortilis*, and *Commiphora* spp., especially *C. madagascariensis* and *C. merkeri*. In Kenya, the transition to semidesert vegetation is marked by *Acacia reficiens* ss. *misera*. Other species commonly include *Boscia* spp. and *Grewia* spp. The most common emergent trees include *Adansonia digitata* and *Euphorbia* spp. Within the wooded areas, the ground cover is of scattered tussocks of grass, often dominated by *Aristida* spp., and occasionally *Sporobolus robustus*.

Growing stock for the region is 359 million tonnes and sustainable yield is 5.5 million tonnes a year. Migrant pastoralism is the dominant land use, with burning of grasses to encourage fresh, palatable growth and, locally, concentrated browsing of trees, in

addition to cutting for fuelwood. In consequence, some degradation of the quality of the biomass occurs.

Because the woody vegetation is often spiny, this vegetation class provides problems, both of access to fuelwood and its use. Both the growing stock and sustainable yield of the woody biomass resource are small, and considerable clearance for fuelwood and charcoal already has taken place—even in protected areas like Serengeti National Park. Although biomass destruction by elephants and other game has occurred in recent years, this problem is more serious in the bushlands of Kenya than in Tanzania.

Class 43—Moist *Acacia-Commiphora* Bushland and Thicket

Moist *Acacia-Commiphora* Bushland and Thicket covers a large triangular area of central Tanzania, extending from Lake Victoria in the north almost as far south as Lake Nyasa, and reaching the Usambara Mountains in the east. Elsewhere in Tanzania, outliers exist in the southeast and around Lake Rukwa in the southwest, giving an area of 174,052 square kilometers, or 18.8 percent of the country. In Kenya, where Dry *Acacia-Commiphora* Bushland and Thicket (Class 41) predominates, this moister variant grows in areas of greater rainfall—notably along the coast and in the southern and western highlands.

Moist *Acacia-Commiphora* Bushland and Thicket is much less extensive in Kenya than in Tanzania, covering only 6.6 percent of the country (38,836 square kilometers). Other important areas exist in central and southern Rwanda (5,796 square kilometers, or 30 percent of the country) and on the central plateau of Burundi (2,319 square kilometers, or 8.7 percent of the land area).

Values of NDVI are invariably greater than those for Dry *Acacia-Commiphora* Bushland and Thicket (Class 41) and reach 0.4 during the wet seasons (figure 12-1). In Tanzania, extensive areas of Class 43 occur between 1,000 and 2,000 meters, but the class also exists at low elevation on the Kenyan coastal plain.

In Tanzania, the vegetation is mostly dense bushland of 3 to 7 meters height, with occasional emergent trees to 20 meters, and is characteristic of the semiarid regions of the country. The trees are generally fire resistant, with evergreen species making up only 2.5 to 10 percent of the trees and shrubs. Grass cover is thinner than in more open bushland, but may still reach a height of 1.5 meters. It is dominated by tall *Hyparrhenia* spp. with shorter *Panicum*, *Setaria*, and *Themeda* species.

The principal tree species are *Acacia*, notably *A. gerrardii*, *A. hockii*, *A. mellifera*, *A. nilotica*, *A. seyal*, and *A. tortilis*, and *Commiphora*, especially *C. africana*, *C. caerulea*, *C. mollis*, and *C. schimperi*. Other common tree species include *Adansonia digitata, Boscia coriacea, Cadaba farinosa, Cadia* spp., *Delonix elata, Lannea* spp., *Sterculia* spp., and *Terminalia* spp. A shrub layer also exists, dominated by *Combretum aculeatum, Grewia* spp., and *Maerua* spp. These regions also contain succulents and climbers.

In areas where branches interlace and impede passage, a thicket is formed within the bushland. These may be small patches around old termite mounds or may extend over hundreds of square kilometers. In the drier areas of Tanzania, thickets often are fully deciduous and contain a rich variety of plant species. The best such example is the "Itigi Thicket" covering 620 square kilometers of the central plateau. The principal species here are *Baphia massaiensis, B. burttii, Bussea massaiensis,* and *Pseudoprosopsis fischeri,* forming a dense thicket 3 to 6 meters in height, occasionally broken by emergent evergreen and semi-evergreen trees to 8 meters.

Other species associations within the thicket include taller stands of *Craibia brevicaudata* ss. *burtii, Combretum trothae, Grewia burtii,* and *Tapiphyllum floribundum.* Other important types of thicket include:

- The "*Commiphora-Cordyla* Thickets" of eastern Tanzania, dominated by *Croton* spp., *Hippocratea* spp., *Lannea* spp., and *Strychnos* spp.
- The "*Commiphora-Cordyla* Thickets" of central Tanzania, with species including *Acacia circummarginata, Commiphora caerulea, C. hornbyi, C. merkeri,* and *Cordyla densiflora*
- "*Euphorbia* Thicket"
- A thicket occurring on rocky hills that is dominated by *Dalbergia* spp., *Diospyros* spp., *Dombeya* spp., *Markhamia* spp., *Strychnos* spp., and *Teclea* spp.

In drier upland areas, the transition between Moist *Acacia-Commiphora* Bushland and Thicket and Montane Forest (Class 84) is marked by an evergreen to semi-evergreen bushland 3 to 7 meters in height. This rather dense vegetation is dominated by *Acokanthera* spp., *Carissa edulis, Dodonaea viscosa, Euclea* spp., *Olea africana, Tarchonanthus camphoratus,* and *Teclea* spp.

The presence of this vegetation class around Lake Rukwa in southwestern Tanzania is the result of *miombo* woodland clearance. Repeated clearance for cultivation here has led to a decline in the original *miombo* species, which was replaced by fire-resistant species common to *Combretum* savanna, especially *Brachystegia spiciformis, Combretum mechowianum, Diplorhynchus condylocarpon,* and *Syzygium guineense.*

In Rwanda, this class consists mostly of "Herbaceous Savanna," but also includes areas of "Wooded Savanna," especially in the north (AID 1987). The "Herbaceous Savanna" is dominated by *Themeda triandra,* but includes scattered individuals of *Acacia cam-*

pylacantha, *A. hebecladoides, A. seyal* ss. *multijuga,* and *A. sieberana.* The areas of "Wooded Savanna" are dominated by *Combretum elaeagnifolium,* and include *Acacia campylacantha, A. hebecladoides,* and, more occasionally, *Dombeya madiensis* and *Gardenia jovis-tonantis.*

In central Burundi, where this class also occurs, most of the area is taken up by coffee, tea, and tobacco cultivation. However, the vegetation also may include *Exotheca abyssinica* savanna, replacing the original humid forest, and *Ericaceous* and Afro-alpine vegetation dominated by *Agrostis* spp., *Deschampsia* spp., *Festuca* spp., *Koeleria* spp., *Pentaschistis* spp., and *Poa* spp.

This is regionally one of the more important productive classes; growing stock is 378 million tonnes and sustainable yield is 4.5 million tonnes a year.

The greater precipitation associated with Class 43 allows a wider range of land uses than in Class 41. Bush fallowing allows the regeneration of tree cover, but with a bias toward fire-resistant species. Sedentary agriculture entails almost complete loss of trees, with a tendency for deforestation to spread outward from settlements.

Class 51—*Acacia* Woodland Mosaic

This class, which covers 3.4 percent of the region, is almost entirely restricted to Kenya (43,105 square kilometers) and Tanzania (18,970 square kilometers).

In Kenya, the main occurrence is a large arc to the south and east of the Central Highlands at an altitude of about 1,000 meters, with Dry *Acacia-Commiphora* Bushland and Thicket (Class 41) adjacent to it at lower altitudes. Another area forms an arc on the southern margin of Tsavo National Park. The third Kenyan area is around the equator on the Somalian border.

Northern Tanzania has two main occurrences: an area running east-west around Arusha, and a smaller arc on the eastern side of Serengeti National Park. Growing stock for this class is 138 million tonnes, with a sustainable yield 3.9 million tonnes a year.

Rainfall may be as high as 750 millimeters, but to the south of Tsavo National Park and on the Kenya-Somalia border, it is below 500 millimeters. These two areas in Kenya have two wet seasons, as does the main Kenyan area. The two areas in Tanzania have one wet season.

Land use has contributed significantly to the characteristics of the vegetation in much of the area, with pastoralism dominating in the drier areas, more or less intensive smallholder farming in moister areas, but with sections in Tanzania unused. Small parts of Tsavo and Serengeti national parks lie in this class. The main modification to the biomass in Class 51 has been through clearance of trees for agriculture, grazing, browsing, and burning of grassland. Considerable tree cover has been lost in the main Kenyan extent of this class, so that cultivation and grazing land dominate except in the less-favorable areas of higher, steeper, rockier slopes. Charcoal burning also has caused conversion to grassland (White 1983).

Acacia spp. dominate the tree cover of Class 51, with some *Commiphora* spp. Evergreen species such as *Carissa edulis, Dodonaea viscosa, Euclea divinorum,* and *E. racemosa* may be present, with a tendency for *Acacia* spp. to invade areas degraded by concentrated browsing. Under severe browsing, woody plants may be absent or reduced to bushes, such as *Boscia* spp. (White 1983). South of Nairobi, cultivation and pastoralism have completely removed the tree cover over extensive areas.

Class 52—East African Low Woody Biomass Mosaic

The East African Low Woody Biomass Mosaic covers much of the northeastern quarter of Uganda, extending eastward over the highlands of southwestern Kenya. It is of considerable importance in Uganda, covering 13.7 percent of the country, or 32,671 square kilometers. In Kenya, it occurs largely in inaccessible highlands and covers 51,905 square kilometers, or 8.8 percent of the country. The class is associated with a 6-month dry season.

In Uganda and Kenya, much of the mosaic is likely to consist of "*Combretum* Small Tree Savanna," dominated by *C. binderanum, C. ghasalense, C. molle,* and *C. zeyheri,* among others. In Uganda, this "*Combretum* Small Tree Savanna" is combined with occurrences of *Acacia* spp. and *Albizia zygia* and the "*Terminalia* Savanna" in the north of Uganda. This latter community is dominated by *T. glaucescens* along with *Albizia zygia, Combretum molle,* and other deciduous species; it probably is derived from denser woodland.

In the highlands of southwestern Kenya, it is likely that the natural vegetation in the mosaic consists of evergreen and semi-evergreen bushland. This vegetation type occurs on drier mountain slopes throughout East Africa, often as a transition between "Montane Forest" (Class 84) and Moist *Acacia-Commiphora* Bushland and Thicket (Class 43). Although the species structure is rather variable, a number nearly always are present, including *Acokanthera* spp., *Carissa edulis, Dodonaea viscosa, Euclea* spp., *Olea africana, Sansevieria* spp., *Tarchonanthus camphoratus,* and *Teclea* spp., as well as succulents such as *Aloe* spp. and *Euphorbia* spp.

Smallholder cultivation with some pastoralism are the dominant land uses, although the land is of poor quality and little anthropogenic pressure threatens the biomass. Regional growing stock is 181 million tonnes and sustainable yield is 5.3 million tonnes a year.

Class 61—Open Woodland

Open Woodland exists throughout Tanzania, where it occurs as small patches scattered among larger blocks of Seasonal and Wet *Miombo* Woodland (Classes 66 and 67) and Dry and Moist *Acacia-Commiphora* Bushland and Thicket (Classes 41 and 43). Together these blocks add up to an area of 21,025 square kilometers, covering 2.3 percent of the country. The remainder of this woodland exists in Burundi, with 1,212 square kilometers along the northwestern border and in the southeast. This covers 4.5 percent of Burundi.

The rather random distribution of this vegetation in Tanzania makes accurate description difficult. It is probably true that, in many cases, the species structure of small patches of Open Woodland is similar to that of the surrounding vegetation, the major difference being in density. In central Burundi, this class is probably similar to the savanna already described under Moist *Acacia-Commiphora* Bushland and Thicket (Class 43). The extensive area in northwestern Burundi includes expanses of "Wooded Savanna," dominated by *Acacia albida* and *A. gerrardii.* This region also includes "Transitional Rain Forest," containing *Albizia* spp., *Hyphaene benguelensis,* and *Newtonia buchananii* (M'Hirit 1986).

Growing stock of Class 61 is 68.9 million tonnes and sustainable yield is 1.0 million tonnes per year.

Class 65—Moist Sudanian Woodland

Moist Sudanian woodland in the East African region is restricted to northern Uganda, where it forms the eastern edge of a land cover belt extending as far west as Mali. It is classified by White (1983) as "Sudanian *Isoberlinia* and Related Woodlands." This land cover class covers 15,334 square kilometers (6.4 percent) of Uganda. Climatically, the area has a dry season of about 6 months and annual precipitation of about 1,000 millimeters.

The vegetation may be described as Open Woodland savanna. It seems to be a degraded *miombo* woodland, although it does not possess either of the two genera most characteristic of *miombo, Brachystegia* and *Julbernardia.* There occurs, however, a single species in each of the other two common *miombo* genera, *Monotes* and *Uapaca.* This class also is shorter than true *miombo* woodland, rarely exceeding 15 meters in height. At the southern edge of the main *Isoberlinia* belt, including Uganda, *Isoberlinia doka* becomes more important, although it is scattered.

Isoberlinia spp. have the ability to regrow from underground suckers and thus are able to withstand fire. An understory of climbers and shrubs may be present, with a herbaceous carpet of perennial grasses including *Andropogon* spp., *Eragrostis* spp., *Pennisetum* spp., and *Schizachyrium* spp.

To a considerable extent, the form of Class 65 has been determined by human activity such as rotational bush fallowing (which entails clearing and burning), sedentary smallholder cultivation, and grazing and browsing by domestic animals. Because population densities are low, fuelwood is in good supply. Growing stock is 48 million tonnes and sustainable yield is just less than 1 million tonnes a year.

Class 66—Seasonal *Miombo* Woodland

Seasonal *Miombo* Woodland is very extensive in the west of Tanzania, where it occurs from Kagera in the north to Lake Malawi in the south, covering much of the regions of Shinyanga, Kigoma, Tabora, Rukwa, Mbeya, and Iringa. Large areas of Seasonal *Miombo* Woodland occur along escarpments flanking the Lake Tanganyika trough, with smaller fragments extending north along the border with Rwanda and Burundi, and into the extreme southwest of Uganda. In Burundi itself, this class accounts for much of the northeastern and southern regions.

Apart from the northern tip of Zanzibar, the remaining areas of Seasonal *Miombo* Woodland are scattered through Tanzania, especially along the border with Mozambique, east of Lake Victoria, in the Central Highlands, and in the eastern districts of Lindi and Pwani. Scattered outliers also exist within the Wet *Miombo* Woodland (Class 67) of southeastern Tanzania.

The figures for area and percentage cover of Seasonal *Miombo* Woodland are: Tanzania, 336,565 square kilometers (36.4 percent); Burundi, 10,855 square kilometers (40.6 percent); Uganda, 5,638 square kilometers (2.4 percent); and Rwanda, 2,108 square kilometers (10.9 percent).

Values of NDVI reach 0.5 in the wet season and show the marked seasonality characteristic of these woodlands (Millington and others 1989), with dry-season minima of about 0.2 (figure 12-2). It is interesting to note the close similarity of the NDVI temporal profiles for Seasonal *Miombo* Woodland in Tanzania and Zaire (figure 11-3). The dry season in this class lasts about 6 months.

As White (1983) points out, "where climate changes rapidly, as on the escarpments flanking the Lake Tanganyika and Lake Malawi troughs . . . it has not always been possible to map wetter and drier *miombo* separately." Much of this vegetation class is distributed on the steeply sloping shores of Lake Tanganyika, along the coast of Zanzibar, and as scattered patches in other blocks of fairly uniform vegetation, suggesting that it often is a product of local conditions. Nonetheless, in many cases it is possible to suggest a probable composition for this vegetation class.

Along Lake Tanganyika and the border regions of Rwanda, Burundi, and southwestern Uganda, an area

Figure 12-2. NDVI **Profile, Seasonal** *Miombo* **Woodland (Class 66)**

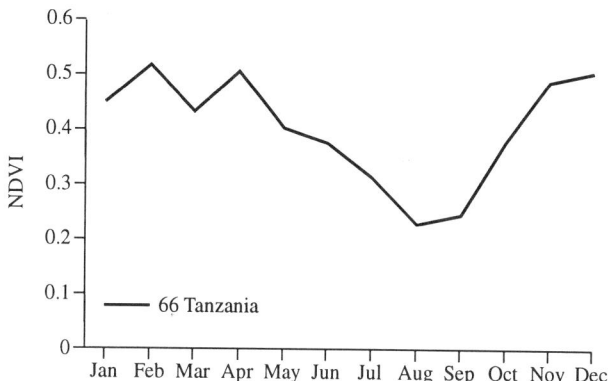

where mean annual rainfall exceeds 1,000 millimeters, the Seasonal *Miombo* Woodland tends to be fairly wet and therefore floristically richer than drier *miombo* variants. Evergreen species are more common and strips of riparian forest may develop around permanent streams.

The woodland canopy usually is between 15 and 20 meters in height, with dominant species including *Isoberlinia angolensis*, *I. paniculata*, and *Terminalia mollis*. Also common are *Brachystegia* spp., although their occurrence varies with local conditions and mean annual precipitation. Much of the Seasonal *Miombo* Woodland along the border with Mozambique also is rather wet, and in addition to the species just mentioned, *Julbernardia magnistipulata* is common. Similarly, the former species also appear with *Isoberlinia tomentosa* where Seasonal *Miombo* Woodland coincides with the greater rainfall areas of the eastern Plateau.

Although slightly open, the woodlands just described generally lack a shrub layer, having instead a ground layer of the grasses *Andropogon* spp. and *Panicum maximum*, and saplings of the main tree species. In Burundi, the principal species in this class are *Brachystegia* spp. and *Maesopsis* spp., reflecting the transition toward forest vegetation.

The largest block of Seasonal *Miombo* Woodland extends south from Lake Victoria to just north of Lake Malawi. It is drier than the Seasonal *Miombo* Woodland on the western border of Tanzania and is similarly dominated by *Brachystegia* spp. (notably *B. boehmii* and *B. spiciformis*) and *Julbernardia globiflora*. However, some species associated with drier vegetation also are present, for example, *Acacia brevispica*, *Acalypha fruticosa*, *Euclea divinorum*, *Grewia bicolor*, *Lantana camara*, *Ormocarpum trichocarpum*, and *Terminalia mollis*. In general, however, the woodlands are floristically poor, with the canopy rarely attaining 15 meters. In especially poor conditions, the canopy may be as low as 3 meters; in such cases, it often is dominated by *Monotes* spp. and *Uapaca kirkiana*.

Regionally, this is one of the most significant classes. Growing stock is 1969 million tonnes and sustainable yield is 31.6 million tonnes a year.

Human population density throughout Class 66 is generally low, partly because tsetse limits the keeping of cattle. Bush fallowing entails some renewal of biomass, but the intensity of land use is low and vegetation is substantially intact. Few problems exist with the fuelwood supply.

Class 67—Wet *Miombo* Woodland

Wet *Miombo* Woodland dominates the large block of woodland in the southeast of Tanzania, which extends over a roughly triangular area from Tanga District in the north to Lake Malawi and the Mozambique border in the south. Smaller patches are found in the western half of the country, especially along Lake Tanganyika and the borders with Rwanda and Burundi, and to the north and east of Lake Rukwa. In all, Wet *Miombo* Woodland covers 11.6 percent of Tanzania, an area of 106,919 square kilometers.

Elsewhere, significant outliers occur in the northeast and southwest of Rwanda (2,319 square kilometers, or 12 percent), the southwestern corner of Uganda (10,644 square kilometers, or 4.5 percent), and to a much lesser extent in western Burundi (1,844 square kilometers, or 6.9 percent). In Tanzania, Wet *Miombo* Woodland is the equivalent at lower altitude of Seasonal *Miombo* Woodland, with extensive areas below about 500 meters.

Large areas of southeastern Tanzania covered by this class were previously labeled "Dry *Miombo* Woodland" (White 1983). However, the evergreen nature of the entire vegetation community is more apparent here, and some areas appear to have virtually no period without growth. The woodland is dominated by *Brachystegia boehmii*, *B. spiciformis*, *Julbernardia globiflora*, and *J. magnistipulata*, forming a canopy that rarely exceeds 15 meters in height.

In many eastern districts, including areas of the coastal lowlands, the Wet *Miombo* Woodland is probably derived from the "Dry Lowland Evergreen Forest" that once may have covered wide expanses of the eastern plateau (Polhill 1968). Hence, Wet *Miombo* Woodland contains a number of tree species from this older vegetation community, notably *Dalbergia* spp., *Ostryoderris* spp., *Pleurostylia* spp., *Sclerocarya* spp., and *Tamarindus* spp. In the extreme southeast, the *Brachystegia-Julbernardia* woodlands are replaced by structurally similar woodlands dominated by other leguminous tree species, especially *Dalbergia* spp., *Lonchocarpus* spp., and *Millettia* spp.

Some areas of Wet *Miombo* Woodland in the southwest of Tanzania have been affected by clearance for shifting agriculture, especially near the south of Lake Tanganyika. Here a mosaic of woodland and cultivation plots exists, where woodland is cleared and burned prior to a cultivation cycle of 2 to 4 years. In a study of the regrowth of abandoned cultivation plots in Zambia, Stomgaard (1985, 1986) identified three successional stages in the regrowth of cleared *miombo* woodland:

1. In the first year after clearance, crops still dominate, although plots are beginning to be invaded by shrubs, especially *Euphorbia tirucalli* and *Smilax kraussiana*.

2. In the 2 to 6 years after clearance, further invasion by shrub and grass species occurs. The woody vegetation at this time is dominated by *Euphorbia tirucalli* and *Smilax kraussiana*. By the end of this phase, the ground cover is dominated by grasses, notably *Rynchelytrum repens*.

3. From 6 to 25 years after clearance, the reinvasion of woody species is sufficient to form a canopy woodland, although it usually lacks a shrub layer.

However, the original *Brachystegia-Julbernardia* woodland is replaced by fire-resistant and fire-tolerant trees of the "*Combretum* Savanna," most notably *Combretum mechowianum*, *Diplorhynchus condylocarpon*, and *Syzygium guineense*. The tree canopy enables *Uapaca* spp. to invade, and the main components of the *Brachystegia-Julbernardia* woodland may reappear toward the end of this stage. Generally, however, the increasing pressure on land has shortened recovery time of abandoned plots, so the cultivation savanna is spreading at the expense of *miombo* woodlands.

The remaining areas of Wet *Miombo* Woodland have a varied composition. Along the Tanzania-Burundi border, this class appears in a fairly high rainfall area at an altitude of more than 1,200 meters, indicating a moist type of montane forest vegetation. In this case, the dominant genera include *Aningeria*, *Cassipourea*, *Chrysophyllum*, *Macaranga*, *Neoboutonia*, *Parinari*, *Polyscias*, and *Tabernaemontana*. Shrub and herb layers tend to be fairly discrete and dominance among the tree species is low.

In the moister climate of Rwanda, Wet *Miombo* Woodland often consists of dense stands of "*Acacia* Savanna." The dominant species are *A. campylacantha*, *A. hebecladoides*, *A. seyal* spp. *multijuga*, *A. senegal*, and *A. sieberana*, often with an understory of *A. seyal*. In the southwest of Rwanda, extending down the western side of Burundi, this class also is represented by "*Acacia* Savanna," but the dominant species here are *A. albida* and *A. gerrardii*. On the eastern edge of this zone, toward the wetter highlands of Burundi, the vegetation may take on the character of Open Woodland (Class 61) rather than savanna, with *Brachystegia* spp. becoming more common (M'Hirit 1986).

Growing stock for the region is 1,555 million tonnes and sustainable yield is 16.2 million tonnes a year.

Land use in the area of Wet *Miombo* Woodland varies from sedentary smallholder cultivation in more densely populated areas to extensive rotational fallow. In general, few problems of overuse occur.

Class 71—Evergreen Woodland Mosaic

Despite its considerable extent, the Evergreen Woodland Mosaic is difficult to describe due to its scattered distribution. The area and percentage cover in each country are: Tanzania, 23,291 square kilometers (2.5 percent); Kenya, 9,485 square kilometers (1.6 percent); Uganda, 1,423 square kilometers (0.6 percent); Rwanda, 843 square kilometers (4.4 percent); and Burundi, 790 square kilometers (3 percent).

The carpet areas of Evergreen Woodland Mosaic occur on the coast and up to about 500 meters altitude in northeastern Tanzania, scattered within larger blocks of vegetation, especially the *Miombo* Woodlands (Classes 66 and 67). This mosaic frequently exists in mountainous areas and near lakes.

The word "evergreen" in the class name needs to be qualified because the tree species in this vegetation, mainly common *miombo* species such as *Brachystegia* spp., *Isoberlinia* spp., and *Julbernardia* spp., are deciduous. The evergreen nature of the woodland mosaic appears to result from an evergreen understory that is recorded by the AVHRR sensor whether or not the canopy tree species are in leaf. This suggests small patches of vegetation with a tree flora similar to the surrounding woodland, but with a lush understory resulting from favorable local conditions. These conditions undoubtedly are related to moisture availability, as indicated by the more humid climate of lakeside and mountainous locations.

Growing stock in this class is 18.3 million tonnes and sustainable yield is 680,000 tonnes per year. Population density is relatively sparse and no serious shortage of fuelwood exists at present.

Class 72—Cultivation and Forest/Woodland Mosaic

This land cover class occurs between sea level and 2,000 meters altitude in all five countries of the East African region. In Kenya and Uganda, it is restricted to the Lake Victoria Basin, where much of the original rain forest has been cleared for cultivation. The mosaic covers 2.9 percent (6,850 square kilometers) of Uganda, but only 0.3 percent (2,002 square kilometers) of Kenya.

In Tanzania, blocks exist around Lake Victoria, especially to the west, and on the coast around Dar es Salaam. The area covered is 9,327 square kilometers, or 1 percent of the country. The mosaic also occurs on the eastern side of Zanzibar.

There are fragments along the northern border of Burundi and larger areas in north and central Rwanda. The class is important in both countries, covering 17.4 percent (3,372 square kilometers) of Rwanda and 7.7 percent (2,055 square kilometers) of Burundi.

The greater productivity of this class is confirmed by NDVI curves. An example from west Tanzania shows values consistently above 0.3, with no lengthy dry season (figure 12-3).

In the Lake Victoria Basin, where virtually no dry season occurs, the mosaic is made up of wooded grassland with relics of the original peripheral "Guineo-Congolian Forest." A description of this vegetation is presented under the Forest classes (82, 85, 86, and 87). The area to the north and west of Lake Victoria is the largest area of anthropogenic savanna in the region, stretching from the humid areas near the Kakamega Forest in Kenya around the lake through southern Uganda, and into the northwest of Tanzania.

Near Dar es Salaam, the mosaic is similar to the East African Low Woody Biomass Mosaic (Class 52) described for the same area. However, both this area and those on Zanzibar probably have suffered more degradation due to overuse than the high woody biomass mosaics such as those typified by this class. Conversely, the woodland patches within the High Woody Biomass Mosaic in Tanga District probably consist of Wet *Miombo* Woodland (Class 67), the principal species being *Brachystegia boehmii, B. spiciformis, Julbernardia globiflora,* and *J. magnistipulata,* and possibly older forest remnants (see Class 67).

In Rwanda, the mosaic area in the north includes "Montane Forest" and "Wooded Savanna with Forest Regrowth" (AID 1987). The "Montane Forests" are characterized by *Podocarpus usambarensis* and *Syzygium parvifolium,* among other species. The open areas of the mosaic consist of "Wooded *Combretum elaeagnifolium* Savanna," including *Acacia* spp. In southeastern Rwanda, and across the border into Burundi, this

Figure 12-3. NDVI Profiles, High Woody Biomass Mosaic (Classes 72 and 73)

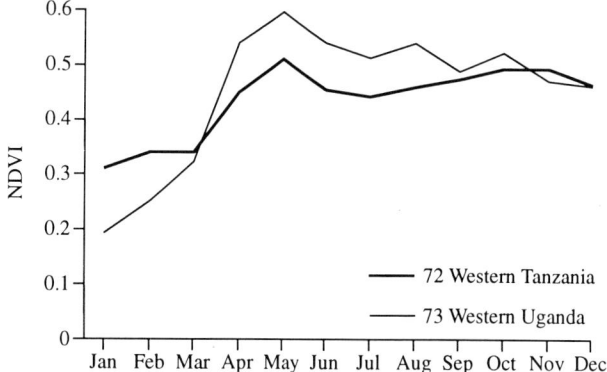

class consists largely of "*Brachystegia* Wooded Savanna" and areas of "Gallery Forest." In Buyenzi and Mugamba districts in northern Burundi, this class contains "Ombrophilous Mountain Forest," with *Albizia gummifera, Prunus africana, Polyscias* spp., and *Symphonia* spp. The associated altimontane prairies are dominated by *Agrostis* spp. and *Erica* spp.

This class is to a large extent the product of deforestation for smallholder settlement, either as rotational fallow, or increasingly for permanent settlement. As population increases, it is likely that the woody biomass in this class will be further reduced.

Growing stock for Class 72 is 40 million tonnes and sustainable yield is 448,000 tonnes per year. The fuelwood supply is threatened by the removal of woodland as agriculture is extended.

Class 73—Cultivation and Forest Regrowth Mosaic

Cultivation and Forest Regrowth Mosaic occurs extensively around the north of the Lake Victoria Basin, from central and western Uganda to southwestern Kenya. It covers 18 percent (43,105 square kilometers) of Uganda and 2.9 percent (16,915 square kilometers) of Kenya. Regional growing stock is 101 million tonnes and sustainable yield is 1.1 million tonnes a year.

Some uncultivated area of this mosaic includes Guinean Woodland (Class 74), described in the next section. However, this land cover class is most extensive on the northern fringe of the degraded "Peripheral Guineo-Congolian Humid Tropical Forests" north of Lake Victoria. Therefore, small areas of secondary forest regrowth might be expected on formerly cultivated land. As an example of forest regrowth in the drier and floristically poorer margins of the Guineo-Congolian rain forest belt, the following description is adapted from White's account (1983) of the invasion of secondary wooded grassland by forest species at Olokemeji in Nigeria.

After 6 years without fire, the site had been invaded by *Antiaris toxicaria, Ceiba pentandra, Celtis brownii, Diospyros mespiliformis, Hildegardia barteri, Holarrhena floribunda, Malacantha alnifolia, Manilkara obovata,* and *Zanthoxylum xanthoxyloides*. After another 25 years, a canopy 8 to 11 meters tall had formed, dominated by forest species such as *Afzelia africana, Diospyros mespiliformis, Hildegardia barteri,* and *Manilkara obovata*. Even so, more than a dozen savanna species still persisted. However, the pressure on agricultural land in Uganda makes it unlikely that succession would be allowed to continue undisturbed for such a long period.

The NDVI curve shows values in excess of 0.5 for the wet season but reduces to 0.2 in January for a site in western Uganda (figure 12-3). These relatively large values indicate both the ability of the remaining tree species to "green up" in response to the wet-season conditions and the growth of crops within this mosaic.

Growing stock is 101 million tonnes and sustainable yield is 1.1 million tonnes per year. Fuelwood supplies will become more difficult to obtain as forest is removed for agriculture.

Class 74—Guinean Woodland

In the East African region, Guinean Woodland is restricted to Uganda, where it occupies much of the northwestern sector of the country. It is similar to the "*Combretum* Small-Tree Savanna" that grows extensively over East Africa. This is characterized by a number of broad-leaved *Combretum* species, notably *C. binderanum, C. ghasalense, C. molle,* and *C. zeyheri*. In Uganda, however, the *Combretum* savanna combines with *Acacia* spp. and *Albizia zygia*. In wetter areas of Uganda, the vegetation also contains *Terminalia mollis* and *T. glaucescens,* replaced by the smaller-leaved *T. brownii* in drier regions (Trapnell and Langdale-Brown 1972). In northern Uganda, the *Combretum* species are replaced by *Butyrospermum paradoxum* ss. *niloticum*.

Generally, the trees of the savanna are rather scattered, reaching about 5 to 10 meters in height. They grow in a sea of tall perennial grasses of 1 to 2 meters, often dominated by *Hyparrhenia* spp. Guinean Woodland covers 7.9 percent of Uganda, an area of 18,918 square kilometers.

Growing stock in Class 74 is 35 million tonnes and sustainable yield is 394,000 tonnes per year. Because the population is relatively low, no fuelwood shortage exists yet.

Class 82—Evergreen Forest

Evergreen Forest is scattered throughout the region and, with the exception of some coastal areas of Tanzania, occurs in highland areas between 1,000 and 2,000 meters. Generally, it occurs along the Lake Tanganyika escarpment, in the highlands of north Tanzania and south Kenya, and in the highlands and coastal areas of southern and eastern Tanzania between Lake Malawi, the Mozambican border, and Tanga District.

The area and percentage cover for each country are: Tanzania, 109,975 square kilometers (11.9 percent); Kenya, 6,580 square kilometers (1.2 percent); Burundi, 5,111 square kilometers (19.1 percent); Uganda, 3,056 square kilometers (1.3 percent); and Rwanda, 2,846 square kilometers (14.7 percent). This is one of the most significant classes for woody biomass.

It is difficult to describe a typical vegetation for such a wide-ranging group of montane forest areas, given the diversity of Afromontane vegetation. It is described for a number of these areas by Lucas (1968), Osmaston (1968), and Polhill (1968). The following generally describes Afromontane vegetation types in the region.

Afromontane rain forests grow in a fairly wide range of local conditions, but rainfall generally exceeds 1,250 millimeters a year, and often is more than 2,500 millimeters. Most forest lies between 1,200 and 2,500 meters altitude, although this too is variable. Typical tree species include *Aningeria adolfi–friedericii, Ocotea usambarensis, Olea capensis, Parinari excelsa, Podocarpus latifolius, Prunus africana, Syzygium guineense* ss. *afromontanum*, and *Tabernaemontana johnstonii*.

"Afromontane Bamboo" (*Arundinaria alpina*) is widespread in the highlands of East Africa. Occurring mostly between 2,380 and 3,000 meters, it is extensive on the Aberdare Mountains and Mount Kenya, and on the gentler slopes of the Ruwenzori Mountains. It is almost absent on Mount Kilimanjaro. Tree species scattered among the bamboo include *Dombeya gvetzenii, Hagenia abyssinica, Ilex mitis, Juniperus procera, Nuxia congesta, Podocarpus latifolius, Prunus africana,* and *Tabernaemontana johnstonii*.

The phenology for this vegetation class shows consistently high values of NDVI (above 0.4) and no marked seasonality (figure 12-4).

Another feature of most high mountains, and the summits of many smaller mountains, is "Evergreen Bushland and Thicket"—particularly near the coast and large lakes (White 1983). Unburned thicket often is between 3 and 13 meters tall, and typical genera include *Blaeria, Erica, Philippia,* and *Vaccinium*. The most extensive vegetation in the mountains of East Africa is, however, grassland; the vast majority of it is secondary, resulting from destructive human activity. Secondary grassland in the montane forest belt typically includes *Exotheca abyssinica, Loudetia simplex, Themeda triandra,* and a number of other species including

Andropogon spp., *Hyparrhenia* spp., *Pennisetum* spp., and *Setaria* spp. In the "Heathland" and "Afro-alpine" zones, the most common genera in secondary grassland are *Agrostis, Deschampsia, Festuca,* and *Poa*.

Along the coast of Tanzania, the Evergreen Forest class is probably derived from the "Dry Lowland Evergreen Forest" that formerly covered much of the coastal plateau. The vegetation consists of *Brachystegia-Julbernardia miombo* woodland containing a number of forest species from the old lowland forest. It is more fully described under Wet *Miombo* Woodland (Class 67).

The likelihood that fast-disappearing forest will be protected by law makes this class an unpromising source of fuel. In fact, much of the Evergreen Forest exists in protected areas, including the Mount Kenya and Nairobi national parks in Kenya, the Ruwenzori National Park in Uganda, and the Kilimanjaro Forest Reserve in Tanzania. Those areas lacking protection are rapidly being cleared for smallholdings.

Growing stock in Class 82 is 767 million tonnes and sustainable yield is 64 million tonnes per year. The fuelwood supply is likely to become a problem in areas being cleared for agriculture.

Class 83—Coastal and Gallery Forest

Coastal and Gallery Forest is most extensive along the Kenyan coast, although significant areas are dotted along the northern coast of Tanzania. Elsewhere, smaller areas are on Mount Kilimanjaro and in the Crater Highlands and west Usambara Mountains of northern Tanzania. This class accounts for 1.8 percent of Kenya (10,539 square kilometers) and 0.6 percent of Tanzania (5,586 square kilometers).

The southernmost block of this forest lies between Kisiji and Kilwa Kivinje, in the Tanzanian district of Pwani. Here, the forest is made up of extensive stands of mangrove. The main genera are *Avicennia* spp. and *Rhizophora* spp., although distinct mangrove zones have developed along the East African coast, and other species, especially *Bruguiera gymnorrhiza* and *Heritiera littoralis*, also are present.

Along the coast of Kenya and northern Tanzania, as far south as Dar es Salaam, this class is a mosaic of "Moist Forest" with "Transitional Evergreen Bushland and Scrub Forest." This is a diverse forest vegetation, its 15 to 20-meter canopy composed mainly of *Afzelia quanzensis* and *Erythrina sacleuxii*. Emergent species, some reaching 30 to 35 meters, include *Albizia adianthifolia, Balanites wilsoniana, Combretum schumannii, Julbernardia magnistipulata, Lannea* spp., and *Manilkara sansibarensis*. The coastal fringe of the mosaic, especially along the northern Kenyan coast, consists of mangrove swamp dominated by *Avicennia* spp. and *Rhizophora* spp.

Figure 12-4. NDVI **Profile, Evergreen Forest (Class 82)**

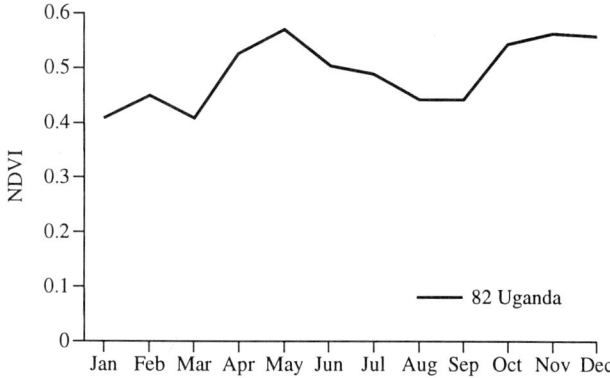

In the remaining mountainous regions, the gallery forest component of this class covers small areas at high altitudes. The greater rainfall of Mount Kilimanjaro and the Crater Highlands gives rise to "Upland Rain Forest," dominated by *Aningeria* spp., *Parinari* spp., and *Ocotea* spp., together with *Cassipourea* spp., *Chrysophyllum* spp., *Macaranga* spp., *Neoboutonia* spp., *Podocarpus* spp., *Polyscias* spp., and *Tabernaemontana* spp. A similar vegetation community could be expected for this class on the slopes of the Usambara Mountains in Tanga District, although here it probably merges with "Moist Lowland Forest," containing genera such as *Allanblackia* spp., *Cephalosphaera* spp., *Isoberlinia* spp., and *Newtonia* spp.

Growing stock in Class 83 is 145.5 million tonnes and sustainable yield is 8.1 million tonnes per year. The fuelwood supply is good generally, although mangrove cutting is causing undesirable environmental effects in the coastal zone.

Class 85—Mesophilous Humid Tropical Forest

This land cover class occurs along the northern margin of Lake Victoria in an area that has no dry season. Most of the class occurs in Uganda, where the area of 13,490 square kilometers constitutes 5.6 percent of the country. Here it probably represents White's "Drier Peripheral Semi-Evergreen Guineo-Congolian Rain Forest" (1983). A block of Mesophilous Humid Tropical Forest lies in the extreme southwest of Uganda, extending into both Tanzania and Rwanda. This is an area of highlands exceeding 1,400 meters in elevation, with a vegetation similar to that of the Wet *Miombo* Woodland (Class 67) on the Burundi border—that is, dominated by *Aningeria* spp., *Parinari* spp., and *Ocotea* spp. At the upper limit other genera tend to dominate, notably *Hagenia* spp., *Myrica* spp., *Nuxia* spp., and *Rapanea* spp., especially on soil derived from volcanic rock.

In Kenya, this class makes up a single block of 843 square kilometers, corresponding in part to Kakamega Forest on the Nandi escarpment east of Lake Victoria. The forest contains both lowland species (including *Aningeria altissima*, *Cordia millenii*, *Entandrophragma angolense*, *Maesopsis eminii*, and *Monodora myristica*) and Afromontane species (such as *Apodytes dimidiata*, *Macaranga kilimandscharica*, *Neoboutonia macrocalyx*, *Prunus africana*, *Strombosia schefflleri*, and *Turraea holstii*).

A number of types of land use occur in this class: permanent smallholdings, rotational fallow, and mixed farming. Considerable areas of the forest have been cleared for agriculture; this threatens the long-term fuelwood supply. Growing stock of Class 85 is 183.1 million tonnes and the sustainable yield is 21.8 million tonnes per year.

Class 84—Montane Forest; Class 86—Humid Tropical Swamp Forest; Class 87—Ombrophilous Humid Tropical Forest

Much of this region's forested area is concentrated around the northern and western shores of Lake Victoria, in southern Uganda. Much smaller areas exist in the north of Uganda and in the highlands of Kenya and northern Tanzania. Ombrophilous Humid Tropical Forest (Class 87) is the most extensive of the forest classes, covering 33,145 square kilometers (13.4 percent) of Uganda, 3,531 square kilometers (0.6 percent) of Kenya, and 1,370 square kilometers (0.2 percent) of Tanzania.

Two other related classes, Montane Forest (Class 84) and Humid Tropical Swamp Forest (Class 86), are far smaller, with an area of less than 6,000 square kilometers for the entire region. Because of the difficulty in distinguishing among these three forest types, the following description covers all three classes.

The area around Lake Victoria has been extensively cultivated, so the forests of south Uganda exist, for the most part, in a mosaic with other vegetation and cropland. This takes the form of a grassland about 2 meters tall, with a varying density of fire-resistant trees. The principal grasses include *Andropogon* spp., *Hyparrhenia* spp., and *Loudetia* spp. Tree species include *Burkea africana*, *Combretum collinum*, *Strychnos* spp., and *Terminalia* spp.

This region is at the eastern edge of the "Guineo-Congolian Rain Forest" zone and is thus a drier variant of this rain forest type, containing most of the important species. The dominant trees include *Albizia* spp., *Aningeria altissima*, *Celtis* spp., *Chrysophyllum albidum*, *Entandrophragma* spp., and *Khaya* spp.

The shores of Lake Victoria host extensive swamp forests. Around the mouth of the Kagera River on the western shore, these are dominated by *Baikiaea insignis* (*eminii*) and *Podocarpus falcatus*. Western Uganda also contains forest dominated by *Parinari excelsa*. Another important group of forests in this region are those dominated by *Celtis* spp., especially *C. mildbraedii*, with a variety of other species. Related forests in the Mount Kenya area are dominated by *Newtonia buchananii*. In Tanzania, forests mapped in this class include *Allanblackia stuhlmannii* and *Isoberlinia scheffleri*, with *Newtonia* spp. and *Parinari* spp.

The montane forests of East Africa mapped in Class 84 have a variety of dominant species, *Aningeria adolfi–friedericii*, *Entandrophragma excelsum*, *Ocotea usambarensis*, and *Podocarpus milanjianus* being the most common. Kenya hosts some related forests that include *Albizia* spp., *Polyscias* spp., and *Olea welwitschii* as dominants. A general description of the Afromontane vegetation in the region is in the Evergreen Forest (Class 82) section.

In Lake Victoria Basin, a few fragments remain of the "scrub Forest" that once covered much of the area. These relicts exist in south Uganda, notably around Lake Edward, and in Burundi. Scrub forest usually is dominated by *Euphorbia dawei*, forming a canopy 12 to 15 meters in height. Often associated with this canopy is *Cynometra alexandri*, a canopy and emergent species of rain forests that here rarely exceeds 10 meters in height.

A common characteristic of these three high-biomass classes is that they are threatened by exploitation for cultivation and this jeopardizes the fuelwood supply, particularly in Uganda. The growing stock of Class 87, which is the most extensive of the three classes, is 399 million tonnes. The sustainable yield is 56.3 million tonnes.

Land Cover Class Tables

Tables 12-1 through 12-5, beginning on page 132, present summaries for each land cover class of the area, showing growing stock and sustainable yield for the East African nations of Burundi, Kenya, Rwanda, Tanzania, and Uganda.

References

Every effort has been made to facilitate access to the documents listed here. Some documents, however, lack full bibliographic information because it was unavailable; also, some documents are of limited circulation.

AID (U.S. Agency for International Development). 1987. "Draft Environmental Profile on Rwanda." AID Office of Forestry, Environment and Natural Resources, RSSA A/TOA 1–77. Prepared in cooperation with U.S. Man and Biosphere (MAB) Program of the Department of State.

Berry, L., T. Taurus, and R. Ford. 1980. *East Africa Country Profiles—Somalia*. Worcester, Mass.: Clark University, Program for International Development.

Lucas, G. L. 1968. "Kenya." In I. Hedberg and O. Hedberg, eds., "Conservation of Vegetation in Africa South of the Sahara." *Acta Phytogeographica Suecica* 54:152–66.

M'Hirit, O. 1986. *Besoins en matière d'éducation et de formation forestière au Burundi*. Résumé introductoire à la réunion de la commission tripartite. Rome: FAO.

Moore, J. E. 1971. "Vegetation." In L. Berry, ed., *Tanzania in Maps*. London: English Universities Press.

Morgan, W. T. W. 1973. *East Africa*. London: Longman.

Osmaston, H. A. 1968. "Uganda." In I. Hedberg and O. Hedberg, eds., "Conservation of Vegetation in Africa South of the Sahara." *Acta Phytogeographica Suecica* 54:148–52.

Polhill, R. 1968. "Tanzania." In I. Hedberg and O. Hedberg, eds., "Conservation of Vegetation in Africa South of the Sahara." *Acta Phytogeographica Suecica* 54:166–79.

Stomgaard, Peter. 1985. "Biomass, Growth and Burning of Woodland in a Shifting Cultivation Area of South Central Africa." *Forest Ecology and Management* 12:163–78.

Stomgaard, Peter. 1986. "Early Secondary Succession in Abandoned Shifting Cultivator's Plots in the *Miombo* of South Central Africa." *Biotropica* 18(2): 97–106.

Trapnell, C. G., and I. Langdale-Brown. 1972. "Natural Vegetation." In W. T. W. Morgan, ed., *East Africa, Its Peoples and Resources*. Rev. ed. Nairobi: Oxford University Press.

White, F. 1983. "The Vegetation of Africa." *Natural Resources Research Series* 20. Paris: UNESCO/AETFAT/UNSO (United Nations Educational, Scientific and Cultural Organization/Association pour l'Etude Taxonomique de la Flore de l'Afrique Tropicale/United Nations Sudano-Sahelian Office).

Table 12-1. Land Cover Classes—Burundi (East Africa Region)

Land cover class	Area km²	Area Percent	Growing stock Thousand tonnes	Growing stock Percent	Sustainable yield Thousand tonnes per year	Sustainable yield Percent
24	685	2.56	42.47	0.03	6.85	0.17
2	685	2.56	42.47	0.03	6.85	0.17
43	2,319	8.66	3,923.75	3.14	48.70	1.23
4	2,319	8.66	3,923.75	3.14	48.70	1.23
61	1,212	4.53	3,563.28	2.85	53.33	1.35
66	10,855	40.55	60,386.36	48.36	966.09	24.48
67	1,844	6.89	21,943.60	17.57	245.25	6.22
6	13,911	51.97	85,893.24	68.78	1,264.67	32.05
71	709	2.95	405.27	0.32	15.01	0.38
72	2,055	7.68	3,460.62	2.77	39.05	0.99
7	2,845	10.63	3,865.89	3.09	54.06	1.37
82	5,111	19.09	30,666.00	24.56	2,545.28	64.50
83	53	0.20	470.53	0.38	26.39	0.67
8	5,164	19.29	31,136.53	24.94	2,571.67	65.17
Lakes	1,844	6.89	0.00	0.00	0.00	0.00
Total	26,768	100.00	124,861.89	100.00	3,945.95	100.00
(Percentage of region)	(1.49)		(1.93)		(1.72)	

Note: In the following tables, details may not add to totals because of rounding.
Source: Authors' calculations from data bases derived from land cover classification and table 4-1.

Table 12-2. Land Cover Classes—Kenya (East Africa Region)

Land cover class	Area km²	Area Percent	Growing stock Thousand tonnes	Growing stock Percent	Sustainable yield Thousand tonnes per year	Sustainable yield Percent
0	14,439	2.45	0.00	0.00	0.00	0.00
11	2,002	0.34	454.45	0.05	20.02	0.06
1	2,002	0.34	454.45	0.05	20.02	0.06
21	104,600	17.74	34,518.00	3.68	1,046.00	3.34
22	1,001	0.17	330.33	0.04	10.01	0.03
24	6,376	1.08	395.31	0.04	63.76	0.20
25	3,320	0.56	1,095.60	0.12	33.20	0.11
2	115,297	19.55	36,339.24	3.88	1,152.97	3.68
31	1,528	0.26	1,306.44	0.14	76.40	0.24
33	21,552	3.66	21,552.00	2.29	1,077.60	3.44
3	23,080	3.92	22,858.44	2.43	1,154.00	3.68
41	214,312	36.35	297,893.68	31.72	4,500.55	14.35
42	211	0.04	390.56	0.04	4.43	0.01
43	38,836	6.59	65,710.51	7.00	815.56	2.60
44	316	0.05	44.56	0.00	11.06	0.04
45	6,271	1.06	11,607.62	1.24	131.69	0.42
4	259,946	44.09	375,646.93	40.00	5,463.29	17.42
51	43,105	7.31	96,124.15	10.23	2,715.62	8.66
52	51,905	8.80	111,595.75	11.88	3,270.01	10.43
5	95,010	16.11	207,719.90	22.11	5,985.63	19.09

(Table continues on the following page.)

Table 12-2 (*continued*)

Land cover class	Area km²	Area Percent	Growing stock Thousand tonnes	Growing stock Percent	Sustainable yield Thousand tonnes per year	Sustainable yield Percent
61	580	0.10	1,705.20	0.18	25.52	0.08
64	2,055	0.35	5,486.85	0.58	94.53	0.30
65	1,475	0.25	3,938.25	0.42	70.80	0.23
66	4,479	0.76	24,916.68	2.65	398.63	1.27
67	685	0.12	8,151.50	0.87	91.11	0.29
6	9,274	1.57	44,198.48	4.70	680.59	2.17
71	9,485	1.61	4,865.81	0.52	180.22	0.57
72	2,002	0.34	3,371.37	0.36	38.04	0.12
73	16,915	2.87	28,484.86	3.03	321.38	1.02
74	1,844	0.31	3,105.30	0.33	35.04	0.11
7	30,246	5.13	39,827.33	4.24	574.67	1.83
82	6,850	1.16	41,100.00	4.38	3,411.30	10.88
83	10,539	1.79	93,565.24	9.96	5,248.42	16.74
84	3,004	0.51	29,739.60	3.17	1,192.59	3.80
85	843	0.14	10,771.85	1.15	1,247.64	3.98
87	3,531	0.60	37,036.66	3.94	5,225.88	16.67
8	24,767	4.20	212,213.35	22.59	16,325.83	52.06
Lakes	15,545	2.64	0.00	0.00	0.00	0.00
Total	589,606	100.00	939,258.13	100.00	31,357.00	100.00
(Percentage of region)	(32.78)		(14.50)		(13.69)	

Source: Authors' calculations from data bases derived from land cover classification and table 4-1.

Table 12-3. Land Cover Classes—Rwanda (East Africa Region)

Land cover class	Area km²	Area Percent	Growing stock Thousand tonnes	Growing stock Percent	Sustainable yield Thousand tonnes per year	Sustainable yield Percent
24	158	0.82	9.80	0.01	1.58	0.07
2	158	0.82	9.80	0.01	1.58	0.07
42	105	0.54	194.35	0.27	2.21	0.10
43	5,796	29.97	9,806.83	13.14	121.72	5.72
4	5,901	30.51	10,001.18	13.68	123.93	5.82
61	211	1.09	620.34	0.85	9.28	0.44
66	2,108	10.90	11,726.80	16.03	187.61	8.82
67	2,319	11.99	27,596.10	37.73	308.43	14.49
6	4,638	23.98	39,943.24	54.61	505.32	23.74
71	843	4.36	432.46	0.59	16.02	0.75
72	3,372	17.44	5,678.45	7.76	64.07	3.01
7	4,215	21.80	6,110.91	8.35	80.08	3.76
82	2,846	14.72	17,076.00	23.35	1,417.31	66.60
8	2,846	14.72	17,076.00	23.35	1,417.31	66.60
Lakes	1,581	8.18	0.00	0.00	0.00	0.00
Total	19,339	100.00	73,141.13	100.00	2,128.22	100.00
(Percentage of region)	(1.08)		(1.13)		(0.93)	

Source: Authors' calculations from data bases derived from land cover classification and table 4-1.

Table 12-4. Land Cover Classes—Tanzania (East Africa Region)

Land cover class	Area km²	Percent	Growing stock Thousand tonnes	Percent	Sustainable yield Thousand tonnes per year	Percent
0	474	0.05	0.00	0.00	0.00	0.00
11	3,952	0.43	897.10	0.02	39.52	0.04
14	105	0.01	23.84	0.00	1.05	0.00
1	4,057	0.44	920.94	0.02	40.57	0.04
24	23,607	2.56	1,463.63	0.03	236.07	0.21
25	105	0.01	34.65	0.00	1.05	0.00
2	23,712	2.57	1,498.28	0.03	237.12	0.21
33	10,065	1.09	10,065.00	0.23	503.25	0.45
3	10,065	1.09	10,065.00	0.23	503.25	0.45
41	36,044	3.90	50,101.16	1.15	756.92	0.68
42	158	0.02	292.46	0.01	3.32	0.00
43	174,052	18.84	294,495.98	6.75	3,655.09	3.27
4	210,254	22.76	344,889.60	7.91	4,415.33	3.95
51	18,970	2.05	42,303.10	0.97	1,195.11	1.07
5	18,970	2.05	42,303.10	0.97	1,195.11	1.07
61	21,025	2.28	61,813.50	1.42	925.10	0.83
65	1,212	0.13	3,236.04	0.07	58.18	0.05
66	336,565	36.43	1,872,311.09	42.93	29,954.28	26.81
67	106,919	11.57	1,272,336.10	29.18	14,220.23	12.73
6	465,721	50.41	3,209,696.73	73.60	45,157.79	40.41
71	23,291	2.52	11,948.28	0.27	442.53	0.40
72	9,327	1.01	15,706.67	0.36	177.21	0.16
7	32,618	3.53	27,654.95	0.63	619.74	0.55
82	109,975	11.90	659,850.00	15.13	54,767.55	49.01
83	5,586	0.60	49,592.51	1.14	2,781.83	2.49
87	1,370	0.15	14,369.93	0.33	2,027.60	1.81
8	116,931	12.65	723,812.44	16.60	59,576.98	53.31
Lakes	41,050	4.44	0.00	0.00	0.00	0.00
Total	923,852	100.00	4,360,841.05	100.00	111,745.89	100.00
(Percentage of region)	(51.36)		(67.30)		(48.80)	

Source: Authors' calculations from data bases derived from land cover classification and table 4-1.

Table 12-5. Land Cover Classes—Uganda (East Africa Region)

Land cover class	Area km²	Percent	Growing stock Thousand tonnes	Percent	Sustainable yield Thousand tonnes per year	Percent
23	2,213	0.93	730.29	0.07	22.13	0.03
24	105	0.04	6.51	0.00	1.05	0.00
25	738	0.31	243.54	0.02	7.38	0.01
2	3,056	1.28	980.34	0.09	30.56	0.04
33	1,317	0.55	1,317.00	0.13	65.85	0.08
3	1,317	0.55	1,317.00	0.13	65.85	0.08
41	8,115	3.39	11,279.85	1.15	170.42	0.21
42	527	0.22	975.48	0.10	11.07	0.01
43	2,213	0.93	3,744.40	0.38	46.47	0.06
45	843	0.35	1,560.39	0.16	17.70	0.02
4	11,698	4.89	17,560.12	1.79	245.66	0.31
51	263	0.11	586.49	0.06	16.57	0.02
52	32,671	13.67	70,242.65	7.16	2,058.27	2.58
5	32,934	13.78	70,829.14	7.22	2,074.84	2.60
61	422	0.18	1,240.68	0.13	18.57	0.02
62	158	0.07	421.86	0.04	12.17	0.02
64	263	0.11	702.21	0.07	12.10	0.02
65	15,334	6.41	40,941.78	4.17	736.03	0.92
66	5,638	2.36	31,364.19	3.20	501.78	0.63
67	10,644	4.45	126,663.60	12.91	1,415.65	1.77
6	32,459	13.58	201,334.32	20.52	2,696.30	3.38
71	1,423	0.60	730.00	0.07	27.04	0.03
72	6,850	2.87	11,535.40	1.18	130.15	0.16
73	43,105	18.03	72,588.82	7.40	819.00	1.03
74	18,918	7.91	31,857.91	3.25	359.44	0.45
7	70,296	29.40	116,712.13	11.90	1,335.63	1.67
82	3,056	1.28	18,336.00	1.87	1,521.89	1.91
83	211	0.09	1,873.26	0.19	105.08	0.13
84	1,370	0.57	13,563.00	1.38	543.89	0.68
85	13,490	5.64	172,375.22	17.56	19,965.20	25.01
86	1,475	0.62	18,847.55	1.92	2,183.00	2.73
87	33,145	13.86	347,657.91	35.43	49,054.60	61.45
8	52,747	22.06	572,652.93	58.35	73,373.66	91.92
Lakes	34,568	14.46	0.00	0.00	0.00	0.00
Total	239,075	100.00	981,385.98	100.00	79,822.49	100.00
(Percentage of region)	(13.29)		(15.15)		(34.86)	

Source: Authors' calculations from data bases derived from land cover classification and table 4-1.

13

Southern Africa

Roger Bevan

This chapter presents a detailed description of the most important land cover classes in this region. Helpful figures in other chapters include figure 3-1 (cloud cover); figures 3-2, 3-3, and 3-4 (NDVI summary land cover profiles); figure 3-5 (regional summary map of land cover classes); figures 7-1 and 7-2 (continental maps of growing stock and sustainable yield); and the "Regional Land Cover Class Map of Southern Africa" at the end of this volume.

Helpful tables in other chapters include table 3-2 (land cover classes); table 4-1 (data and sources for growing stock and sustainable yield); and table 6-6 (southern Africa estimated woody biomass by summary class).

Class 0—Desert

This class occupies approximately 6 percent of the southern Africa region, including 20,446 square kilometers in Angola, 29,457 square kilometers in Botswana, 221,478 square kilometers in Namibia, and 79,148 square kilometers in South Africa. It comprises most of the Namib Desert and its immediate environs, a large triangular area stretching along the Atlantic coast from near Namibe in southern Angola to south of Alexander Bay in South Africa. It stretches inland about 100 kilometers in southern Angola, but widens southward to about 500 kilometers in Cape Province. It is broken only by the slightly better vegetated Namaland Mountains in southern Namibia. Outliers occur in the Great Karroo in Cape Province of South Africa, the Etosha Salt Pans in north Namibia, the Makgadikgadi Pan in northeastern Botswana, and in degraded parts of the Kalahari Desert of south-central and southwestern Botswana.

The phenology of this class shows a consistently low NDVI, mostly between 0.02 and 0.06 for the entire year.

It peaks at 0.06 in northern Namibia and southern Angola in March and July, but has a slightly greater peak of 0.08 in September in southern Namibia. Precipitation is mostly less than 100 millimeters in the Namib Desert at the core of this class, but may exceed 250 millimeters in the wettest parts.

The coastal belt is largely frost-free, but there may be only 5 or 6 frost-free months in southern Namibia and interior Cape Province. The slightly wetter areas of this class largely coincide with White's "Bushy Karroo–Namib Shrubland" and "semidesert" vegetation (1983) and the drier parts of the "Dwarf Karroo Shrubland," together with areas of the "Kalahari/Karroo-Namib Transition." These classes all represent transitions at the desert edge to areas of greater biomass.

The Namib Desert occupies a coastal peneplain, an extensive area of bare rock often covered with mobile sand. In southwestern Angola, the eastern part of the Desert of Moçâmedes may rarely and locally support scattered low bushes and dwarf trees in depressions that receive water. In Namibia, true desert areas support very little woody biomass. However, on the desert fringes in the Namaqualand foothills, in the lower Orange River Valley on the Namibian–South African border, and on higher ground to the east, this class may contain shrubland with widely scattered small bushy trees and larger shrubs. Succulents are abundant, especially in the south.

Welwitschia bainesii is widespread from central Namib northward, especially in the transition zone between the inner and outer Namib. It is particularly abundant in the shade of riparian bushland dominated by *Colophospermum mopane* and *Terminalia prunioides*. On the stony flats between watercourses, *Welwitschia bainesii* is the only woody plant, associated with grasses. On the margins of larger rivers, taller shrubs and small trees such as *Acacia albida, A. erioloba,*

Euclea pseudebenus, Ficus sycomorus, Rhus lancea, and *Salvadora persica* occur locally.

In the outliers of this class in Cape Province, a "Dwarf Karroo" vegetation dominates, including dwarf shrubs, mostly Compositae, and larger shrubs, such as *Rhigozum trichotomum*. Bushes and trees are almost entirely absent and succulents are less prominent than grasses. Soils often are slightly saline, so that halophytes such as *Salsola tuberculata* are widespread. In the very arid Karroo, large, brackish flats or *vloere* exist, sometimes covered by *Salsola aphylla* and other halophytes, but often these areas are bare.

The Etosha Salt Pans of Namibia and the Makgadikgadi Pan contain no woody vegetation. Outliers of this class in the Kalahari may contain limited woody biomass stocks. Southwestern Botswana's rolling sand dune areas contain a few widely scattered trees such as *Boscia albitrunca* and *Acacia mellifera* ss. *detinens*, occurring mainly on dune crests, with shrubs such as *Rhigozum trichotomum* confined to the interdune troughs. Farther northeast, the rolling sandy country with wide plains, depressions, and pans includes trees such as *Acacia erioloba, A. mellifera* (detinens), *Boscia albitrunca, Dichrostachys cinerea,* and *Terminalia sericea.* Low shrubs include *Acacia hebeclada, Bauhinia macrantha, Boscia albitrunca, Dichrostachys cinerea, Grewia* spp., and *Ziziphus mucronata.*

This class is clearly of almost no value as woody biomass. The very limited standing stock would not be replaced for many years if it were removed as a result of severe environmental constraints on biological production.

Class 11—*Veld* Grassland

High *Veld* Grassland is typical of the interior plateau areas of western Lesotho, Orange Free State, parts of southwestern Cape Province, and southern Transvaal. It constitutes nearly 6 percent of southern Africa and occurs largely in South Africa (214,364 square kilometers). It coincides largely with White and Moll's "High *Veld* Grassland" (1978) and Werger and Coetzee's "Moist Cool Temperate Grassland" (1978).

Small outliers occur in Bophuthatswana, just north of Pretoria, and in small tongues within the coastal mountain ranges north of Port Elizabeth. Other small outliers occur in the extreme southwest and northeast of Angola (22,922 square kilometers) and in south-central Mozambique (6,271 square kilometers). The largest outlier of this class occurs in western and northwestern Botswana (53,380 square kilometers) and northeastern Namibia (15,282 square kilometers). It covers large areas of central Ghanzi, Ngamiland, and North East districts in Botswana and coincides with the "Kaukau *Veld*" (White 1983) on the Namibian-Botswana border.

Although this latter area was previously classified as "Woodland-Bushland" (ETC Foundation 1987; Millington and others 1989), it can have a more-or-less continuous grass sward and has a sufficiently similar NDVI phenology to be classed with the *veld* grasslands of South Africa. In semiarid areas, classes may vary from year to year because of fluctuations in rainfall.

Although somewhat ecologically dissimilar, the various regions of this class all have moderately high productivity in the wet season, with NDVI values greater than 0.35, reaching peaks of 0.43 in December and January. The values are sustained at greater than 0.35 from February to April, but decline to approximately 0.2 from June to September. A quick transition from greater summer NDVI values to the smaller values of the winter months largely reflects seasonality in precipitation and evapotranspiration.

The high *veld* grasslands of South Africa are a typical continental interior high plateau ranging from 1,000 to 1,800 meters altitude. The grassland vegetation usually is considered the result of hot, wet summers and cold, dry winters, which restrict tree growth (Taylor 1972). The growth of woody vegetation also is restricted by other factors, such as the more frequent incidence of frost, which can be expected at least 80 days a year for most of this class and up to 120 days in the highest areas. Rainfall increases from approximately 600 millimeters in the west to more than 800 millimeters in the east. The topography is a plateau, flat to gently rolling. It has predominantly black montmorillonitic soils in some of the more northern parts. In more southerly areas, soils are mainly acid to neutral, yellow and grey, sandy and loamy, and overlie sandstones (Scheepers 1975).

Most authors—for example, Werger and Coetzee (1978) and Acocks (1975)—distinguish "true" grasslands from "false" grasslands, the latter being largely anthropogenically induced. The former include the so-called sour grasses, tall grasses with wide, coarse leaves that have less nutritive value in winter than "sweet" grasses. They usually occur at slightly higher altitudes and form uniform-looking sward of *Andropogon* spp., *Chloris virgata, Eragrostis* spp., and *Panicum coloratum. Themeda triandra* dominates the less-acid "mixed" and "sweet" grasses, which have short, narrow leaves and often intermingle with forbs.

The "false" grasslands occur in potentially more woody areas that have suffered excessive burning, overgrazing, or cultivation (Werger and Coetzee 1978). Overgrazing often results in replacement of some of these species by tough and unpalatable *Aristida* spp. and *Chloris virgata.*

Riparian woodland commonly exists along watercourses, although the trees usually are stunted as a result of climatic severity and exploitation. Partially wooded are the Mohokare and Sengu valleys in

Lesotho, the Orange, Vaal, Madder, and smaller tributary valleys in South Africa, plus small valleys in the coastal ranges of Cape Province. The deep, sandy levees fringing the Orange River support riparian forest 6 to 10 meters tall, composed principally of *Acacia karroo*, *Celtis africana*, and *Diospyros lycioides*. Elsewhere, these trees have a bushlike form and may include *Rhus laucea* and *Ziziphus mucronata*, as well as *A. karroo*, *C. africana*, and *D. lycioides*.

In dry, low-lying, undisturbed or protected areas, low bushland and scrubland less than 5 meters high occurs. This is dominated by *A. karroo, Buddleja saligna, Celtis africana, Cussonia* spp., *Diospyros* spp., *Ehretia rigida, Euclea crispa, Grewia occidentalis, Olea africana, Osyris* spp., *Rhus* spp., and *Ziziphus mucronata*. Southward from Lesotho, *Aloe ferox* occurs on north-facing slopes. Much of the southwestern Cape Province is classified as "High *Veld*/Karroo Transition" by White (1983), with overgrazing having converted much of this area to a secondary dwarf shrubland with grasses such as *Aristida congesta, Cynodon hirsutus, Eragrostis* spp., and *Themeda triandra*.

Other areas of greater woody biomass occur in the higher parts of eastern Orange Free State and adjacent areas of Lesotho. These are probably transitional to Transitional Wooded Grassland (Class 24) and Open Woodland (Class 61). Northwest of Kimberley, this class occurs over coarse, stony soils where trees have been cleared for mining and where grazing has replaced "sweeter" *Themeda*-dominant vegetation by the more "sour" *Aristida*-dominant communities.

Over much of South Africa, this class coincides with good ranching country, and in more temperate areas is used extensively for maize and wheat cultivation. The southern part of the so-called Maize Triangle coincides with the northern part of this land cover class in South Africa. Maize cultivation is extensive, with fairly small yields of 0.9 to 3.2 tonnes per hectare (Christopher 1982). Wheat is grown in southeastern Orange Free State as a winter crop. Farther south, in the "High *Veld*/Karroo Transition" of the southwestern Cape Province, the land is largely devoted to pastoralism and subsistence agriculture.

In northwestern Botswana and northeastern Namibia, this class is represented largely by the "Kalahari Thornveld/Zambezian Broad-Leaved Transition" (White 1983) and is included as "Plains Bushveld" class by Werger and Coetzee (1978). It is an area of coarse-textured soils on tropical plains with 500 to 700 millimeters rainfall a year. This wooded grassland is the characteristic vegetation of the thick mantle of the Kalahari Sands (White 1983).

In Botswana, the almost continuous grass sward is less than 1 meter high and consists of *Anthephora argentea, A. pubescens, Eragrostis* spp., *Panicum kalaharense*, and *Schmidtia* spp. In the south, the principal trees and bushes include *Acacia erioloba, A. fleckii, A. hebeclada, A. leuderitzii, A. mellifera, A. tortilis, Boscia albitrunca, Dichrostachys cinerea*, and *Terminalia sericea*. In the north, broad-leaved trees are more common, including *Combretum collinum, Commiphora africana, C. angolensis, Ochna pulchra*, and *Ziziphus mucronata*, but with *Acacia* spp. still dominant. In both areas, the trees are always less than 7 meters tall (usually much smaller) and are quite widely spaced, with dominant grassy sward beneath them.

An outlier of this class occurs in northwestern Angola and extends into Zaire. It is described by Belgian writers as "steppes" and probably is the result of degradation of woodland (see Chapter 11, Central Africa, Class 11). The other outlier in south-central Mozambique takes the form of an "Open Shrubby Savanna Grassland." This was mapped by ETC Foundation (1987) and Millington and others (1989) as Dry *Miombo* Woodland but is, in fact, a degraded woodland area, consisting of scattered low trees and bushes, including *Uapaca* spp., *Monotes* spp., and *Protea* spp. in a well-developed grass sward about 0.5 to 1.5 meters high.

The fuelwood resource in this land cover class is small, both in growing stock and sustainable yield. More important, it is very patchy in distribution. In Lesotho and South Africa, fuelwood supply is further restricted by the land tenure pattern. Consequently, the potential fuelwood supply is poor in these areas, which include a number of important settlements such as Teyateyaneng, Maseru, Mafeteng, and Quthing in Lesotho; Virginia, Bethlehem, and Bloemfontein in the Orange Free State; Cradock and Middelburg in southwestern Cape Province; quite densely settled parts of Bophuthatswana, both north of Pretoria and west of Lesotho; the north of Transkei near Matatiele; and the north of both Ciskei and Transkei near Queenstown.

The pressure on limited woody biomass stock is particularly marked near black townships and in the densely settled rural areas of the homelands. Here, pastoralism and subsistence agriculture are more common than the commercial agriculture which dominates much of the heart of this land cover class.

Class 14—Montane Grassland and Heathland

Montane Grassland and Heathland covers only 15,440 square kilometers of southern Africa, which is less than 1 percent of the land area. Nevertheless, this class is locally significant in the Drakensberg Mountains and in Lesotho. True subalpine and alpine grassland and heathland occur mainly in three areas in Lesotho (Mokhotlong, Qacha's Nek, and Thaba-Tseka districts) which total 7,535 square kilometers and four small areas in adjacent South Africa and Transkei, mostly in the Drakensberg range, between Kopshorn and Scobell's Kop near Barclay East, which total 6,587 square kilometers. Small outliers also occur to the west above Rouxville and Sterkstrom in eastern Cape Province, also at high altitude. All area in this class is above 1,500 meters, and the majority of it is above 2,000 meters.

The phenology of this class exhibits a fairly marked seasonal pattern, with a moderately high NDVI of 0.35 in the warmer period of December to April, rapidly giving way to values of about 0.19 in the cold period from June to October.

The subalpine belt in Lesotho exists between 1,830 meters and 2,895 meters. Soils are very thin and stony and the climate is severe. Rainfall is variable (500 to 1,200 millimeters) and much of it falls as snow. Temperatures colder than 0°C are common in winter, but in January mean temperatures of about 15°C are common. The frost-free period is about 185 days.

Subalpine vegetation is dominated by fire-controlled *Themeda-Festuca* grassland, the species of which differ with aspect and altitude. *Chrysocoma* dominates the grassland over about 13 percent of this zone, and this is believed indicative of overgrazing. The scant woody vegetation in the subalpine belt indicates a very small fuelwood potential.

The alpine vegetation occurs in more severe conditions than the subalpine grasslands and is dominated by homogeneous low, woody heathlands. The dominant species, *Erica* spp. and *Helichrysum* spp., are interspersed with grassland. All are evergreen woody plants adapted to the cold, dry climate and all show a reduction in height with increasing altitude. Various vegetation types correspond to different environmental situations, such as bogs, pools, stream banks, wet and dry meadows, and cliffs, but almost all are dominated by grasses, mosses, and herbs.

Only the heath communities have a large proportion of woody species, but even in the best developed of these communities, which are restricted to the summits of the Drakensbergs, the woody species rarely attain more than 1 meter in height. Isolated patches of scrub, often only up to 2 meters in height, and dominated by *Buddleja corrugata, Leucosidea sericea,* and *Passerina montana,* exist in undisturbed areas on the lower mountain slopes (Millington and others 1989). However, no fuelwood resource exists over much of this part of the class.

Class 24—Transitional Wooded Grassland

Two large areas of this class, together with a number of smaller areas, cover more than 650,000 square kilometers in southern Africa. One area is in eastern Botswana (240,607 square kilometers), southwestern Zimbabwe (25,610 square kilometers), and northern Transvaal. Outliers extend along the South African–Mozambican border into southern Mozambique (16,388 square kilometers are in Mozambique) and into northern Orange Free State.

The other area stretches north-northwest to south-southeast adjacent to the drier classes of land cover that border the Namib Desert, from southwestern Angola (15,967 square kilometers) across central Namibia (161,458 square kilometers), into northern Cape Province. Outliers occur in South Africa in the Langeberg Mountains, Asbestos Mountains, Mount Gakarosa, and Kaap Plateau of northern Cape Province, near Kimberley on the border between Orange Free State and Cape Province, and in a few small areas in the coastal ranges of southern Cape Province, on the north-facing slopes of the Langeberg and Kougaberge ranges. Altogether this class covers 171,365 square kilometers in South Africa. Small areas totaling 25,294 square kilometers also occur in the extreme west of Zambia.

This class coincides largely with areas receiving 200 to 500 millimeters rainfall a year, although a minority of areas receive substantially more. The phenology varies slightly between eastern and western representatives, as follows.

In eastern Botswana, the seasonality is less marked because of a low productivity in wet season. The greatest NDVI values occur in April, May, and November, when levels of 0.21 to 0.25 are recorded. For the rest of the year, a low level of between 0.12 and 0.18 is maintained (figure 13-1).

In northwestern Namibia, seasonality is somewhat greater, but the overall annual levels are similar. Peak NDVI values of 0.33 and 0.30 are experienced in March and April, declining to 0.12 in December and January (figure 13-1).

In eastern Botswana, this land cover class is floristically similar to the *Acacia* Woodland Mosaic (Class 51) and the Open Woodland (Class 61) areas to the north, dominated by *Colophospermum mopane*, with *Acacia*

Figure 13-1. NDVI **Profiles, Transitional Wooded Grassland (Class 24)**

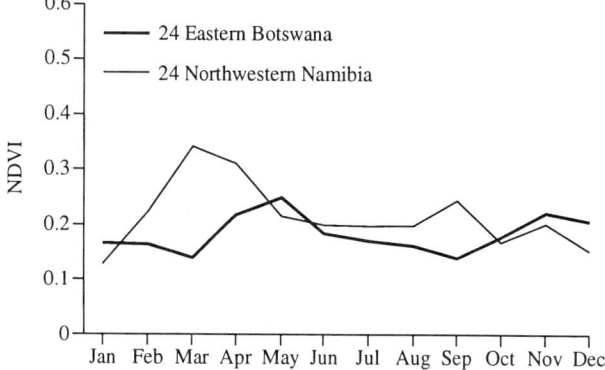

spp., *Combretum* spp., *Terminalia prunioides*, and *Ziziphus mucronata* and including shrubby *Colophospermum mopane*, *Grewia flava*, and *Terminalia sericea*. A strong ecological gradient exists to the west, and once the sandier soils developed on the Kalahari Sands are encountered, it grades rapidly into bushland and shrubby bushland. Floristically and structurally, it is therefore quite diverse, but nearly always has a well-developed grassy sward.

Eastward, the class continues into southwestern Zimbabwe and northern Transvaal, mostly on the flanks of the Limpopo Valley, being most extensive in Matabeleland South. It occurs along the hot, dry Limpopo Valley and along the northern tributary valleys of the Mwenezi, Shashe, Tuli, and Umzingwani rivers (as well as in the middle of the Sabi Valley in the east) in Zimbabwe, and the southern tributary valleys of the Mogol, Palala, Mogalakwena, and Sand valleys in the Transvaal. The class also includes a large part of the lower Limpopo Valley in northern Venda.

In these regions, it also contains a variety of similar vegetation types—"Arid and Dry Mountain Bushvelds," "Arid Spiny and Thorn Plains *Veld*," "*Terminalia* Sandveld," and low "*Colophospermum mopane* Bushland" (Werger and Coetzee 1978). All the vegetation types are dominated by low trees and bushes, some of which are emergent trees above a well-developed ground layer of grasses and herbs. They are similar in phenology, biomass productivity, and growing stock.

Especially in the dry valleys of southern Zimbabwe, a deficit in soil moisture exists during the long dry season. This seasonal drought is accentuated by two factors. First, soils have low infiltration capacities and high runoff rates; consequently, little rainfall during the wet season infiltrates and replenishes the soil moisture reserve. Second, rooting depth is very limited in many cases. The vegetation is adjusted to these inhospitable dry-season conditions.

Slight variation in soil properties is reflected in slight variation in the vegetation communities. On the alluvial soils of the river valleys, low shrubby vegetation is dominant. Variation in the vegetation communities reflects variation in both soil texture and fertility, as well as disturbance. The latter can be quite extensive, as many of these soils are quite fertile and suitable for grazing or cultivation. A few areas of badly degraded savanna do occur in these valleys, and the land grades into low, unproductive Wooded Shrubland (Class 35).

Vegetation in the Limpopo Valley is open grassy savanna with a smaller woody biomass component. This is mainly restricted to scattered emergent trees such as *Acacia* spp., *Dichrostachys cinerea*, and *Sclerocarya caffra*, which attain heights of about 10 meters, and scattered shrubs varying from 1 to 3 meters, dominated by *Grewia flava*, *Ormocarpum trichocarpum*, *Rhus pyroides*, and *Ziziphus mucronata*.

In other areas, vegetation is restricted to thorny and spiny shrubs scattered among grasslands. Woody shrubs and stunted trees reach about 5 meters. Typical woody constituents are *Boscia* spp., *Commiphora* spp., *Dichrostachys cinerea*, *Grewia flava*, *Lycium* spp., *Rhigozum brevispinosum*, *Terminalia prunioides*, and *Ziziphus mucronata*. The only common emergent is the low, shrubby *Colophospermum mopane*, which here rarely exceeds 5 meters in height and often reaches only 2 meters. It occurs as scattered individuals in a shrubby ground cover interspersed with grasses.

On escarpments overlooking the river valleys is a very open savanna of small trees and bushes. The structure varies from dense woody stands reaching 4 to 8 meters in height, through a variety of shrub types, to the most arid form, in which trees rarely exceed 3 meters and vegetation is very open with grass and fern layers dominant. The species are related to those in the lowlands, and the main woody species are *Acacia* spp., *Androstachys johnsonii*, *Boscia albitrunca*, *Colophospermum mopane*, *Combretum apiculatum*, *Commiphora glandulosa*, *Kirkia acuminata*, *Pterocarpus rotundifolius*, *Terminalia prunioides*, and *Ximenia americana*.

Distribution of the various species is mainly dependent on soil properties. On sandy parts of the Highveld, *Terminalia sericea*-dominated savanna occurs, with an open canopy and with slightly taller trees, usually 8 to 10 meters high, underlain by a well-developed grass layer with few shrubs. These areas may be extensively grazed (Millington and others 1989).

In Mozambique, this class coincides largely with White's "Halophytic Communities" (1983). Rainfall of 400 to 600 millimeters a year with moderately saline soils produces grassland with *Acacia nilotica* ss. *kraussiana* with extensive bare areas. However, much of this area is agricultural land or degraded land, characterized by low productivity and marked seasonality. These areas are dominated by agriculture, which has destroyed the vegetation to such an extent that occasionally only very small patches of thicket and grass savanna are left between farms. These usually are areas either of scrubby vegetation less than 3 meters high or of open, grass-dominated savanna with isolated trees which attain 5 to 10 meters.

In northern Transvaal, this class occupies a significant area along the Limpopo Valley, as described earlier, but it also extends south of the Soutpansberg Mountains onto higher areas northwest of Pietersburg and southward along the Mozambican border in Kruger National Park. It largely consists of a dry savanna type of vegetation—"South Zambezian Undifferentiated Woodland and Scrub Woodland," according to White (1983). It usually occurs on alkaline soils of hotter, drier areas of lowland valleys and rejuvenated upland regions. Calcretes are common. Rainfall varies from 325 to 1,000 millimeters and altitude from 150 to 1,525 meters.

Well-developed woodland more than 9 meters tall is localized; elsewhere, it is mostly scrub woodland. Most of the larger woody plants are about 7 meters high with occasional taller emergents. In its natural state, the vegetation can be quite dense, or even closed, but never impenetrable. Some of this is now preserved in the Kruger National Park, but elsewhere in northern Transvaal the area is extensively grazed. A wide variety of species includes *Acacia* spp., *Combretum* spp., *Commiphora mollis, Dichrostachys cinerea, Diospyros mespiliformis, Sclerocarya caffra, Terminalia prunioides, T. sericea,* and *Ziziphus mucronata.* Commonly referred to as the "Transvaal Bushveld" (Christopher 1982), the drier parts are suited only to extensive cattle grazing, and cultivation depends on irrigation.

The other significant area of this class stretches north-northwest to south-southeast, from southwestern Angola to the north of Cape Province. The most northerly of these areas is on the lower slopes of the coastal ranges inland from the Desert of Moçâmedes. Examples occur along a distinct north-south-oriented altitudinal zone along the coast between Namibe and Tombua, and on the high ridges of the Serra da Chela. Here, as farther south, this class is largely adjacent to the Desert (Class 0) and Bushy Shrubland (Class 33).

The vegetation is mainly low, open, and shrubby, with some taller shrubs to 7 meters in height (for example, *Acacia mellifera, A. reficiens,* and *Commiphora angolensis*), passing into *Colophospermum mopane*-dominated communities with grasses on the higher parts. Around the Angolan towns of N'Giva, Anhanca, Evale, Namacunde, Nehone, and Moygua, this land cover class is a degraded form of Open Woodland (Class 61) to the north. This is a secondary vegetation dominated by stunted tree thickets on grassland, caused by clearance near settlements and, in part, by military activities and rural population movement.

Very extensive areas of this class occur in the high plateaus of Namibia, including large areas in northern Ovamboland, around Windhoek in central Namibia, and in the higher parts of Namaland, including parts of the Hanam Plateau. The communities often are dominated by *Colophospermum mopane,* with *Ceraria longepedunculata, Commiphora* spp., and *Sesamothamnus* spp. occurring as trees to 15 meters height over a grassy sward. Farther south through Namibia, the principal bushes and trees gradually become *Acacia* spp., *Combretum apiculatum, Dombeya rotundifolia, Euclea undulata, Ficus* spp., *Rhus marlotii,* and *Tarchonanthus camphoratus.*

In the more eastern parts on the fine red Kalahari Sands, *Acacia mellifera* ss. *detinens, Catophractes alexandri, Combretum apiculatum, Dichrostachys cinerea,* and *Terminalia sericea* become more frequent. This also is the case in the small outliers of this class in the west of Botswana and the extreme northwest of Cape Province. Occurrences of this class in Namibia largely coincide with the more densely populated areas, including Windhoek, and with areas of privately owned farmland, which is principally devoted to cattle ranching (van der Merwe 1983).

Other outliers occur farther south in Cape Province. The area just east of Springbok coincides with the "False Succulent Karroo" of Acocks (1953) and is very similar to the areas in the north in Namibia, except that succulents are more abundant. The large areas northwest of Kimberley in the Langeberg and Asbestos mountains, and northern Kaap Plateau, as well as those southeast of Kimberley in Orange Free State, correspond with the "Kalahari Thornveld and Shrub Bushveld" of Acocks (1953) and White's "Bushy Karroo Shrubland" (1983). A similar vegetation ecology occurs in the coastal range outliers, described by Acocks (1953) as "Succulent Mountain Scrub" and "Karroid Broken *Veld.*"

All these areas have a distinctive shrub layer with a scrubby succulent vegetation. Small trees and shrubs include *Aloe dichotoma, Ceraria namaquensis, Euclea tomentosa,* and *Ficus ingens,* with smaller shrubs often of *Euphorbia* spp. Shrubs usually are less than 2 meters tall, but larger woody plants, either arborescent succulents or nonsucculent bushy trees, are scattered over the landscape. These rarely exceed 5 meters in height, the most notable being *Acacia erioloba, A. mellifera* ss. *detinens, Aloe dichotoma, Boscia* spp., *Ceraria* spp., *Cotyledon paniculata, Ehretia rigida, Euphorbia* spp., and *Grewia flava.*

This vegetation is floristically similar to Wooded Shrubland (Class 35), but includes areas of commercial agriculture, such as the irrigated zone along the Orange River near Upington.

Although the growing stock is greater than in many adjacent semiarid areas, the annual productivity is quite small, which limits fuelwood extraction from this land cover class. It is a transitional class between woodland and grassland, but more the latter, and commonly is used for extensive grazing. Although some floristic overlap with other land cover classes probably exists, this class is phenologically distinct in the 1986 AVHRR NDVI data.

Many areas in this class already have been extensively exploited, as in southern Angola, in the Limpopo Valley in Zimbabwe, and in the Changane Valley in Mozambique. In many more-open areas, growing stock not only is small but is scattered. In southern Zimbabwe, some of these areas coincide with communal areas, where deforestation is common because of fire and land needed for grazing, cultivation, construction timber, and fuelwood. In other areas, such as Kruger National Park, woody vegetation is protected from exploitation. Overall, it is a land cover class of limited fuelwood potential. Its woody biomass growing stock is estimated at 41 million tonnes for the whole of southern Africa, with a sustainable yield of 6.1 million tonnes.

Class 31—*Veld* Shrubland and Cultivation

This class occurs only in South Africa. It covers 4,005 square kilometers of western Cape Province, partly in lower, flatter land around the Great Berg River, southeastward from St. Helena Bay and partly inland from Piketberg to Malmesbury, including some of the low hills up to about 800 meters. It occurs mostly on the Malmesbury Shales and part of the Quaternary and Tertiary Sands nearer the coast. It coincides with the "Coastal Renosterbosveld" of Acocks (1953), now referred to as "Renosterveld" by Taylor (1978) and as "Coastal Macchia" by Moll and Bossi (1984). It also overlaps in the north with the "strandveld" of Acocks (1953).

This class has a phenology distinctive from the "Bushy Karroo–Namib Shrubland" that mostly surrounds it, having a much more marked seasonal pattern. Precipitation occurs almost entirely in winter, totaling 570 millimeters a year. A low summer NDVI of about 0.12 occurs from November to May. This increases quite rapidly to approximately 0.58 in July, August, and September in response to the winter rains.

This class coincides with an area of shales that weather to form fine-grained soils, denser and more fertile than the sands of the mountains or the coast on which Hill Shrubland and Bushy Shrubland (Classes 32 and 33) occur. This is the so-called swartveld, a gently undulating landscape which has been extensively plowed for wheat cultivation, so that only traces of natural vegetation remain. This has been one of the most important wheat-growing areas of South Africa since early white settlement and, although yields are small, it remains a prime producer of South Africa's wheat. Wheat is grown in rotation with oats and other cereals, with some sheep farming, using stubble or fallow land. These virtually continuous wheatlands (Christopher 1982) contrast markedly with the surrounding shrubland.

Natural communities that do remain are dominated by the shrub *Elytropappus rhinocerotis* (renosterbos), usually densely branched and about 0.5 to 1.5 meters high. Grasses often form a significant component, and scattered patches of tall scrub 3 to 4 meters high may occur, often introduced plants such as the Australian *Acacia cyclops*, *A. saligna*, and *Leptospermum laevigatum*. Semisucculent and broad sclerophyllous plants also occur, as do species of tropical affinity, principally

Diospyros austro-africana, *Euclea* spp., *Maytenus heterophylla*, *Osyris* spp., and *Rhus* spp. These occur as scattered individuals 2 to 3 meters high, or locally as small thickets.

This densely settled and intensively cultivated area is not a significant source of woodfuel, and access to the limited stock of woody biomass undoubtedly is restricted by patterns of land ownership.

Class 32—Hill Shrubland

This land cover class is relatively limited, present only in South Africa, covering 24,978 square kilometers. It is limited largely to the lower slopes of western Cape Province, occurring mainly at the foot of the north-northwest-trending subparallel coastal ranges, including the Cedarberg Ranges in the south, the Bokkeveldberge and the Hardeveld in the center, and the slopes around Springbok, Nababeep, and Steinkop in the north.

The ranges are composed of the Table Mountain Sandstone or the Witteberg Quartzites of the Cape System, while the Cape Granites frequently form the foothills and lower slopes, leading down onto the Bokkeveld Shales and sandstone of the Cape System on the lower ground. The soils have little water-retaining capability, are acidic, and generally are infertile. Deeper, brown or reddish sandy loams that are more fertile occur over granites and shales.

Precipitation varies considerably with relief from 150 to 250 millimeters in the north to more than 500 millimeters in the south, with a distinct winter maximum. This class includes "Mountain fynbos" areas in the south. The term *fynbos* has been used to describe the Mediterranean-type evergreen sclerophyllous shrubland, but it equally may be classed as a form of heathland (Moll and Jarman 1984). Acocks (1975) described these areas as "Macchia"; others, such as Moll and Jarman (1984), refer to some parts as "Mountain Renosterveld." Farther north, the class includes the more southerly parts of the "Karroo-Namib Domain" of Werger (1978), also referred to by White (1983) as "Bushy and Succulent Karroo Shrubland."

The NDVI phenology is similar to Bushy Shrubland (Class 33), reflecting a Mediterranean-type seasonal pattern, but with a slightly higher NDVI level, especially in winter as a result of the greater precipitation. Between December and June, NDVI levels are 0.12 to 0.15, increasing to 0.22 to 0.3 from July to October, peaking at about 0.3 in September.

In the southern true Mediterranean area of this class, the "Mountain fynbos" is very complex floristically. It changes from tall, proteoid shrubs 1.5 to 2.5 meters high on the lower slopes to shorter, ericoid forms on the upper slopes, with restioids often dominating the exposed ridges. *Protea nereifolia* is the most common shrub on the lower slopes. Only rarely are small trees observed, although *Heeria argentea* and *Maytenus oleoides* occur on rocky hillsides. Where surface rock is extensive, *Maytenus acuminata*, *Olea africana*, *Olinia ventosa*, and *Podocarpus elongatus* may form a closed scrub forest 7 to 8 meters in height. *Olea capensis* and *Widdringtonia cupressoides* also can occur as scrub forest, usually less than 7 meters tall.

Above the proteoid zone, the shrubs are smaller, usually only 1.5 meters high, and are dominated by *Erica* spp. and *Phylica* spp. Trees are absent except in the Cedarberg ranges, where *Widdringtonia cedarbergensis* persists on rocky scarps and screes. Upper parts of these mountains exhibit a low restioid vegetation, usually less than 0.5 meter high, such as that in the northern Cedarberg Mountains. There, *Restio curviramis* can form dense, low, rounded, cushionlike tufts 20 centimeters high, with *Cannamois nitida* growing in dense tufts 1 meter high.

Farther north in western Cape Province, this montane vegetation grades into the southern end of Werger's "Western Cape Domain" (1978) of the Karroo-Namib region. This area was classified as "Western Mountain Karroo" by Acocks (1953). At its northern limit, succulent dwarf shrubs are fairly common, but farther south, as precipitation increases, succulents are less common and the dwarf shrubs grow only 1 meter high. These include *Asparagus capensis*, *Cotyledon wallichii*, *Euphorbia mauritanica*, *Galenia africana*, *Pentzia* spp., *Pteronia glauca*, *Pterothrix spinescens*, *Ruschia ferox*, *Zygophyllum gilfillani*, and others. Higher up the mountain ridges, between Calvinia and the southern Cape Mountains, this karroid vegetation merges into "Mountain Renosterveld," with dwarf shrubs such as *Elytropappus rhinocerotis*, *Eriocephalus africanus*, *Pentzia incana*, *Relhania* spp., and *Ruschia multiflora*.

Woody vegetation decreases northward throughout this class, but nowhere is it of great significance. It is dominated again by small growing stock and poor sustainable yield.

Class 33—Bushy Shrubland

This class covers an extensive lowland area along the western coast of Cape Province from Cape Town to the Namibian border, stretching along the southern coast as far as Cape Barracouta, and reaching into interior Cape Province as far as the Roggeveld. It covers 117,405 square kilometers, with 70,032 square kilometers in South Africa, 33,725 square kilometers in Angola, and 7,483 square kilometers in Namibia.

It coincides largely with the "Western Cape Domain" of the "Karroo-Namib Realm" of Werger (1978), although it also includes some areas classified as "Cape Shrubland" (part of the *fynbos*) by Werger (1978), some classified as "Capensis Realm" by White

(1983), and some classified as "Coastal Renosterveld" and "strandveld" by Acocks (1975). Much of this latter area has been disturbed and degraded to a "Coastal Macchia," according to Moll and Bossi (1984).

Precipitation varies between 150 and 250 millimeters in the north but exceeds 300 millimeters in the south of this area; most falls in the winter months. Mean daily temperatures vary from about 13°C in July to about 22°C in January. The topography trends parallel to the Atlantic coast (northwest to southeast) and southern coast (west to east) with the two series of folds meeting near Ceres. This class dominates the lower land around the coast and between these ranges.

In the valleys, this class overlies Bokkeveld Shales and sandstones of the Cape System and the Malmesbury Shales of the Archaean Complex. Along the coast, it overlies sands, conglomerates, and limestones of Tertiary and Recent age. Soils from the Bokkeveld series are fairly fertile, but coastal soils are often shallow and clayey.

The other significant occurrence of this class is at the northern edge of the Desert of Moçâmedes in Angola. It extends along much of the Angolan coastal plain and along the foot of the escarpment inland from the desert areas in Benguela and Namibe provinces. This area coincides with the northern end of the "Namaland (or Namaqualand) Domain" of Werger (1978) and the northern extreme of the "Bushy Karroo Shrubland" of White (1983). Patches also occur in northern Namibia, including an area on the fringe of the Etosha Pan. Precipitation levels are less in this area, often less than 100 millimeters, and mean daily temperatures reach 15°C in July but only 20°C in January.

The phenology of this class in western Cape Province demonstrates the effect of the winter rainfall maximum, with NDVI values of about 0.19 from July to October (late winter to spring), declining to slightly lower values of approximately 0.12 from December to May. Although of similar level, the phenology curve for the Angolan extent of this class shows a near reversal of this pattern, with values of 0.04 between July and January, and slightly greater values above 0.12 between February and May. The generally lower NDVI values in Angola probably reflect less precipitation. The different seasonal pattern reflects the slight summer maximum of rainfall in this northern part of the class.

Floristically, the southern area of the class is one of extreme diversity (Taylor 1972), but with few woody plants or trees. Typically, the *fynbos* occurs in the form of sclerophyllous shrubland of 1 to 3 meters height, with some scattered taller bushes and, rarely, widely spaced trees. In many parts of the Cape lowlands, this class has been replaced by secondary shrubland dominated by *Elytropappus rhinocerotis* (renosterbos). Streams may be fringed with riparian thicket and scrub forest. The *Protea* spp. bushes and other taller plants that occur in bushier *fynbos* may thicken to form thickets 4 to 6 meters high, but large areas are degraded by fire, overgrazing, and cultivation.

The varied flora is characterized physiognomically by three elements: restioid, ericoid, and proteoid. Restioid usually is tufted, 20 centimeters to 2 meters high, with nearly leafless tubular or nonwoody stems. Ericoid is characterized by small, narrow, rolled leaves. Proteoid is represented by taller bushes to 6 meters, with leaves moderate in size, but still with relatively small woody elements. Species common to both western and southern coast areas include the ericoids *Anthospermum aethiopicum*, *Eriocephalus racemosus*, *E. umbellulatus*, and *Metalasia muricata*. A conspicuous restioid is *Thamnochortus erectus*. Proteoids are represented by *Leucadendron coniferum*, *L. muirii*, *Protea obtusifolia*, and *P. susannae*.

In typical *fynbos*, true trees are virtually absent. The only species with well-defined boles are *Leucadendron argenteum*, *Widdringtonia cedarbergensis*, and *W. schwarzii*. *Olea capensis* and *W. cupressoides*, which often occur as trees elsewhere, here exhibit a bushy habit and are less than 7 meters tall. Dense thickets up to 6 meters sometimes occur on stream banks and are dominated by *Berzelia lanuginosa* or *Leucadendron salicifolium*. The latter protects its seeds from fire, and many other *fynbos* plants have many protective characteristics against fire.

Because of intensive land use in areas covered by this class, many areas have been invaded by alien species, especially *Acacia cyanophylla*, *A. cyclops*, *A. melanoxylon*, *Hakea acicularis*, and *Pinus pinaster*. The lower and more gradual water courses are fringed with dense thickets, 5 to 7 meters tall, of *Brabeium stellatifolium*, *Freylinia oppositifolia*, and *Metrosideros angustifolia*, with some *Cunonia capensis*, *Hartogia capensis*, *Ilex mitis*, and *Maytenus oleoides*.

In the northern part of this class, in southwestern Angola, the vegetation forms part of Werger's "Namaland Domain" (1978), a low, open, shrubby vegetation with occasional succulents, and with some taller shrub communities including, for example, *Acacia mellifera*, *A. reficiens*, and *Commiphora angolensis*. On higher areas inland, this shrub vegetation grades into *Colophospermum mopane* communities, with some grasses on sites that are less rocky and that have deeper soils. Along streams and valleys, the communities may contain several species of *Commiphora* spp. and *Euphorbia* spp. as well as *Catophractes alexandri*, *Phaeoptilum spinosum*, and *Rhigozum virgatum*.

Much of the coastal vegetation in this part of Angola was classified by Millington and others (1989) as "Dry Coastal Savanna" and "Arid Coastal Thicket." A thin band of desertlike vegetation is included in this class, stretching along the coast north of Lobito, including

areas of bare saline flats and grassy sand dunes where the only woody species is the shrubby *Strychnos spinosa*. However, it is similar to the South African occurrences in form and growing stock, and the woody component rarely exceeds 2 meters in height except near the escarpment.

Land use in this class is mostly limited to grazing, except around Cape Town where cultivation, olive-growing, and commercial stock-rearing occur. This land cover class is of very restricted fuelwood value, dominated by a small growing stock, a small wood-to-grass ratio, and a fairly small sustainable yield. Little potential for fuelwood exploitation exists except locally. The woody biomass growing stock is estimated at 116 million tonnes, with a sustainable yield of 5.7 million tonnes.

Class 34—Kalahari Shrubland

This class covers 4 percent of southern Africa, an area of nearly a quarter of a million square kilometers. It mostly coincides with the Kalahari Desert of southwestern Botswana (52,643 square kilometers), southeastern Namibia (63,603 square kilometers), and northern Cape Province (129,367 square kilometers), including parts of the arid Karroo. Outliers occur in the higher areas to the east of the Namib Desert, in the southern Great Karroo of Cape Province, and in small areas in central Botswana. It mostly coincides with White's "Kalahari Deciduous *Acacia* Bushland and Wooded Grassland" (1983) and with the "Kalahari/Karroo-Namib Transition."

Precipitation in these areas ranges from 80 to 250 millimeters, with a summer maximum. This class shows only a slight seasonality, with a fairly uniform NDVI throughout the year that varies from 0.12 to 0.17 for every month except February, when it peaks at 0.21 (figure 13-2). This is somewhat lower than for the Wooded Shrubland (Class 35), which occurs on its

Figure 13-2. NDVI Profiles, Shrublands (Classes 34 and 35)

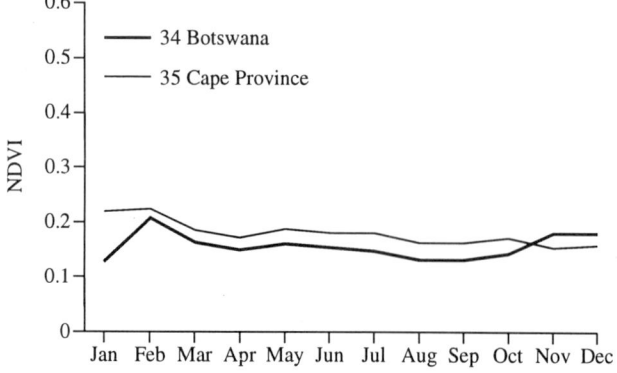

slightly wetter margins (figure 13-2). On its drier margins, this class often is adjacent to the Namib Desert.

In southwestern Botswana and southeastern Namibia, the vegetation is very sparse on the rolling sand dunes with few, widely scattered trees, always less than 7 meters tall and normally much shorter. They mostly occur on dune crests and include *Boscia albitrunca* and *Acacia mellifera* ss. *detinens*. Shrubs such as *Rhigozum trichotomum* are confined to the troughs between the dunes. Shrubs restricted to this area include *Acacia haematoxylon* and *Monechma* spp. Farther to the northeast, in Botswana, the small outliers of this class may contain a slightly greater tree component with *Acacia* spp., *Burkea africana*, *Combretum* spp., *Commiphora* spp., *Peltophorum africanum*, and *Terminalia sericea*, but again they are mostly confined to sand ridges and have limited annual productivity. Similar parallel ridges of sand dunes occur in northern Cape Province, where the commonest scattered trees are *Acacia erioloba*, *A. reficiens*, *Albizia anthelmintica*, *Boscia albitrunca*, and *Terminalia sericea*. Similar species occur in the Great Karroo outlier of this class, but the woody component is limited to bushes and small trees; most are less than 3 or 4 meters high.

In the outliers bordering the hills above the Namib Desert, the vegetation is a very dry form of White's "Bushy Karroo Shrubland" (1983). Arborescent succulents or nonsucculent bushy trees, usually less than 5 meters tall, are scattered above an open shrub layer 2 meters high. Succulent species include *Aloe dichotoma*, *Ceraria* spp., *Cotyledon paniculata*, *Euphorbia* spp., and *Pachypodium* spp., with some water-storing species, such as *Cyphostemma currorii*. A wide variety of nonsucculent bushes, bushy trees, and tall shrubs also is present, including *Acacia erioloba*, *A. mellifera* ss. *detinens*, *Boscia* spp., *Ehretia rigida*, and *Grewia flava*.

Land use in this class is mostly very extensive grazing. It includes the Gemsbok National Park of southwestern Botswana and northern Cape Province, as well as other game reserves.

Although the fuelwood resource is somewhat better than in the desert that this class fringes, this class has a small sustainable yield. Where extraction takes place, it greatly exceeds the annual rate of wood productivity. Growing stock is estimated at 145 million tonnes, with half of this in South Africa. The sustainable yield is estimated at 12 million tonnes.

Class 35—Wooded Shrubland

Wooded Shrubland covers 4 percent of southern Africa, occurring on 244,644 square kilometers. It is largely adjacent to, and very similar to, Bushy Shrubland (Class 33), but usually occurs on the wetter margins, often at higher altitude, although not always. Its type area is in the central Upper Karroo and Great Karroo areas of Cape Province. Here it occupies large areas to the north of the coastal ranges, covering 123,781 square kilometers. In Namibia, it occurs as a series of broken patches in the higher ranges, trending north-northwest to south-southeast, on the inland side of the coastal desert areas, covering 50,061 square kilometers.

A few small areas occur in southwestern Angola, north and east of the Desert of Moçâmedes covering 2,002 square kilometers. In Botswana, it occurs largely in wetter areas to the east and northeast of the Kalahari Desert, covering 60,547 square kilometers. Small outliers occur on the border with Zimbabwe near Tuli, in northern Transvaal, and in drier parts of interior Mozambique.

This class has fairly uniform NDVI values throughout the year, between 0.15 and 0.18 every month except January and February, when it increases to 0.22 (figure 13-2). This profile reflects the slightly greater potential for woody biomass production in this class than in the drier Kalahari Shrubland (Class 34). Precipitation is mostly 100 to 250 millimeters a year, with a distinct summer maximum, but may exceed 400 millimeters in the northeastern outliers.

Wooded Shrubland mostly coincides with White's "Bushy Karroo Shrubland" (1983) in South Africa and Namibia. This is a shrubland dotted with small bushy trees and large shrubs. It occurs in the Karroo, where soils are mostly derived from Dwyka tillite and dolerite, and often are clayey and have a tendency to accumulate salts. The class also occurs on rocky, well-drained mountain slopes in Namibia.

Succulents are abundant, often dominating nonsucculent smaller shrubs. Shrubs are usually less than 2 meters tall, but larger, woody plants, either arborescent succulents or nonsucculent bushy trees, are thinly scattered throughout the landscape. They rarely exceed 5 meters in height and often are confined to rockier areas where water supply is enhanced by runoff or seepage from surrounding slopes.

Of the arborescent succulents, *Aloe dichotoma* (which grows to 5 meters) is the most abundant, with *Ceraria* spp., *Cotyledon paniculata*, and *Pachypodium* spp., all of which attain 3 to 5 meters in height. Toward the north, the succulents *Euphorphia currorii* and *E. eduardoi* occur; also present may be species with water-storing stems, such as *Cyphostemma currorii*, *Moringa ovalifolia*, and *Sesamothamnus* spp. (White 1983). Succulent arborescents are less common in the south on the Karroo.

A wide variety of nonsucculent bushes, bushy trees, and tall shrubs are likely to occur with varying density, including *Acacia erioloba*, *A. mellifera* ss. *detinens*, *Boscia* spp., *Ehretia rigida*, and *Grewia flava*. In the north, *Acacia montis-usti*, *A. robynsiana*, *Adenolobus pechuelii*, *Colophospermum mopane*, *Commiphora* spp., *Euphorbia guerichiana*, and *Rhigozum virgatum* occur in scattered clusters, 2.5 to 4 meters tall.

In Botswana, this class coincides with White's "Kalahari Deciduous *Acacia* Bushland and Wooded Grassland" (1983) and was classified as "Southern Kalahari Bush Savanna" by Weare and Yalala (1971). This vegetation largely occurs on the rolling sandy country of the slightly wetter parts of the Kalahari Desert, an area of wide plains, depressions, and pans. The main tree species are *Acacia erioloba, A. mellifera* ss. *detinens, Boscia albitrunca, Dichrostachys cinerea,* and *Terminalia sericea.* Low shrubs include *Acacia hebeclada, Bauhinia macrantha, Grewia* spp., and *Ziziphus mucronata.* Generally, the densest tree and shrub growth is on the rises of sandy ridges; the depressions are more open.

Farther north, the vegetation becomes a tree and bush savanna with a slightly greater tree component in the vegetation, including *Acacia* spp., *Boscia albitrunca, Burkea africana, Combretum* spp., *Commiphora* spp., *Croton zambesicus, Lonchocarpus nelsii, Ochna pulchra, Peltophorum africanum, Rhus tenuinervis,* and *Terminalia sericea.* These trees also are more common on sand ridges, whereas low-growing shrubs are common on the plains and include *Croton subgrattissimus, Grewia* spp., *Commiphora pyracanthoides,* and *Ximenia caffra.*

The small outlier in southern Zimbabwe is parallel to and just north of the Limpopo River, stretching across the border between Botswana and Zimbabwe close to Beitridge. It is a degraded shrub savanna. Similar outliers exist in northern Transvaal. Here, the seasonal winter drought is accentuated by soil of low infiltration capability and shallow rooting depths, plus pressure from nearby settlements such as Tuli in Zimbabwe or from land clearance and grazing, as is the case in the Transvaal. The vegetation mainly consists of very low, shrubby *Colophospermum mopane* up to 5 meters in height, occurring as scattered individuals, with a variety of *Acacia* spp., *Boscia* spp., *Dichrostachys cinerea,* and *Sclerocarya caffra.* The outliers in Mozambique can be classed as "Dry Riparian Woodland," with shrubs and small trees up to 5 meters tall scattered in a grass-dominated savanna (Millington and others 1989).

Much of this area, especially in South Africa, is given over to extensive cattle ranching. The carrying capability is low, however, commonly a single cow per 6 to 25 hectares (Christopher 1982).

The growing stock in this land cover class is estimated to be 137 million tonnes for southern Africa. The sustainable yield is estimated at 12 million tonnes a year, seriously limiting the potential for extraction. It is probable that even a small rate of fuelwood extraction would exceed the annual rate of wood production.

Class 42—*Fynbos* Thicket

This land cover class is confined to a few small areas that trend east-west in the southern coastal ranges of Cape Province in South Africa, between Cape Town and Port Elizabeth, 17,021 square kilometers in all. These areas were classified by Acocks (1953) and Moll and Bossi (1984) as "Macchia" or "False Macchia," but they also contain large cultivated areas.

The class occurs on south-facing parts of the coastal ranges, which are composed mostly of Table Mountain sandstones and shales, and on the coastal lowlands. The south-facing aspect promotes greater precipitation and somewhat more even distribution than in the extreme southwestern Cape Province. Precipitation is altitudinally linked across a wide range from 500 to 1,500 millimeters, moving from winter maxima in the west to spring and autumn peaks in the east. This results in a fairly consistent phenology. It is lowest in June, July, and August, at about 0.38, and peaks at about 0.5 in February and March. This is a somewhat greater NDVI value than the other *fynbos* vegetation areas in Hill Shrubland and Bushy Shrubland (Classes 32 and 33), perhaps because in this class *fynbos* enjoys partial protection from fire. Some of the floral and biomass changes with elevation are described by Rutherford (1972, 1978). These areas also are more woody, with somewhat better growth conditions. They seem to be transitional from the Bushy Shrubland (Class 33) to the "Warm Temperate Woodland" in Class 67.

White (1983) suggests that *Protea* spp. bushes and other tall plants that often occur scattered in bushy *fynbos* may produce dense, impenetrable thickets 4 to 6 meters high if they are protected from fire for long periods. This may apply in these areas. A wide variety of species occurs, including restioid (for example, the large *Willdenowia striata)* and ericoid forms, while proteoid shrubs also are common. In these areas of greater NDVI values, certain shrubs and trees that are not true *fynbos* species locally form thicket or scrub forest up to 10 meters tall. These include *Cassine peragua, Euclea tomentosa, E. undulata, Maytenus heterophylla, Myrsine africana, Chionanthus foveolatus, Pterocelastrus tricuspidatus, Putterlickia pyracantha, Rhus* spp., *Tarchonanthus camphoratus,* and *Zygophyllum morgsana.* Some encroachment of *Elytropappus* spp. may occur.

The most westerly part of this land cover class occurs in the vicinity of Stellenbosch, Paarl, and Wellington, where vineyards are a dominant feature, together with some olive growing. The more easterly parts include some sheep and cattle grazing and wheat cultivation. The valley floors support citrus orchards. For example, the Gamtoos Valley, possessing the most valued land of the region, is intensively cultivated.

The thicket areas in this class have some fuelwood potential, but access is limited by land tenure patterns, and areas of commercial cultivation obviously have little fuelwood potential. Thus, despite an estimated 47 million tonnes of growing stock on this small area of South Africa, the sustainable yield is estimated at only about 0.5 million tonnes, so this remains an area of limited woody biomass resource.

Class 43—Moist *Acacia-Commiphora* Bushland and Thicket

This class covers 441,586 square kilometers of southern Africa, more than 7 percent of this part of the continent. It crosses many floristic regions, but they are classified together by phenology and exhibit considerable ecological and structural similarity. The class occurs in patches that form an arc extending from the coastal plain of Angola, through southern Angola, parts of Zambia, Zimbabwe, and Malawi, to a southerly extension in southern Mozambique, southern Transvaal, and Swaziland.

South Africa contains 111,767 square kilometers of this class, Mozambique 75,038 square kilometers, Angola 71,224 square kilometers, Zimbabwe 63,498 square kilometers, Zambia 55,067 square kilometers, and Malawi 32,408 square kilometers, with smaller occurrences in Namibia, Botswana, and Swaziland. All areas exhibit the broad characteristics of an open bushland or thicket, frequently above a dense grassy sward, but often this is considerably degraded or altered by agricultural activity.

The phenology of this class has a distinct seasonality, with summer NDVI maxima (between December and April) of 0.35 to 0.47, followed by a fairly gradual decrease to September (0.12), and a rapid increase from November to December (figure 13-3). Most areas coincide with a mean annual precipitation within the range of 600 to 900 millimeters, usually with a summer maximum.

The largest contiguous area of this class occurs in southern Transvaal, southern Swaziland, and inland Natal. This area coincides with two of White's transitional areas (1983): from "Afromontane Scrub/Forest"

Figure 13-3. NDVI **Profile, Moist** *Acacia-Commiphora* **Bushland and Thicket (Class 43)**

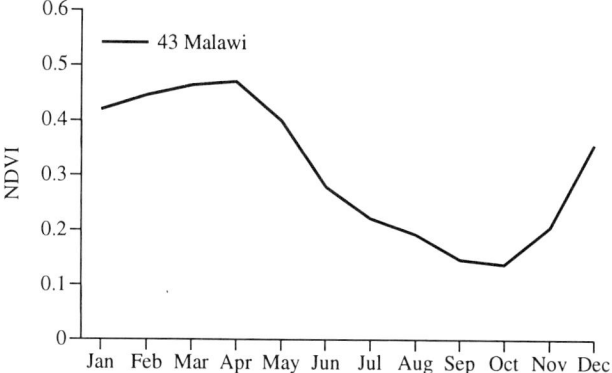

to "Highveld Grassland" and from "Afromontane Scrub/Forest" to "Coastal Forest." The area lies chiefly between 800 and 1,700 meters. The vegetation is mainly grassland, but originally was probably bushland with patches of scrub forest. The relics of this woody vegetation have Afromontane affinities at higher altitudes and include *Apodytes dimidiata* and *Halleria lucida*. The lowland remnants, which are related to coastal forests, include *Acacia* spp., *Aloe arborescens*, *Celtis africana*, *Commiphora harveyi*, *Ficus* spp., and *Syzygium cordatum*.

In South Africa, this land cover class borders the Highveld grasslands in the west and high-altitude woodlands in the east (for example, on the Lubombo Hills and forest plantations on the South African–Swazi border). The grassveld on the higher areas often is an open, grassy savanna with areas of bushes and smaller trees, represented in moister areas by *Acacia karroo*, *Maesa lanceolota*, *Syzygium cordatum*, and *Vernonia ampla*, and in drier areas by *Acacia nilotica*, *Dichrostachys cinerea*, *Ficus capensis*, *Maytenus senegalensis*, and *Sclerocarya birrea*.

This class now is largely an area of intensive commercial agriculture in the west, especially from Ermelo to Harrismith. This is mainly pastoral agriculture, but with some maize and sorghum cultivation. Farther east and south, in Kwazulu and inland Natal, are considerable areas of subsistence agriculture. As a result, bushes and trees are very restricted by cultivation and grazing, and woody vegetation now is confined to watercourses, poorer soils, and steeper slopes.

In many areas, most notably Swaziland, lower *veld* areas have been overgrazed and invaded by *Acacia nigrescens*, with secondary *Combretum zeyheri*, *Dichrostachys cinerea* ss. *africana*, and *Sclerocarya birrea*. The trees can attain 10 meters height, and between them are areas of low bushes, the most common being *Cordia gharaf*, *Ormocarpum trichocarpum*, and *Sclerocarya caffra*. These frequently are shorter than 4 meters, but tree density is great, ranging from 325 to 625 trees per hectare. Eradication of this thorny bushland is difficult and expensive. This sort of bushy savanna is both widespread and currently extending, especially in Swaziland (ETC Foundation 1987).

In the Highveld and Middleveld of Swaziland, grazing and cultivation have led to reduced woody biomass and an increase in the proportion of grasses. This also is true of the more western parts of this class in South Africa, including the small outliers on the margins of the Drakensbergs to the south and east of Lesotho.

This class also occurs in Mozambique, mostly on lower hills facing the coast in Inhambane Province and on the slopes of the upper Save Valley, extending into southeastern Zimbabwe on south-facing slopes in Southern Matabeleland. Most of these areas are a low-woody-biomass thicketlike form of the "Dry *Miombo*

Woodland" areas that surround them. Some are very degraded, whereas others possibly are edaphic climax, dry thicket vegetation. The proportion of shrubs to trees varies from high thicket vegetation, in which trees dominate and attain 10 meters in height, to low thicket vegetation, in which shrubs up to 3 meters are more prevalent than the trees, which occasionally reach 8 meters.

With decreasing height, the canopy becomes more open and grasses and herbs increase. They are floristically related to *miombo* woodland and the main tree species are *Acacia nigrescens*, *Albizia* spp., *Brachystegia boehmii*, and *Julbernardia globiflora*. Some of these assume a shrublike habit, but other shrub species also invade the thicket vegetation. In northern Mozambique, this class largely occurs in the middle Zambezi Valley, following closely the 800-millimeter isohyet, but small outliers also occur on the slopes of the upper and middle Lurio Valley in the northeast.

A similar vegetation occurs in Malawi in the Shire Valley north of Blantyre, on the Phalombe Plains, on the Shire Highlands, and over a large area north and south of Lilongwe. Most of these areas are characterized by very open woodland or a "Cultivation Savanna" (a term commonly used in Malawi to describe a savanna of maize cultivation and grassland with very scattered *miombo* woodland and exotic tree species). It generally grades from very open woodland to a grassy savanna in which the tree component often is represented only by mangoes. It mainly develops after woodland clearance for cultivation and fuelwood extraction.

On very thin, stony soils, a dry, open-canopy *miombo* woodland is the natural vegetation. An example is on the Rift Valley escarpment, where it is known locally as "Msuku Woodland." The main canopy dominants are *Brachystegia* spp., especially *B. boehmii*. In most areas, however, dry, open-canopy *miombo* woodland results from clearing well-developed woodland, and under increased population pressure it grades into "Cultivation Savanna." Typical of this is the Lilongwe Plateau in Malawi, where the only remnants of open-canopy woodland are sacred graveyards among grasslands and maize cultivation. The woody component of the "Cultivation Savanna" away from these remnants is dominated by mangoes (*Mangifera indica*), *Acacia* spp., *Combretum* spp., *Piliostigma* spp., and *Uapaca kirkiana*. It is particularly common on the Lilongwe Plain and in the Namwera and Malindi regions (Millington and others 1989).

In the Southern Region of Malawi, population pressure in this class has been far greater than farther north and "Cultivation Savanna" and woody thickets are common. Fertile soils are characterized by grassland interrupted by copses of relict woodland as well as isolated trees such as *Acacia albida*, *Cordyla africana*,

Adansonia digitata, *Hyphaene ventricosa*, and *Sterculia* spp. On low-fertility soils, either thicket or savanna vegetation forms. Thickets occur on alluvial soils to the south of Lake Malawi; here the species are similar to those in wooded areas, although they are more stunted. On thin, stony soils, a wooded savanna occurs which is dominated by *Brachystegia* spp., *Combretum* spp., *Colophospermum mopane*, and *Pterocarpus* spp.

Edwards (1982) quantitatively analyzed disturbed savanna woodland near Blantyre. Compared with mature *miombo* woodland, this disturbed savanna woodland (a) lacks stratification, (b) has a significantly lower basal area (4.87 square meters per hectare compared with 11.8 square meters per hectare), and (c) has a better-developed grass layer. The main dominants are *Brachystegia boehmii*, *Burkea africana*, *Julbernardia globiflora*, *Monotes africanus*, and *Pseudolachnostylis maprouneifolia*.

Areas of this class that occur in Zimbabwe are again largely on valley slopes, mostly to the north and west of Harare (for example, on the flanks of the Hunyani Range). Patches also exist near the Mozambican border, mainly on the sides of the Sabi Valley. As in other areas, such "Dry Savanna Woodland" (as it is called in Zimbabwe) is floristically poor and dominated by *Brachystegia spiciformis* and *Julbernardia globiflora*, with a number of smaller trees.

At its drier margins and in more disturbed areas, however, the canopy can be as low as 3 meters and the class takes the form of an open shrubby savanna. Here the dominant tree is *Brachystegia boehmii*, with other small, invading trees and shrubs such as *B. spiciformis*, *Colophospermum mopane*, *Julbernardia globiflora*, *Kirkia acuminata*, and *Sclerocarya caffra*. Some of these areas may include "Escarpment Thicket," which is common along the entire length of the Zambezi escarpment. This thicket rarely attains 10 meters height and forms a dense mixture of tree and shrub species, dominated by *Combretum* spp., *Commiphora* spp., and *Pterocarpus antunesii*.

In Zambia, a compact representative area of this class exists in the Luangwa Valley and around Choma in the Southern District, together with scattered patches on the sides of the upper Zambezi Valley. In the Luangwa Valley, the vegetation is largely altered "*Mopane* Woodland." "*Mopane* Woodland" is usually a single-story, open, deciduous woodland attaining 6 to 18 meters in which the dominant tree is *Colophospermum mopane*. Although it often occurs as pure stands, occasionally "Munga Woodland" trees invade, especially *Acacia nigrescens*, *Adansonia digitata*, *Combretum imberbe*, *Kirkia acuminata*, and *Lannea stuhlmannii*. Shorter trees and shrubs usually are absent and the grass and herb cover is only locally dominant. It is restricted to alkaline soils on valley floors which flood in the wet season and are dry in the dry season.

Destruction of *"Mopane* Woodland" leads to a drastic reduction in woody biomass because it is replaced by coarse tussock grassland.

Farther west in Zambia, this class is represented by areas on the slopes of the upper Zambezi Valley and its tributaries. It exhibits a similar *Colophospermum mopane*-dominated open woodland, but stands mixed with *Commiphora* spp. and *Terminalia sericea* are common and sometimes grade into the *"Baikiaea* Woodlands" on deeper alluvial soils near drainage lines.

The other compact outlier of this class around Choma consists partly of "Munga Woodland" (Fanshawe 1969). This is a type of savanna woodland with an open, parklike appearance, with either one or two woody layers, dominated by *Acacia* spp., *Combretum* spp., and *Terminalia* spp. in the upper story and a dense woody understory reaching 4.5 meters. The bushes are deciduous to semideciduous and are again dominated by *Acacia* spp. and *Combretum* spp. In some areas of "Munga Woodland," woody thickets exist. These are related to specific soil and water conditions, and usually include *Commiphora mollis*, *Euphorbia candelabrum*, *Markhamia obtusifolia*, and *Schrebera trichoclada*.

Although it exhibits variations, "Munga Woodland" generally is an invasive secondary woodland. It appears to have developed by the invasion of woodland trees onto alluvial grasslands and all the tree species are fire resistant. Many other types of woodland in Zambia are currently being invaded by "Munga Woodland," in particular the riparian woodlands.

In Angola, small areas occur in the southeast, similar to those of western Zambia. The main areas are in the southwest, between Chiange and Otechinjan; on the west-facing slopes northwest of Lubango; and on the slopes above Benguela and Lobito. These include the driest variants of the Angolan *miombo* woodlands and the *"Baikiaea* and *Mopane* Woodlands." The canopy trees are generally low, about 10 meters, and are mainly *Brachystegia* spp. and *Julbernardia* spp. On higher ground, a variety of shrubby and stunted trees invade to form a bushy thicket. The areas farther north are floristically richer than the *miombo* and *"Mopane* Woodlands," and dominant species become *Adansonia digitata*, *Dichrostachys* spp., *Euphorbia conspicua*, *Setaria welwitschii*, and *Sterculia setigera*.

Land use in this class ranges from commercial cultivation (for example, in parts of Malawi and South Africa) to subsistence farming (for example, Zambia and Swaziland). Pastoralism is inhibited by tsetse in many parts of Zambia and Zimbabwe.

The fuelwood potential of all these types of open woodland and thicket is significant, although limited by the seasonal drought. Woody biomass growing stock for southern Africa is estimated at 747 million tonnes, with a sustainable yield of 9.1 million tonnes.

Class 51—*Acacia* Woodland Mosaic

This class covers 519,892 square kilometers or 8.7 percent of southern Africa. Although best developed in Botswana, this class is distributed in an arc from northwestern Namibia, through Botswana, into southern Cape Province. It occurs in northern Botswana in the Chobe region to the east of the Okavango Delta, extending north into the Caprivi Strip and around to the west of the Okavango Delta near Tsau. In southern Botswana, it is strongly developed along the Botswana–South African border and in the Kalahari, with a number of significant patches south of the Makgadikgadi Pan. It covers 97,012 square kilometers in Botswana.

It also occurs extensively in Namibia (166,464 square kilometers), with patches to the east of the Etosha Pan, in the Great Karas Berg, in the hilly parts of Namaland, in the mountains surrounding Windhoek, and northward as far as the Angolan border.

In South Africa, 142,277 square kilometers are mapped in this class; the principal areas are in northern Cape Province and Bophuthatswana. Patches also occur in the Transvaal near Pretoria, to the north and east near Potgietersrus and Pietersburg, and between the Drakensberg Ranges and Kruger National Park. Further small areas occur scattered on the fringes of *Veld* Grassland (Class 11) in Orange Free State and southwestern Cape Province. The southernmost representatives of the class occur on the slopes of the coastal ranges, including some areas in Transkei.

Other small areas occur in southern Angola (25,768 square kilometers), eastern and southern Zambia along the Zambezi Valley (9,116 square kilometers), and in northern Zimbabwe on the north-facing slopes of the Zambezi Valley. The class covers 47,637 square kilometers in Zimbabwe. Other areas occur in Matabeleland and into eastern Mozambique on the sides of the Limpopo Valley (31,143 square kilometers).

Nearly all of these areas are transitional between Transitional Wooded Grassland (Class 24) and Open Woodland (Class 61), although in the south and west they often are adjacent to Wooded Shrubland (Class 35). This mosaic of *Acacia* woodland, Transitional Wooded Grassland (Class 24), and Wooded Shrubland (Class 35) is not as productive as Open Woodland (Class 61). Its occurrences mainly coincide with those areas that have an annual rainfall of 400 to 700 millimeters, although it occasionally occurs in both drier and wetter areas, depending on local soil-moisture holding conditions.

Values of NDVI are high, more than 0.36 between December and March and peaking at 0.42 in February, with a slow decline from March to September when the minimum of 0.17 occurs (figure 13-4).

Figure 13-4. NDVI **Profile,** *Acacia* **Woodland Mosaic (Class 51)**

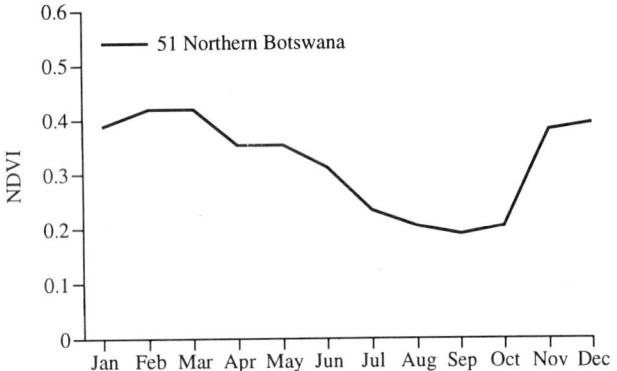

In Botswana, bushland along the eastern side of the country comprises trees of small to medium height, with a well-developed shrub layer. In the south, the dominant tree is *Peltophorum africanum* with *Acacia* spp., *Bauhinia macrantha, Boscia* spp., *Burkea africana, Ochna pulchra,* and *Terminalia sericea.* Farther north, the bushland is dominated by scattered trees of *Acacia* spp., *Combretum imberbe, Peltophorum africanum,* and *Sclerocarya caffra.* Among these trees are numerous smaller ones, such as *Acacia* spp., *Albizia* spp., *Boscia albitrunca, Combretum apiculatum, Commiphora schimperi, Terminalia sericea,* and *Ziziphus mucronata.* The main shrubs are *Dichrostachys cinerea* and *Grewia* spp.

To the north of the Tswapong Hills, the vegetation changes both floristically and structurally as *Colophospermum mopane* becomes more important. This species occurs both as a medium-height tree and in shrub form It usually is the dominant species wherever it occurs. Associated trees include *Acacia nigrescens, Burkea africana, Combretum* spp., *Commiphora mossambicensis, Sclerocarya caffra,* and *Terminalia prunioides.* The most common shrubs are *Acacia* spp., *Commiphora pyracanthoides, Dichrostachys cinerea,* and *Grewia* spp.

The scrub woodland to the north of the Tswapong Hills in Botswana is the vegetation community with the greatest woody biomass potential in this class. It has floristic similarities to Open Woodland (Class 61) and is dominated by *Colophospermum mopane.* The associated trees have variable geographical distributions and may become locally dominant. They include *Acacia nigrescens, Burkea africana, Combretum* spp., *Commiphora mossambicensis, Kirkia acuminata,* and *Sclerocarya caffra. Acacia nigrescens* and *Burkea africana* are the most common over large areas. The shrub layer consists of varying proportions of *Acacia* spp., *C. mopane, Dichrostachys cinerea, Grewia* spp., *Terminalia prunioides,* and *Ziziphus mucronata.*

The portion of this land cover class in eastern Botswana differs from that in the Kalahari Desert and the northwest of the country in that the shrub layer is better developed and tree density is much greater. Nevertheless, the canopy is generally open and, especially in the *Colophospermum mopane*-dominated areas, the grass cover is usually poor. The trees usually are less than 12 meters high, with average heights between 7 and 9 meters.

The more open woodland in the north of Botswana is composed chiefly of *Acacia* spp., *Combretum* spp., *Terminalia prunioides,* and *Ziziphus mucronata,* with shrubby *Colophospermum mopane, Grewia flava,* and *T. sericea.* In the northwest, scrub woodland occurs along the Ghanzi Ridge, which is rich in quartzite and limestone and where vegetation is denser than in the surrounding areas. Grasses are relatively unimportant here, and shrubs are more important. Trees include *Acacia* spp., *Albizia anthelmintica, Boscia rehmanniana, Combretum* spp., *Lonchocarpus nelsii, Montinia caryophyllacea,* and *Peltophorum africanum.* A distinct grassland transition zone between dense and open woodland occurs around the Mabebe Depression; it is dominated by *Cenchrus ciliarus* and *Chloris gayana.*

This land cover class continues into northeastern Namibia, including a large part of what White (1983) classifies as the "Transition to Zambezian Broadleaved Woodland." It is dominated by *Acacia* spp., but also includes *Combretum collinum, Commiphora africana, C. angolensis, Ochna pulchra,* and *Ziziphus mucronata,* forming a widely spaced woodland with trees normally less than 7 meters tall.

Farther west, the largest occurrence of this class in Namibia is between Etosha Pan and Grootfontein, and along the Kaokoland Escarpment in the north (as well as other west-facing escarpments throughout Namibia, which become drier at lower altitudes). Here *Colophospermum mopane* dominates the class at 7 to 10 meters height and forms a light, stunted woodland with a shrubby understory on stony or sandy soils. Here and in Angola, where a number of small outliers occur, the main associated plants are *Acacia* spp., *Boscia microphylla, Combretum* spp., *Commiphora* spp., *Grewia villosa, Rhigozum* spp., *Terminalia* spp., and *Ximenia* spp.

In the Windhoek Mountains, a vegetation with similar structure consists of tall bushes and small trees over a grassy sward. The principal bushes and trees are *Acacia hebeclada, A. hererocusis, A. reficiens, Albizia anthelmintica, Combretum apiculatum, Dombeya rotundifolia, Euclea undulata, Ficus cordata, F. guerichiana, Heeria (Ozorea), Rhus marlotii,* and *Tarchonanthus camphoratus.*

Other areas of this class in Angola with a similar phenology and productivity occur in the upper tributaries of the Zambezi Valley. These continue along the flanks of the Zambezi Valley in Zambia. In Zambia, it is again "*Colophospermum mopane*-dominated Woodland" with *Acacia nigrescens, Adansonia digitata, Combretum imberbe, Kirkia acuminata,* and *Lannea stuhlmannii.*

C. mopane prefers sodium-rich soil in wetter areas such as this, because it disperses the clay particles which then accumulate in the deeper horizons to form an impervious layer, resulting in a small water storage capability and poor depth penetration. These occurrences continue along the Zambezi Valley beyond Maramba and into northern Zimbabwe, south of Lake Kariba. Further outliers occur in Matabeleland as, for example, south of the Matopo Hills facing the Limpopo Valley.

In Mozambique, this class occurs on well-drained sites above the floor of the Limpopo Valley and is represented by a mixture of *"Mopane* Woodland" and *"Dry Miombo* Woodland," often adjacent to Transitional Wooded Grassland (Class 24). It also includes the dry riparian woodland along the Limpopo, where the vegetation is essentially a grass-dominated savanna with shrubs (3 to 5 meters) and small trees (5 to 10 meters). The species are related to the secondary invasive tree and shrub species of the surrounding woodlands.

In South Africa, this class transcends a number of floristic regions, but tends to occur on the margins of the *Veld* Grassland (Class 11). In northern Transvaal, it coincides with the "Sour and Mixed Bushveld" of Acocks (1975) and with White's "South Zambezian Undifferentiated Woodland and Scrub Woodland" (1983). The soils of these flatter areas at altitudes of 500 to 1,000 meters are nutrient-poor, leached, and occasionally waterlogged during the wet season. Precipitation always exceeds 500 millimeters a year.

A mosaic of dense woodland with grass understory is interspersed with more grassy patches in which *Burkea africana*, *Terminalia sericea*, and *Ochna pulchra* dominate the woody component, and *Eragrostis pallens* and *Panicum maximum* dominate the herb layers. The tall perennial grasses are mainly *Andropogoneae* and have poor nutritive value, especially in the winter; hence the term "sour *veld*." Werger and Coetzee (1978) refer to these areas as part of the "Broad-Orthophyll Plains Bushveld," which includes the species just mentioned and *Combretum apiculatum* as dominant on sandy soils derived from granite. They also refer to *"Terminalia* Sandveld" communities in the deep sandy areas of the Transvaal, which are largely on the plateaus and on small outliers in the low *veld* on granite soils.

Patches of this class also occur in northern Cape Province, and these have similar characteristics to occurrences in southern Botswana. Other outliers occur along the south-facing slopes of the Kaap Plateau and in the middle of the Vaal Valley between Prieska and Barclay West. Here, *Acacia erioloba* in particular has been removed from large areas to provide fuel for the mines at Kimberley. Similar vegetation occurs in southern Orange Free State, in the vicinity of Spring-fontein, Bethulie, and Wepener, and in the Orange River and Caledon valleys. These areas reflect mosaics of agricultural land within the natural *veld* vegetation.

Further small outliers occur in southwestern Cape Province and Transkei as a transition zone between the *veld* and the coastal range woodlands. Here, species are more varied than those of Transvaal. In some cases they represent degraded woodland, especially in Transkei.

Most of these areas in South Africa have considerable agricultural activity, mostly cultivation of maize. In addition, these areas include some of the most densely populated parts of the country. The woody component of the vegetation may be an important source of fuelwood for some rural populations. However, this woody biomass often exists on privately owned land, thereby restricting access. Elsewhere, these areas provide a significant fuelwood source, most notably in northern Botswana and northwestern Namibia.

Fuelwood exploitation needs to be carefully monitored and controlled because of small sustainable yield. Cattle grazing is an important consideration in Botswana, where a carrying capability as small as one beast per 25 hectares (Millington and others 1989) makes overgrazing commonplace. The drier southern and western parts of this class are most prone to excessive exploitation.

Class 61—Open Woodland

This land cover class occupies more than 8 percent of southern Africa, nearly a half a million square kilometers. It occurs extensively in southern and southeastern Angola (127,944 square kilometers), southwestern Zambia (72,561 square kilometers) and northeastern Namibia (99,278 square kilometers), including the Caprivi Strip into the north of Botswana (24,767 square kilometers). It then becomes a broken but extensive band through central Zimbabwe (115,455 square kilometers), curving southward into southern Mozambique (26,295 square kilometers) and into small areas in northern Transvaal and Orange Free State (19,023 square kilometers). Significant, although smaller, areas occur in Swaziland (2,319 square kilometers) and Lesotho (369 square kilometers). Although they transcend a number of floristic groups (White 1983), all these areas are classified together by productivity and phenology.

Open Woodland occurs mostly where precipitation ranges from 600 to 800 millimeters a year with a summer maximum. It also occurs in areas outside this precipitation range, depending on soil drainage conditions. In southeastern Angola, the phenology exhibits NDVI peaks of 0.42 in December and 0.38 in April, declining slowly to a minimum of 0.21 in September.

In central Zimbabwe, a similar pattern is displayed, with a peak of 0.36 in March and low of 0.2 in September (figure 13-5).

In the south and east of Angola, the seasonal drought is manifest in scant annual rainfall and the long dry season. This seasonality is accentuated by the soils developed on the Kalahari Sands and the thin, stony soils of the Bié Plateau. The inability of these soils to hold water throughout the dry season subjects the trees to a severe seasonal drought. This woodland differs from woodlands on the Angolan plateaus: it is more open than Seasonal *Miombo* Woodland (Class 66), structurally different from Wet *Miombo* Woodland (Class 67), and often floristically distinct from Evergreen Forest (Class 82).

The canopy trees are generally low (about 10 meters) and are interspersed in a grass layer of up to 2 meters height. On the Bié Plateau, the dominant trees are *Brachystegia boehmii, B. gosswieleri, B. spiciformis, Julbernardia globiflora*, and *J. paniculata*. On the high parts of the plateau, between 1,900 and 2,200 meters, the canopy is much lower (less than 5 meters) and is dominated by *B. floribunda, B. spiciformis*, and *J. paniculata*. However, a variety of other shrubby and stunted trees invades the high ground to form a bushy thicket. Above 2,200 meters, the tree species die out and the open thicket is replaced by "Montane Grassland."

The fuelwood potential of this woodland is far lower than others in Angola, because both growing stock and productivity are restricted. On the highest parts of the plateau, fuelwood potential is extremely small, but fortunately these areas are quite localized in extent. In southeastern Angola, on the Kalahari Sands, where the rainfall varies between 500 and 1,000 millimeters a year, "*Baikiaea* Woodland" and "*Burkea africana* Savanna" occurs.

In a forest feasibility study of Kuando-Kubango District, Coelho (1967) identified seven forest zones in the area covered by this land cover class. The dominant species characteristic of this class are *Baikiaea plurijuga, Burkea africana, Combretum dinteri, Guibourtia coleosperma*, and *Pterocarpus angolensis*. Tree densities of the dominant species ranged from 1.5 to 47 percent for all trees, with 1.4 to 48 percent for trees 20 centimeters dbh (diameter at breast height). *B. plurijuga* forms a canopy which can reach 20 meters, but is commonly lower in drier areas; beneath the canopy is a rich, dense, shrubby understory (Millington and others 1989).

The topography of ancient dunes on the Kalahari Sands produces a characteristic variation in vegetation patterns throughout the border areas of Angola, Namibia, Zambia, Botswana, and Zimbabwe. The ridges are tens of meters high and several kilometers apart and are dominated by *B. plurijuga, P. angolensis*, and *G. coleosperma*. Interdune areas are characterized by *Combretum* spp. and *Terminalia sericea*. A mixture of these occurs on the intervening slopes together with *Erythrophleum* spp. "*Mopane* Woodland" occurs on basaltic areas and on alluvial soils such as those in the Cunene Valley in southwestern Angola.

"*Mopane* Woodland" has the smallest woody biomass potential of all Angolan woodlands, although the open woodland of the Kuando-Kubango District continues into northwestern Namibia and the Caprivi Strip. In these areas, this class is composed chiefly of *Acacia* spp., *Baikiaea* spp., *Burkea africana, Combretum* spp., *Terminalia prunioides*, and *Ziziphus mucronata*, with some *Colophospermum mopane* shrub. It occurs to the north of the *Acacia* Woodland Mosaic (Class 51) and has a somewhat greater productivity, although it is floristically very similar.

In southwestern Zambia, these woodlands are referred to as "Kalahari Woodlands" by Fanshawe (1969), and include several types. "*Guibourtia* Woodland" is a two-story open woodland with a deciduous-to-semideciduous, floristically rich upper canopy 18 to 24 meters high which includes *G. coleosperma*, various deciduous trees, and invasive "*Miombo* Woodland" species. The understory is composed of a high thicket (1.3 to 2.6 meters) of small trees and shrubs. Climbers and scramblers are scarce, and the grass element is of variable density.

"*Burkea-Erythrophleum* Woodland" is a "Kalahari Woodland" that has a more open canopy than "*Guibourtia* Woodland," and an unstratified understory. It is floristically less diverse and is dominated by *Burkea africana* and *Erythrophleum africanum*. Both of these trees also exist in *Guibourtia* woodland, and *G. coleosperma* also is common. Other trees common to both woodland types are *Amblygonocarpus andongensis, Combretum mechowianum, Cryptosepalum exfoliatum, Dialium engleranum*, and *Monotes* spp. "*Miombo* Woodland" elements are rare in "*Burkea-Erythrophleum* Woodland." Other types of "Kalahari Woodland" are

Figure 13-5. NDVI **Profiles, Open Woodland (Class 61)**

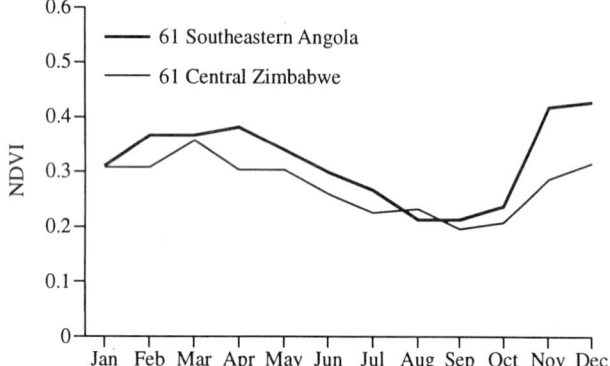

mainly "*Burkea-Diplorhynchus* Woodland" and "*Diplorhynchus* Scrub."

These woodlands are all related, each securing a slightly different ecological situation. In general, they have developed because of the destruction of natural *Baikiaea* forests which now are restricted to the extreme southwest of Zambia. The *Baikiaea* forest area has declined rapidly in recent years due to cultivation, burning, competition from *Cryptosepalum* forest, and, most important, timber extraction. *B. plurijuga* ("Zambezi teak") is a valuable timber tree. The destruction of the forest leads to "Chipya" and other types of "Kalahari Woodland" or to a secondary "*Baikiaea* Woodland" with elements from adjacent "*Munga* and Kalahari Woodlands."

In Zimbabwe, this class includes the same "Open Savanna and *Baikiaea* Woodland" vegetation described above, plus some montane vegetation adjacent to areas of Evergreen Forest (Class 82). Much of this class generally exists on well-drained soils above 1,350 meters, although it occurs as low as 675 meters on the Limpopo Escarpment. The canopy is between 6 and 13 meters tall and is dominated by *Brachystegia spiciformis* and *Julbernardia globiflora*. The tree canopy varies between 50 and 80 percent; shrub cover is open and usually below 50 percent, and grass cover ranges from 50 to 80 percent. An increase in the proportion of *Julbernardia* spp. is associated with increased disturbance, usually after agricultural clearance. Within this woodland are patches of grasslands and savannas of *Acacia* spp. and *Parinari curatellifolia*. All of these divergent vegetation types may be indicative of disturbance or local environmental conditions. The fuelwood potential of these vegetation inclusions is much smaller than the *B. spiciformis–J. globiflora* woodland enclosing them.

In Matabeleland North, west of Hwange on the deep Kalahari Sands, the class is dominated by an open, dry deciduous *Baikiaea* forest. The open nature of the forest is again a function of its exploitation for agriculture, fuelwood, and timber. The *B. plurijuga* woodland canopy is extremely variable in height, ranging from a dwarf form (only 1 to 1.5 meters) to regenerated specimens about 20 meters high. Other canopy trees are rare unless the area has been extensively disturbed.

At its edges, *Baikiaea* forest grades into associated woodlands and bushy savannas which contain a mixture of tree species. Associated with *B. plurijuga* on deep, sandy substrates are *Burkea africana, Entandrophragma caudatum, Erythrophleum africanum, Pterocarpus antunesii*, and *Ricinodendron rautanenii*. The invasive *Acacia eriloba* and *Combretum collinum* also are widespread. A floristically rich, dense, shrubby understory 5 to 8 meters tall exists, in which the main species are *Baphia massaiensis* ss. *obovata, Combretum engleri,*

Dirichletia rogersii, Paropsia barzzeana, Pteleopsis anisoptera, and *Pterocarpus antunesii*. The density of the shrub layer increases with repeated burning, and beneath shrubby understories, grasses are poorly developed.

The montane vegetation in Zimbabwe is complex. True montane forest, similar to that in Malawi and in East Africa, is rare, for two reasons. First, the Zimbabwean mountains are generally lower. Second, many areas of southwestern Zimbabwe are dry leeward slopes of mountains that form the border with Mozambique. Consequently, Zimbabwean montane forests suffer greater drought stress than the equivalent forest in Mozambique. Because of this, savanna woodland occurs to 2,100 meters. The canopy rarely exceeds 6 meters in height and is underlain by sparse shrub, fern, and grass layers. The canopy is dominated by *Brachystegia spiciformis*, with occasional areas of *B. glaucescens, B. taxifolia*, and *Uapaca kirkiana*.

In southern Mozambique, this class occurs in the Limpopo Valley where it is a "Dry *Miombo* Woodland" with a maximum canopy height of about 8 to 10 meters. As it becomes drier it can be restricted to only 3 meters height and forms an open, shrubby savanna. This latter form is the more common. "Dry *Miombo* Woodland" is floristically poor and dominated by *Brachystegia boehmii, B. spiciformis*, and *Julbernardia globiflora*. Scattered among the canopy are smaller trees of *Monotes* spp., *Protea* spp. and *Uapaca* spp. Where the canopy is very disrupted, shrubs commonly invade and a grass layer 0.6 to 1.2 meters high develops. In these cases, the dominant tree is *B. boehmii*, but other small trees and shrubs invade the canopy, such as *Burkea africana, Faurea speciosa, Hymenocardia acida, Ochna schweinfurthiana, Parinari curatellifolia, Swartzia madagascariensis, Syzygium guineense* ss. *guineense, Terminalia brachystemma*, and *Vangueriopsis lancifolia*.

In South Africa, this land cover class is a version of *Acacia* Woodland Mosaic (Class 51) having slightly greater productivity. It represents the denser woodland areas, with dominant species being *Burkea africana, Ochna pulchra*, and *Terminalia sericea*, along with *Combretum apiculatum* on· sandy soils. Much of this class includes a mosaic of agricultural activity and irrigation developments. Examples include intensive orchards north of Nelspruit and around Zebediela in the Transvaal and irrigation developments along the Vaal River in Orange Free State.

In Lesotho, this land cover class is represented by pockets of bushy woodland along escarpments and riparian areas. Along escarpments, the dominant trees are evergreens such as *Diospyros whyteana, Euclea* spp., *Halleria lucida, Ilex mitis, Maytenus* spp., *Olinia emarginata*, and *Podocarpus latifolius*, with a few deciduous species. In the riparian woodland component, the main woody species are *Acacia karroo, Celtis africana, Diospyros lycioides, Populus* spp., *Rhus lancea, Salix* spp., and *Ziziphus mucronata.* Generally, these are poor-quality scrubby woodlands, known as *shallahalla*, and are a product of overgrazing; however, they are an important source of fuelwood in Lesotho.

In Lesotho, this land cover class contains the largest woody biomass stock, but much of it is on escarpments or along rivers and is inaccessible or has restricted access. In Mozambique, the sustainable yield is restricted by seasonal drought. Although the class provides readily accessible woody biomass in areas with otherwise poor potential, the class is not a significant long-term resource.

In most of the other areas—Angola, Botswana, Namibia, Zambia, and Zimbabwe—these woodlands have moderately high growing stock but smaller sustainable yield, so that only slow rates of exploitation are realistic if widespread vegetation destruction is to be avoided. Open Woodland does provide a potentially important fuelwood source, however, in many areas where surrounding lands are less endowed. The most favored sites are the "Montane Woodlands" of Zimbabwe, although many of these areas are reserved. A potential conflict also exists between fuelwood and timber extraction in many areas on the Kalahari Sands. This class has an estimated 1,446 million tonnes of woody growing stock throughout southern Africa, with a sustainable yield of 21 million tonnes.

Class 66—Seasonal *Miombo* Woodland (including Tropical Coastal Woodland)

This is the largest of all land cover classes of southern Africa, accounting for more than 1.3 million square kilometers, or about 22 percent of southern Africa. It consists of well-developed seasonal forest to the south of the Mesophilous Humid Tropical Forest (Class 85) and Ombrophilous Humid Tropical Forest (Class 87) of the equatorial areas. It is characterized by strong seasonality and a semi-evergreen or deciduous nature.

Seasonal *Miombo* Woodland covers nearly the northern half of Angola and along the Angolan escarpment to the south of Huambo. Consequently, it occurs in all districts except Namibe and Kuando-Kubango, covering 450,070 square kilometers. In Zambia, it is widespread, although more scattered, occurring in every province and covering 344,891 square kilometers. It occurs in Malawi (46,425 square kilometers), across the north of Zimbabwe (78,252 square kilometers), and large areas of Mozambique (360,646 square kilometers).

Smaller areas of mostly tropical coastal woodland in southern Mozambique, Natal, eastern Cape Province (54,065 square kilometers total for South Africa) and small parts of Swaziland (4,479 square kilometers)

are ecologically somewhat different, but have a similar phenology and productivity to Seasonal *Miombo* Woodland. These coastal areas in South Africa and Mozambique coincide with White's "Eastern Africa Coastal Mosaic" (1983) and with White and Moll's "Tongaland-Pondoland Regional Mosaic" (1978).

These areas occur mostly where rainfall ranges from 800 to 1,200 millimeters a year, with a summer maximum. The phenology reflects the rainfall seasonality: minimum winter NDVI values occur in August in Angola (0.19), September in Zambia (0.22), and October in northern Mozambique (0.18). All three curves (figure 13-6) show a rapid increase in NDVI values through the spring, staying mostly above 0.40 through the summer months. Peaks are 0.54 in Angola (December), 0.55 in Zambia (December and January), and 0.52 in northern Mozambique (January).

Tropical coastal woodland areas of South Africa and Mozambique have a peak NDVI value of 0.45 in March. For the rest of the period from November to June, NDVI values range from 0.38 to 0.45. After June, levels diminish to approximately 0.25 from August through October, followed by a rapid rise through November-December (figure 13-6).

In Angola, Seasonal *Miombo* Woodland resembles Wet *Miombo* Woodland (Class 67), but has a more marked seasonality, a more open structure, and a slightly lower productivity. Undisturbed Seasonal *Miombo* Woodland has a clear structure, a closed canopy to 30 meters in height, and is floristically rich. The main species are *Brachystegia boehmii, B. gosswieleri, B. spiciformis, B. wangermeeana, Combretum* spp., *Isoberlinia angolensis*, and *Julbernardia paniculata*. A well-developed tree layer at 5 to 10 meters occurs above a dense herbaceous undergrowth. Apart from rainfall amount and seasonality, the strongest controls on *miombo* woodland distribution in Angola appear to be vegetation disturbance, depth of soil, and altitude.

In the dissected river valleys of the north, which trend north-south, a rapidly regenerating riparian for-

Figure 13-6. NDVI **Profiles, Seasonal** *Miombo* **Woodland (Class 66)**

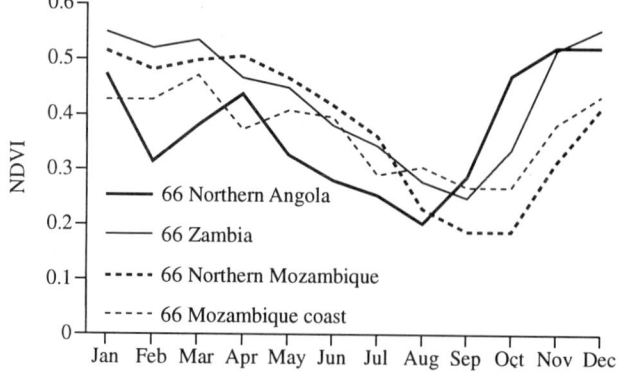

est occurs, 35 to 40 meters tall. On the drier interior plateau, however, the forest is less productive and more sensitive to disturbance. Existing in northwestern Lunda Norte Province is a semi-evergreen woodland of *Marquesia acuminata–Pteleopsis diptera,* 35 to 40 meters tall, although *Marquesia macroura–Brachystegia taxifolia* woodland is more common. This is a successional *miombo* woodland with a dense evergreen canopy of 30 meters height, dominated by the two trees, plus *Brachystegia* spp., *Daniellia alsteeniana, Pterocarpus angolensis,* and *Uapaca* spp. A small tree and shrub layer also occurs, dominated by *Bridelia* spp., *Erythrophleum* spp., *Faurea saligna, Parinari curatellifolia,* and *Uapaca* spp. It is likely that the *miombo* woodland in Bié and Huíla districts are similar to those in Lunda Norte District.

In western parts of Bié, Cuanza Norte, Cuanza Sul, Malanje, Uige, and Zaire districts, wooded savanna-type vegetation is more common than woodland, with patches of bushy thicket and grassland, depending on the environment. This is altogether drier and more deciduous than the surrounding woodlands. The dominant species are *Cochlospermum angolense, Diplorhynchus condylocarpon, Piliostigma thonningii,* and *Terminalia sericea.* The shrub layer can develop into a thicketlike savanna, but usually only a well-developed tall grass layer exists.

To the east, much of this land cover class is characterized by an open-canopy *miombo* woodland that is floristically poorer than wetter *miombo* woodland areas. In particular, *Brachystegia floribunda* is far less common or even absent locally. Other species of *Brachystegia* that occurs in the wetter *miombo* woodlands are less important. In most areas, particularly on thin or freely draining soils, *Brachystegia gosswieleri, B. spiciformis,* and *Julbernardia paniculata* dominate.

The stunted, shrubby understory of trees also increases because more light penetrates through the canopy. The grass and herbaceous ground layer also is better developed. Nevertheless, in wetter areas, the canopy trees are very similar to the wetter *miombo* woodlands.

In northern Bié District, Seasonal *Miombo* Woodland is very poorly developed. The canopy varies between 6 and 30 meters height and the tree canopy cover varies from 15 to 70 percent. The canopy is dominated by *Brachystegia tamarindoides, Marquesia loadensis, Syzygium guineense,* and *Uapaca* spp. The open nature of the canopy leads to a low tree and shrub layer (reaching 5 meters) and an open, grassy undergrowth.

In Zambia, Seasonal *Miombo* Woodland is again closely related to Wet *Miombo* Woodland (Class 67) (Chidumayo 1987). The woodland is similar in structure, but it is more open and leaf loss in the dry season is far greater. Species such as *Brachystegia allenii, B. bussei, Burkea africana, Isoberlinia angolensis, Julbernardia*

globiflora, and *Terminalia sericea* appear with greater frequency in the canopy. The shrub and grass layers also may be better developed. In the hills along the border of Zaire, in Northwestern Province, along the Mozambican border, and on the Zambian extension of the Nyika Plateau, the trees come under severe drought stress due to the smaller water-holding capacities of the thin, stony soils. Here the canopy is dominated by *Brachystegia glaucescens, B. microphylla, B. taxifolia,* and *Cryptosepalum exfoliatum.*

The drier end of this class in Zambia is a "Dry *Miombo* Woodland" featuring both natural and disturbed elements. It is common on alluvial sands and flats, along the Zambezi Valley, the Kafue Flats, and, in part, on the Kalahari Sands. In the first two areas it is closely associated with "Munga Woodland." The canopy of such woodlands is dominated by *Erythrophleum africanum,* commonly with *Brachystegia allenii, B. bussei, Burkea africana, Isoberlinia angolensis, Julbernardia globiflora,* and *Terminalia sericea.* The woodland has a more open canopy of widely spaced deciduous trees, enabling a diverse understory to develop, of which the common elements are *Baphia massaiensis, Combretum elaeagnoides, Crossopteryx febrifuga, Dalbergiella nyasae, Diospyros kirkii, Diplorhynchus condylocarpon,* and *Pseudolachnostylis maprouneifolia.*

"Munga Woodland" is common on the Kafue Flats, in Southern District, and along the Lunsemjwa, lower Luangera, and Zambezi valleys. It is a type of savanna woodland with an open, parklike appearance. One or two woody layers exist, both deciduous, with emergents reaching 18 meters in height. Particularly important in the canopy are *Acacia* spp., *Combretum* spp., and *Terminalia* spp. trees, although the woodland can be floristically diverse. Sometimes a dense understory reaching 4.5 meters occurs. The bushes are deciduous or semideciduous and floristically diverse; but nonetheless, they are dominated by *Acacia* spp. and *Combretum* spp. Climbers occur occasionally, and a tall, thick grass layer always exists. It often is burned annually.

This land cover class also includes a degraded *miombo* woodland, notably in the Kafue Flats and in Luapala, Northern, and Northwestern provinces. In these areas, woodland is destroyed by a type of shifting agriculture practiced by the Bemba in Zambia known as *chitemene.* It usually occurs in "*Miombo* or *Combretum* Woodland." Mainly cassava, cowpea, maize, and millet are cultivated, with groundnuts planted in the second year.

The regrowth of abandoned plots has been studied by Stomgaard (1985, 1986), and is discussed in detail in Chapter 12 (East Africa). Overall, the situation is one of declining *miombo* woodland species as they are replaced by fire-resistant and fire-tolerant trees characteristic of "*Combretum* Savanna." The dominant trees are *Brachystegia spiciformis, Combretum mechowianum, Diplorhynchus condylocarpon,* and *Syzygium guineense.* The canopy these form enables *Uapaca* spp. to invade, and, much later in the succession, the main components of the *Brachystegia-Julbernardia* woodland may reappear. Other vegetation types are common in the degraded Seasonal *Miombo* Woodland; in northern Zambia, these include "Itigi Forest" and "Lake Basin Chipya Woodland."

"Itigi Forest" exists along the Zairian border near the Mweru Swamp. It is composed of deciduous trees 6 to 12 meters high, mainly *Baphia massaiensis, Boscia angustifolia, Burttia prunoides, Bussea massaiensis, Diospyros mweroensis,* and succulents. Underneath is a deciduous-to-evergreen thicket 3 to 4.5 meters high, dominated by *Boscia mossambicensis* and *Teclea isheri.* It is a preclimax forest unable to attain dry deciduous forest forms because of poor drainage conditions.

Disturbance of "Itigi Forest" results in the formation of "Lake Basin Chipya Woodland and Scrubland." This is a three-story woodland with an open deciduous canopy which can attain 27 meters height. The main canopy dominants are *Albizia antunesiana, Burkea africana, Combretum collinum, Erythrophleum africanum, Parinari curatellifolia, Pterocarpus angolensis,* and *Terminalia sericea.* This is underlain by a floristically diverse, evergreen-to-semideciduous understory, 6 to 12 meters high. The main dominants at this level are *Combretum* spp., *Diplorhynchus condylocarpon, Markhamia obtusifolia, Piliostigma thonningii, Pseudolachnostylis maprouneifolia,* and *Syzygium guineense.* Under the middle story are a shrub layer, 2 to 3 meters high, and a rich ground flora, 0.6 to 2 meters high. It occurs extensively in Luapula Province along the Zairian border and around the Bangweulu Swamp.

In the north of Zimbabwe, this class coincides with a "Seasonal Savanna Woodland." In less-degraded and moister areas, the canopy trees reach 8 to 10 meters and are dominated by a mixture of typical Zimbabwean savanna woodland species, mainly *Brachystegia boehmii, B. spiciformis,* and *Julbernardia globiflora,* although other trees are invasive. Tree canopy cover varies between about 50 and 80 percent, and the shrub and grass cover is better developed. The understory is dominated by shrubs and smaller trees such as *Diospyros kirkii, Faurea saligna, Protea gaguedi, Pseudolachnostylis maprouneifolia,* and *Psorospermum febrifugum.* Many of these areas include tobacco cultivation, often grown in rotation with maize, especially in the Harare region.

In the north of Malawi, Seasonal *Miombo* Woodland is dominated by a denser, open-canopy woodland which exhibits a strong seasonality caused by climatic or pedological factors. In the south, the woodland is far more open and commonly grades into thicket and wooded savanna, although some of it can be termed open-canopy woodland. This poorer vegetation in the south is largely a result of extensive clearance for fuelwood and cultivation.

Undisturbed seasonal open-canopy *miombo* woodland occurs in northern and central Malawi, with an open canopy dominated by *Brachystegia boehmii, B. manga, B. stipulata, B. spiciformis, Julbernardia globiflora, J. paniculata,* and *Isoberlinia tomentosa.* Once cleared, the nature and composition of the woodland changes drastically. In the south of the country, three different types of vegetation can be seen, which are described in the following paragraphs.

Slightly denser wooded areas occur on previously cultivated soils that remain somewhat fertile, so that relict *miombo* woodland and isolated trees occur in patches. These contain mainly *Acacia albida, Adansonia digitata, Cordyla africana, Sterculia* spp., and palms, particularly *Hyphaene ventricosa,* between grassland areas.

In less-fertile areas, the vegetation degrades into either a wooded or thicket savanna. Thicket savanna occurs on alluvial soils in the lower Bwanje and Shire valleys and around the southern shores of Lake Malawi. The tree species are similar to those in wooded areas, but all are much more stunted.

On thin or stony soils, a very open wooded savanna forms, dominated by *Brachystegia* spp., *Colophospermum mopane, Combretum* spp., and *Pterocarpus* spp. Considerable areas in this class are devoted to cultivation, including cotton, which is grown mainly along the Lower Shire Valley below 600 meters, and tobacco which is generally grown at altitudes of 500 to 1,000 meters in the Shire Highlands around Blantyre and Limbe and on the Lilongwe Highlands. Other subsistence crops include maize, cassava, millet, and rice, the last along the lake shores.

In Mozambique, this land cover class is an intermediate *miombo* woodland type, between the wet and dry phases. This relation is clearly reflected in its geographical distribution in the Zambezi Valley. In northern Mozambique, it is representative of slightly drier conditions than in the surrounding Wet *Miombo* Woodland (Class 67), and in the south it represents wetter conditions than the surrounding "Dry *Miombo* Woodland" and scrubby vegetation.

In less-degraded areas, and in moister areas, the canopy trees attain 8 to 10 meters height, and, because of seasonality, are more open than Wet *Miombo* Woodland (Class 67). Small trees, shrub layers, and grass layers are better developed. In more drought-prone areas, this woodland degrades into an open savanna with a well-developed herb and grass layer and widely spaced trees to 8 meters; the canopy is dominated by *Brachystegia spiciformis* and *Julbernardia globiflora.* This class also includes the extensive tea plantations of Zambezia.

In southern Mozambique and South Africa, areas of woodland along the coast and in some inland areas have a different floristic composition but similar productivity and phenology. These are tropical coastal woodlands, divided into five types by White and Moll (1978). Four of these—"Sand Forest," "Dune Forest," "Swamp Forest," and "Fringing Forest"—occupy edaphically controlled locations. The fifth, "Undifferentiated Lowland Forest," is more widespread and consists of mixed evergreen and semideciduous species of a height and stratification depending on local site factors.

Canopy heights range from 10 to 30 meters with a large number of woody lianas. Characteristic canopy trees include *Albizia adianthifolia, Brachylaena* spp., *Cassipourea* spp., *Celtis* spp., *Chaetacme aristata, Combretum kraussii, Drypetes gerrardii, Ficus* spp., *Millettia grandis, Mimusops obovata, Protorhus longifolia,* and *Trichilia dregeana.* A widespread subcanopy layer exists, with a wide variety of species; tall subwoody plants up to 23 meters may occur. Many of these areas have been cleared, especially for sugarcane cultivation along the coast of Natal.

"Sand Forest," which largely occurs to the south of Maputo, has a deciduous habit occurring on sandy soils which are pale orange to grey, with rainfall between 700 and 900 millimeters a year. Here the forest is 10 to 25 meters high, with a canopy of *Balanites maughamii, Cleistanthus schlechteri, Newtonia hildebrandtii,* and *Ptaeroxylon obliquum,* with a well-developed subcanopy with *Croton* spp., *Grewia* spp., and *Ochna* spp.

"Dune Forest" is well developed north of Cape St. Lucia, with common pioneering species such as *Scaevola thunbergii* giving way to a full-canopy forest which includes *Diospyros rotundifolia, Euclea natalensis,* and *Mimusops caffra.*

"Swamp Forest" occurs on wet sites near streams and lakes, but is now very limited in extent. The general structure is a closed canopy of even height (about 30 meters) with a sparse woody understory and a well-developed herbaceous layer.

"Fringing forests" follow the Limpopo Valley in southern Mozambique. Common species include *Ekebergia capensis, Ficus* spp., *Ranolfia caffra, Syzygium* spp., *Trichilia emetica,* and *Xanthocercis zambesiaca* (White and Moll 1978).

Other patches of tropical coastal woodland occur southwest of Durban along the Umkomaas Valley inland of the Drakensbergs; around the coast near Port Shepstone, extending inland to the Transkei border; along the "Wild Coast" at the mouth of the Umtata river in Transkei; and along the high ridge that forms the border between Orange Free State and Natal, between Ladysmith and Harrismith. These represent some of the better-preserved forest areas of the Tongaland-Pondoland floristic area recognized by White (1983).

In the mist belt forests of parts of the coastal and interior highlands of Natal, the main canopy can reach 18 to 25 meters and emergents may attain 37 meters. Lianas and epiphytes are rare, but a bush and herbaceous layer encroaches in more open woodland areas. On the coast, the wind-trimmed coastal thicket soon gives way to a scrub forest with *Cordia caffra, Ekebergia capensis, Euclea racemosa, Mimusops caffra, Sideroxylon inerme,* and *Trichilia dregeana* forming an open canopy over a dense, bushy understory.

In places on the rocky Transkei coast, *Euphorbia triangularis* forms a narrow fringe of scrub forest 10 meters tall at the mouths of rivers. Scrub forest in the interior valleys is sometimes dominated by *Aloe bainesii* (12 to 15 meters tall) associated with a wide variety of other species. The original forests would have extended along this entire coastal belt of South Africa, with the most luxuriant stands approaching rain forest in stature and structure. The canopy would have been semi-evergreen to evergreen, with up to 30 species represented in any one place out of approximately 120 species for the entire region (White 1983). However, the remnants of such forests are few, and those that exist are classified as Wet *Miombo* Woodland (Class 67) and Evergreen Woodland Mosaic (Class 71).

Much of this class is a mixture of cultivation and woodland. Cultivation ranges from commercial, as in the sugar plantations in Natal and tobacco estates of northern Zimbabwe, to subsistence, most notably in parts of Angola and Zambia. In Malawi, much of the woodland is owned by tobacco estates, providing an important accessible fuel source for curing tobacco. Of course, this limits access to fuelwood for others.

In South Africa, many forest reserves exist in the tropical coastal woodland, both on the coast and inland. Some small areas have been given over to commercial forestry. Elsewhere, Seasonal *Miombo* Woodland contains extensive growing stock, despite the strong lull in dry-season productivity. Where Seasonal *Miombo* Woodlands occur in association with drier woodland areas, they are important source of fuelwood; but in wetter regions they may be less important than the surrounding Wet *Miombo* Woodland.

Significant variations in growing stock occur between different regions; for example, between eastern and western Angola; between the smaller-potential "Munga Woodland" in Zambia and the wetter Seasonal *Miombo* Woodland; and between the more degraded areas of southern Malawi and the greater potential in northern Malawi. Estimated woody growing stock of this class throughout southern Africa is 7,463 million tonnes, with an estimated sustainable yield of 119 million tonnes, making it a very important resource, but one in need of careful management. It is about one-third of the woody resource of southern Africa.

Class 67—Wet *Miombo* Woodland (including Warm Temperate Woodland)

In southern Africa, this land cover class occurs over about 7 percent of the land surface, an area of 409,916 square kilometers. Two distinct and widely separated areas exist. Warm Temperate Woodland occurs on the coast of South Africa, covering about 23,500 square kilometers. Wet *Miombo* Woodland stretches across the SADC states and covers more than 385,000 square kilometers. These two areas are discussed in the following sections.

Warm Temperate Woodland

The South African area mainly occurs between Knysna in Cape Province and Kokstad in western Natal, although small areas also occur along the coastal ranges of Natal and in the Drakensbergs of Transvaal. This latter area includes one large expanse stretching from just west of Uitenhage to east of Grahamstown; it also includes significant highland areas in Ciskei and Transkei, between King William's Town and Umtata and east of Umtata, from Mt. Frere down the Umzimvubu Valley to the "Wild Coast" near Port St. Johns.

Annual precipitation in this class varies widely within the range of 500 to 1,500 millimeters a year, but usually is between 650 and 750 millimeters. The NDVI is close to 0.50 in the period from November to February, exhibiting a slow decrease to a minimum of 0.27 in August, followed by a rapid increase in the spring (figure 13-7). The summer maximum is a vegetative growth response to precipitation and warmer temperature in coastal South Africa, and the winter minimum

is a response to cooler temperatures rather than drought.

In South Africa, these areas were classified by Acocks (1953) as "Valley Bushveld" and by White and Moll (1978) as transitional between "Afromontane" and "Undifferentiated Lowland Forests of the Tongaland-Pondoland Mosaic." The natural woodlands vary considerably in luxuriance. Some have closed canopies with many trees at 20 to 30 meters in height, but the majority of these temperate woodland areas have a much more open and lower canopy and have been considerably altered by agricultural and other land uses. At their driest end, which is usually accompanied by warmer temperature, they grade into *Fynbos* Thicket (Class 42).

The tree flora is fairly uniform but has a wide variety of species, including *Apodytes dimidiata, Combretum kraussii, Cryptocarya latifolia, Curtisia dentata, Halleria lucida, Ilex mitis, Kigelia africana, Nuxia floribunda, Ocotea bullata, Podocarpus* spp., *Prunus africana, Ptaeroxylon obliquum, Scolopia mundii,* and *Xymalos monospora.*

These arboreal communities exist today largely on steeper slopes in valleys and in higher areas because much of the land, especially in Transkei, has been severely degraded as a result of great population pressure. In addition, on rocky sites, sandstone outcrops, and unstable slopes, bushland is more common. With increasing altitude, ericaceous forms, especially *Passerina* spp., *Philippia* spp., and *Widdringtonia* spp., dominate in a scrub forest of only 5 to 7 meters height.

The most southwesterly occurrence of this land cover class is the Knysna Forest along the coastal area to the west of Port Elizabeth. This varies greatly in stature and floristic composition in relation to slope, aspect, altitude, and soil moisture (Taylor 1978). At lower altitudes, where temperatures are warmer and droughts are more common, this class grades into scrub forest, bushland, and thicket, with dominant species being *Diospyros dichrophylla, Euclea racemosa, Grewia occidentalis, Maerua caffra, Syderoxylon inerme,* and *Rhus* spp.

At higher altitudes, with decreased temperatures and increased humidity, canopy height again falls off rapidly, and forest gives way to scrub forest and thicket. The main species are *Berzelia intermedia, Diospyros glabra, Leucadendron eucalyptifolium, Protea cynaroides,* and *Virgilia oroboides.* Many areas of this class are mosaics of agriculture, orchards, afforested areas, and natural woodlands. This is true of areas north of Nelspruit, in the coastal ranges of Natal, and near East London, where the mosaic of agricultural land and warm temperate coastal forest extends to the "Wild Coast."

This class is an important source of fuelwood in the densely settled Transkei uplands, where it is extensively exploited. There is a marked contrast between

Figure 13-7. NDVI **Profiles, Wet** *Miombo* **Woodland (Class 67)**

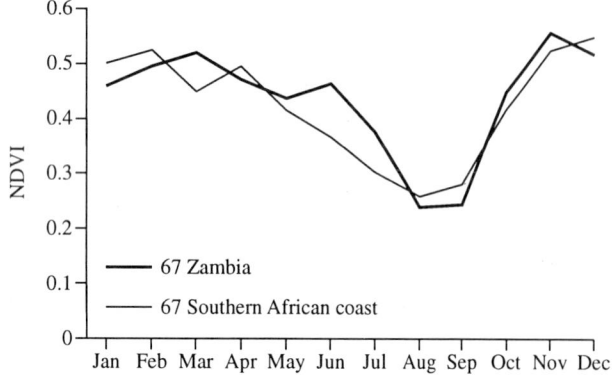

the accessibility of fuelwood in the Transkei and Ciskei compared with the equivalent areas, often commercially managed, in adjacent Cape Province. The woody biomass growing stock in these regions is estimated at about 74 million tonnes, with a sustainable yield of 1.3 million tonnes.

Wet Miombo Woodland

The Wet *Miombo* Woodland part of this class occupies approximately 6 percent of the land area of southern Africa, an area of more than 385,000 square kilometers. It covers large areas in Angola (150,234 square kilometers), especially in northern Angola, where it includes the Dembos cloud forest in the northeast of the country and large areas on the border between Bié and Moxico, in Lunda Sul, southeastern Malanje. It occurs in eastern Moxico and extends across the Zambian border into Northwestern Province.

Elsewhere in Zambia, it occupies a considerable portion of Copperbelt, Central, Laupula, and Lusaka provinces, with smaller areas scattered in Northern, Southern, and Western provinces, totaling 102,967 square kilometers. Malawi has significant occurrences, mainly along the shore of Lake Malawi and at high altitude, totaling 7,694 square kilometers. Important areas also exist on either side of the Zambezi River in Zambia and Zimbabwe. Smaller patches occur in western Zimbabwe, in Matabeleland, and along the higher slopes of the Sabi catchment, extending into Mozambique.

The class covers 22,606 square kilometers of Zimbabwe. In Mozambique, it occurs extensively in the north, particularly along the Luganda River and along the northern coastal areas of Cabo Delgado and Nampula provinces. It also occurs on the east-facing slopes in Manica and Sofala, with one small occurrence on the coast near Inhambane. It covers 102,861 square kilometers in Mozambique.

The phenology indicates that fairly high NDVI values (0.42 to 0.54) are sustained from October to June with a peak of approximately 0.54 in November. The lowest NDVI values (0.26) occur in August and September, increasing rapidly in October (figure 13-7). This reflects the high rainfall, usually greater than 1,000 millimeters with a summer maximum, and a limited dry period in the winter. Chidumayo (1987) divides the *miombo* woodland of Zambia into wetter and drier types at about 1,100 millimeters rainfall a year.

In Angola, these areas were classed by ETC Foundation (1987) and Millington and others (1989) as "Dense Medium Height *Miombo* Woodland." It differs from Evergreen Woodland Mosaic (Class 71) slightly in seasonality and in less overall productivity, although it is denser and well wooded, with similar species composition. It has a well-defined structure and is floristically rich, with a dense, closed canopy, often exceeding 15 meters but capable of attaining 30 meters. It is dominated by *Brachystegia boehmii*, *B. gosswieleri*, *B. spiciformis*, *B. wangermeeana*, *Combretum* spp., *Cussonia angolensis*, *Isoberlinia angolensis*, and *Julbernardia paniculata*.

Other canopy trees are less common, but a well-developed stunted tree layer exists at between 5 and 10 meters. A herbaceous undergrowth also is well developed. Within the woodland, stands of *Uapaca* forest occur, 4 to 10 meters high. These indicate a successional vegetation following agricultural clearance. On the higher ground in the Bié Plateau, the woodland canopy is lower (6 to 25 meters) and more open (cover varies from 30 to 90 percent). Here it is dominated by *Brachystegia bakerana*, *B. longifolia*, *B. spiciformis*, *Copaifera baumiana*, and *Guibourtia coleosperma*. There is a low (0.5 to 2 meters) woody layer with 50 to 60 percent ground cover and a dense, tall grass layer (2 to 3 meters).

The Dembos cloud forest on the northeastern escarpment of the central plateau of Angola between 6°30' N and 9°30' N also is included in this land cover class. Cloud forest vegetation exhibits less moisture stress than that on the adjacent lowlands and plateau due to the frequent mists and increased precipitation (1,100 to 1,500 millimeters a year) from the moisture-laden onshore winds. The forest canopy reaches between 30 and 50 meters and is floristically very rich. The main canopy trees are *Albizia* spp., *Celtis* spp., *Ficus* spp., and *Morus* spp. The area is used extensively for coffee cultivation, with the coffee grown under the shade of the large canopy trees.

The woodlands in this category in eastern Moxico Province are ecologically and floristically related to those of southern Zaire and western Zambia. These woodlands are highly variable and responsive to soil depth and quality. On well-drained soils, various types of "*Miombo* and *Cryptosepalum* Woodland" develop. On the thinnest soils, *Brachystegia macrophylla* and *B. utilis* dominate. Closed woodlands of *B. spiciformis* also exist, and thicker, well-drained soils exhibit either open "*Marquesia calonerus–Uapaca pilosa* Woodland" or, more commonly, dense "*Cryptosepalum maraviense* Woodland." All of these woodlands have high canopies of 15 to 20 meters, but they vary considerably in canopy openness and understory vegetation. On the poorly drained soils, the height of the woodland is less. It is dominated by *Brachystegia boehmii*, *Marquesia katangensis*, and *Uapaca* spp. In many places, however, this low woodland develops into a grassland with very poor fuelwood potential.

In Zambia, Wet *Miombo* Woodland is generally a two-story, closed, semi-evergreen woodland. The main upper-story dominants are *Brachystegia* spp., *Isoberlinia* spp., *Julbernardia* spp., and *Pterocarpus angolen-*

sis. The lower story is less defined structurally but is floristically more diverse. Underneath this exists either a grass and suffrutex layer 0.6 to 1.3 meters high or a dense, evergreen thicket reaching heights of 3.5 meters. A variety of suffrutices are common in *miombo* woodlands, and the grass cover varies both in density and by the season. The dense, evergreen thicket understory indicates that *miombo* woodlands have replaced dry Evergreen Forest in many parts of Zambia (Fanshawe 1969). The main woody thicket species are *Canthium burttii, Cassipourea* spp., and *Chrysophyllum megalismontanum,* as well as many shrubs.

Two main variants of Wet *Miombo* Woodland are apparent in Zambia. On the deeper soils of the Zambian Plateau, the main canopy dominants are *Brachystegia boehmii, B. floribunda, B. spiciformis, B. utilis, Isoberlinia angolensis,* and *Julbernardia paniculata.* As the shallower soils on hills and escarpments are encountered, or in extensive pockets of sand (*isengas*), the canopy dominants change. Initially they change to *B. glaucescens* in the south or *B. microphylla* in the north, and eventually to *B. taxifolia* and *Cryptosepalum exfoliatum,* respectively.

In Malawi, on the high plateau and in mountainous areas, wet Evergreen Forest and Wet *Miombo* Woodland undoubtedly constitute the climax vegetation. Much destruction of these evergreen forests and woodlands provided timber, fuelwood, and land for agriculture. This is particularly so in central and southern Malawi. Climax montane forest now exists only in isolated areas, and the only extensive areas of Wet *Miombo* Woodland are in the Northern Region; many of these areas are now reserved.

Because this is the most productive wood resource in Malawi, the implications of its restricted access are far-reaching for most of the population. Semi-evergreen forests exist on the Kandoli Mountains, Thyolo Mountain, the Mulanje Massif, and Zomba Plateau. These are closely associated with both the montane forests and the Wet *Miombo* Woodlands on the uplands, and are an intermediate type. They are dominated by *Brachystegia spiciformis,* but, unlike the *miombo* woodlands on the lowlands and plateaus, they have a thick, evergreen, shrubby understory, and grasses are poorly developed.

In much of northern Malawi, and on the Viphya Massif, Thyolo Mountains, and the Shire Highlands, there exist dense, closed-canopy *miombo* woodlands with high biomass and little seasonality. These woodlands take on two forms: (a) closed-canopy woodland in areas with precipitation exceeding 1,300 millimeters, and (b) a more open canopy woodland in medium-to-high rainfall regions. The canopy can reach 15 meters, and in some cases two canopies are present, possibly indicating previous woodland clearance. The closed-canopy woodlands are dominated by *Brachystegia longifolia* and *B. spiciformis.* In the open wood-

lands, *B. longifolia* is less important and *B. boehmii, B. spiciformis, B. manga, B. stipulata, Isoberlinia tomentosa, Julbernardia globiflora,* and *J. paniculata* dominate the canopy; the understory is very mixed.

In Mozambique, Wet *Miombo* Woodland is again dominated by *Brachystegia spiciformis* and *Julbernardia globiflora* and has similar structures to those previously described. However, much of this land cover class includes deciduous coastal forests along the Cabo Delgado coast, which are floristically very rich. The most important tree species are *Afzelia quanzensis, Chlorophora excelsa,* and *Sterculia appendiculata.* Other important canopy species and emergents are *Albizia gummifera, Ekebergia capensis, Erythrophleum suaveolens, Khaya nyasica, Millettia stuhlmannii, Newtonia buchananii, Pachystela brevipes, Syzygium guineense,* and *Xylopia aethiopica.* The canopy is about 12.5 meters high with emergents attaining 25 meters. In addition, a well-developed, thick, closed subcanopy exists, consisting mainly of *Albizia adianthifolia, Bauhinia* spp., *Harungana madagascariensis, Macaranga capensis,* and *Trema orientalis.*

The woody biomass reserve of the Wet *Miombo* Woodlands is great in most areas, both in growing stock and productivity. It is an aggressive vegetation type that can withstand extensive exploitation. The fuelwood potential of the Bié Plateau in Angola is generally high, both in growing stock and in sustainable yield. In the areas of this class in eastern Angola, however, the fuelwood potential is quite variable and dependent on local environmental conditions.

In the Dembos cloud forest, much of the understory has been destroyed to plant coffee bushes in the shade of the canopy trees, effectively reducing both the woody biomass and access to fuelwood. For the most part, however, Wet *Miombo* Woodland is a very important fuelwood source in all the countries where it occurs. It has an estimated woody growing stock of 4,875 million tonnes, with a sustainable yield of 54 million tonnes.

Class 71—Evergreen Woodland Mosaic

This class covers 165,779 square kilometers in southern Africa, nearly 3 percent of the land surface. It occurs in two areas: in south-central Angola, where it covers 96,222 square kilometers, and along the coast of central Mozambique, where it covers 59,071 square kilometers. In Angola, it occurs mainly in the high, less-accessible areas of northern Kuando-Kubango and southern Moxico provinces, but also extends into parts of Bié and Huíla provinces. In Mozambique, it is represented by a large area extending through much of Zambezia and Sofala; it is best developed along the coast but extends inland almost to the border with Malawi.

Outliers of the class exist in Mozambique, in Tete and Niassa Provinces in the north, in the extreme northeast, and along the coast of Inhambane and Gaza, especially near the mouth of the Limpopo. Other outliers occur on the border between Mozambique and Zimbabwe in Inyanga National Park. Still other small outliers occur in Malawi (2,635 square kilometers) and in parts of Northern Province in Zambia (6,060 square kilometers).

Evergreen Woodland Mosaic suffers little, if any, seasonal moisture stress. This may be the result of reliable precipitation and local soil moisture conditions. Its "evergreen" nature, however, is more likely the result of understory leaf growth during the short periods when upper-canopy trees are relatively leafless. This "evergreen" behavior is illustrated in the phenology. Photosynthetic activity is high throughout the year, mostly varying between NDVI values of approximately 0.35 and 0.55. High productivity is sustained between November and July with a brief period of low productivity between August and October (figure 13-8).

In Angola, this class was termed Dense High *Miombo* (ETC Foundation 1987; Millington and others 1989). It has a very well defined structure and is floristically rich. The canopy, which often extends to 25 meters, is dense and often closed, although not always. It is dominated by *Brachystegia bakerana*, *Cryptosepalum exfoliatum*, *Dialium engleranum*, *Guibourtia coleosperma* ss. *pseudotaxa*, and *Julbernardia paniculata*. Few other canopy trees occur, but stunted trees grow to between 5 and 10 meters. This lower layer is dominated by *Baphia massaiensis*, *Copaifera baumiana*, *Diospyros* spp., *Paropsia barzzeana*, *Parinari* spp., *Trichilia quadrensis*, and *Xylopia odoratissima*.

In addition to these two tree layers, a very poorly developed grass and herbaceous ground layer exists. Detailed data on the composition of Evergreen Woodland Mosaic in Kuando-Kubango District are available in the forestry feasibility study carried out by Coelho

Figure 13-8. NDVI Profiles, Evergreen Woodland Mosaic (Class 71)

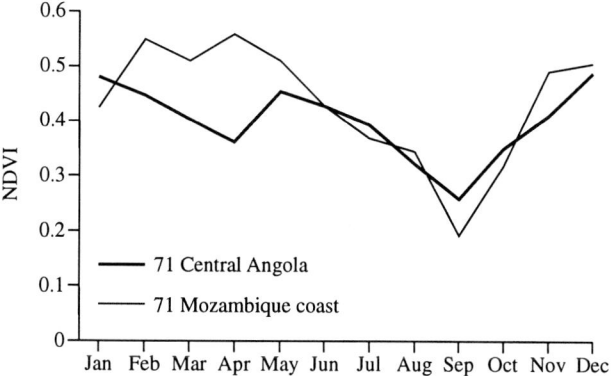

(1967), which is described in more detail in Millington and others (1989). This area differs from that farther north because the trees are related to Open Woodland (Class 61), with which this class interdigitates in the south. The most common tree from Open Woodland (Class 61) is *Burkea africana*, and generally, the Evergreen Woodland Mosaic here is dominated by *Brachystegia bakerana*, *Cryptosepalum pseudotaxus*, *Erythrophleum africanum*, *Guibourtia coleosperma*, and *Isoberlinia* spp., especially *I. baumii*. The canopy becomes more open on the shallower soils of the Bié Plateau and as the Open Woodland (Class 61) to the south is approached.

In Mozambique, the Evergreen Woodland Mosaic is best developed adjacent to the coast in Zambezia and Sofala provinces. The coastal plain is flat and often marshy, with large areas periodically flooded. Little if any seasonal moisture stress occurs. The canopy reaches to more than 15 meters, but more commonly is about 10 meters and is underlain by a layer of low trees (5 to 10 meters high) and a poorly developed herbaceous layer. The canopy is dominated by *Brachystegia spiciformis* and *Julbernardia globiflora*, but the whole woodland is floristically very rich.

The coastal forest belt is characterized mainly by a dry, deciduous forest on the lowland plains, with moister evergreen and semideciduous forest on the windward mountain slopes. The evergreen and semideciduous component of the coastal forest complex is grouped in this land cover class. Much of this forest has now been converted to low woodland (Millington and others 1989) with an upper canopy of 10 to 12 meters and a lower canopy of 5 to 7 meters and is dominated by *Acacia nigrescens* and *Albizia* spp. In many areas it has become so degraded that it now forms a thicketlike vegetation with shrubby species varying between 3 and 7 meters in height and emergents of up to 10 meters.

Not all of the lower forest and thicket can be attributed to degradation, because low, open vegetation occurs naturally on sandy soils of low water-holding capability along the coast (for example, Inhambane Province). Scattered pockets of higher, nondegraded coastal forest still exist, although much is degraded. These forests have a canopy of about 20 meters, with emergents reaching 40 meters. The main species are *Crossonephelis (Melanodiscus) oblongus*, *Lovoa swynnertonii*, and *Maranthes goetzeniana*.

In Manica and Sofala provinces, and on the windward slopes of the Macondes Plateau, the moister forest is semideciduous rather than evergreen. It has a canopy of about 15 to 20 meters with emergents up to 35 meters, the mean height of the larger trees being about 25 meters. The main canopy species are *Erythrophleum suaveolens*, *Newtonia buchananii*, and *Pachystela brevipes*. A well-developed shrub layer contains *Albizia adianthifolia* as the main species. Variations to this general pattern occur, particularly around the Zambezi Delta, where "*Hirtella* Forest" develops as a response to greater precipitation (1,200 to 1,400 millimeters) and a high water table. "*Hirtella* Forest" is a mosaic of *Miombo* Woodland and Semideciduous *Pteleopsis-Erythrophleum* Forest, and is dominated by *Hirtella zanzibarica*. Rice and sugarcane are grown extensively in these coastal plain areas.

In Malawi, this land cover class coincides largely with a mosaic of *miombo* woodland and cultivation along the lower land adjacent to the western shore of Lake Malawi, but it also includes montane areas such as those around Mount Mulanje. Floristically and structurally, it is similar to the *miombo* woodland described in Classes 66 and 67, with the canopy being dominated by *Brachystegia boehmii*, *B. manga*, *B. stipulata*, *B. spiciformis*, *Julbernardia globiflora*, *J. paniculata*, and *Isoberlinia tomentosa*. *B. boehmii* is the dominant tree species in many areas. This land cover class is intermediate between Wet *Miombo* Woodland (Class 67) and Evergreen Forest (Class 82).

This land cover class is also represented in northern Zambia by the vegetation associated with swamps, lakesides, and wet valley floors. Many of these areas are grassland, but some are areas of swamp forest. These are mainly three-story closed evergreen forests with a canopies reaching 27 meters, dominated by *Ilex mitis*, *Syzygium* spp., and *Xylopia* spp. They are underlain by a discontinuous evergreen understory of 9 to 18 meters and a dense evergreen shrub layer which reaches 4.5 meters. The forest floor is either bare or has stands of herbaceous vegetation.

All swamp forests are controlled by high groundwater levels and are small, varying from 1 to 120 hectares. Fanshawe (1969) estimates that only 380 square kilometers of swamp forest exist in Zambia. Other areas of this land cover class occur in valley floor sites, notably the Chambeshi. These are floristically and structurally similar to those described in the adjacent Wet *Miombo* Woodland (Class 67), but with a somewhat different phenology.

For most areas in this class, growing stock is high, as is sustainable yield. Overall, in southern Africa, the estimated growing stock is 83 million tonnes, with a sustainable yield of 3 million tonnes. The high growing stock and sustainable yield mean that, in most places, the fuelwood potential of these forests is great. This is the case for most areas in Angola and Mozambique, but areas in Malawi and Zambia are more variable. In many parts of Malawi, the areas are either associated with lakeside cultivation or are confined to less-accessible montane woodlands. In Zambia, although they have plentiful growing stock, the swamp forests are important only locally as a woody biomass resource. Representatives of the class in the Chambeshi Valley remain an important source of fuelwood.

Class 72—Cultivation and Forest/Woodland Mosaic

This class covers nearly 3 percent of southern Africa, an area of 163,250 square kilometers. It occurs mainly in areas with special drainage characteristics and includes considerable cultivated land. The most distinctive occurrence is the Okavango Delta area in northwestern Botswana, with an associated small area on the Chobe River on the Botswana border, together covering 16,441 square kilometers of Botswana.

Most other areas are less compact and scattered to the north and west. The majority occur in a circle of patches fringing the Evergreen Woodland Mosaic (Class 71) in central Angola (99,172 square kilometers). The largest of these areas is in Kwanyama in southern Angola, with another large area on the southeastern slopes of the Morro de Moco, northwest of Huambo.

Other smaller areas are associated with southward-flowing tributary streams and occur in association with the Zambezi and its tributaries in western Zambia. Small areas exist in the extreme northwest of Zambia around Mwinilunga on the River Lunga, so that altogether this class covers 29,193 square kilometers of Zambia. Small areas occur in northwestern Namibia, associated with the ephemeral drainage of the Omatako (9,801 square kilometers). Other areas are on the Zambian-Zimbabwean border on the shores of Lake Kariba (Zimbabwe has 2,582 square kilometers in this class) and on the border between Botswana and Transvaal on the Limpopo, near Martin's Drift (South Africa has 5,270 square kilometers in this class).

The phenology of this land cover class exhibits a limited seasonality, with NDVI peaks of 0.27 to 0.35 between February and May, followed by a slow decrease to below 0.2 between July and October. The Okavango Delta area shows somewhat greater seasonality, with a marked peak of 0.37 in December and January, slowly decreasing to 0.12 in September. This is followed by a sharp increase through October and November to December (figure 13-9). Moisture availability seems to be the key to this seasonality; this can be influenced by the level of the water table as well as precipitation.

A more compact area of this class occurs in the Okavango Delta, with small outliers in the Chobe Valley on the Namibian border, notably around the large southern meander of the Chobe River. It is partly a fringing forest with floristic similarities to the surrounding woodland types, as well as having species adapted to the edaphic conditions. The class exhibits

Figure 13-9. NDVI **Profiles, Cultivation and Forest/ Woodland Mosaic (Class 72)**

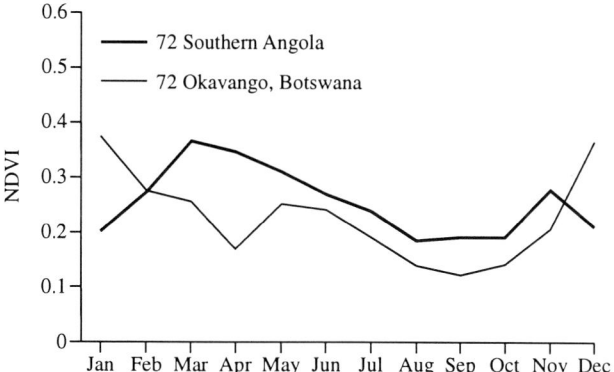

the greatest productivity of any class in Botswana, but is also characterized by some seasonal variation in productivity. Around the Okavango Swamp, grassland is fringed by a belt of large trees such as *Acacia eriolobu, A. galpinii, A. xanthopholoea, Colophospermum mopane, Combretum imberbe, C. megalobotrys, Croton megalobotrys, Diospyros mespiliformis, Ficus sycomorus,* and *Ionchocarpus capassa.* The grassy Chobe floodplain has a similar fringe of woody vegetation, but some areas are given over to cultivation.

Although not as clearly linked to a particular soil moisture regime such as the above, areas with similar biomass and phenology in Angola and western Zambia owe their existence to seasonal flooding. Large areas are periodically flooded by the Zambezi and its tributaries, as well as many other rivers that drain north and south from the central highlands of Angola. During several months of the year, the water table is close to the surface, promoting a short, tufted grassland with thicket on nearby nonflooded sites; the most abundant grass is *Loudetia simplex.* In Kuando-Kubango in southern Angola, such grasslands, interrupted only by thickets, cover large areas.

These areas often are floristically transitional between "*Baikiaea* and *Miombo* Woodlands," but structurally and phenologically they are controlled by seasonal precipitation and flooding. *Baikiaea plurijuga* woodland is dominant in many areas, alternating with grassland and mixed savannas containing *Acacia sieberana, Diospyros mespiliformis,* and sometimes *Acacia erioloba* along the rivers. In some higher areas, however, a woody community of *Combretum imberbe, D. mespiliformis,* and *Ficus sycomorus* occurs, although this is sometimes dominated by *Acacia albida* (an example is along the upper Cunene River). Along drainage lines in Kuando-Kubango, savanna communities occur, dominated by *Acacia* spp., along with island thickets of savanna species, sometimes dominated by the palm *Borassus aethiopum,* on old termitaria.

White (1983) refers to thickets dominated by *Brachystegia bakerana,* sometimes no more than 1.3 meters in height although usually taller, occurring as an ecotone between hydromorphic grasslands and woodlands on better-drained sites. Well-developed forest may develop along the larger water courses, and most of the tree species are deciduous for at least 2 months of the year. The floodplain of the upper Zambezi in Barotseland is flooded each year from mid-February to mid-June to a depth of 2 meters or more. Few tree species can withstand such a pronounced fluctuation in water level, and the outer fringe of riparian forest, 9 to 12 meters tall, is dominated by *Syzygium guineense* ss. *barotsense,* with an understory of *Rhus quartiniana.*

Some thickets in this class occur as regenerative stages after disturbance from overgrazing, shifting cultivation, or fire. Or, they may occur on certain soils, such as loamy sands that have clayey impervious layers at depth.

Woody biomass stock and productivity may appear ample in this class, but accessibility problems exist in many areas. Many areas are remote and for much of the year may be flooded. Some areas are given over to cultivation. Some areas, like Chobe and Okavango, lie partly in national parks. This class is, therefore, of mixed potential for fuelwood.

Class 82—Evergreen Forest

This class covers about 5 percent of southern Africa, an area of 313,906 square kilometers. It consists of two principal types of forest. The larger category consists of high-altitude *miombo* woodland that extends in patches across Angola, Zambia, Zimbabwe, and Mozambique. The smaller category consists of afforested areas and natural woodlands in subtropical South Africa. It includes large areas of west-facing slopes on coastal mountains in Angola, stretching from Zaire Province in the north to the Serra da Chela escarpment in Moçâmedes Province in the south.

Farther inland, this class includes the highest areas of the country on the Bié Plateau to the northwest of Huambo, extensive high areas in Huíla Province, and smaller areas in Kuando-Kubango Province. The coverage in Angola is 82,573 square kilometers. Other occurrences are in Zambia, in the upper Zambezi and tributary areas of Northwestern and Western provinces, including West Lunga National Park, totaling 86,420 square kilometers. It also is an important class in Malawi, where it covers 11,751 square kilometers.

In Mozambique, extensive areas occur on higher land in Niassa Province. A few small high-altitude areas are present in Nampula Province and on Planalto do Mavia in Cabo Delgado Province. The area of this class in Mozambique is 76,039 square kilometers. In Zimbabwe, the most significant areas are in the higher parts of Matabeleland North, totaling 29,931 square kilometers. In South Africa (22,237 square kilometers) and Swaziland (2,687 square kilometers), this class is represented by high-altitude forested areas in south-eastern Transvaal and Swaziland, and at the northern end of the Drakensbergs in northern Transvaal.

In the tropics, this class demonstrates a similar phenology to that of Wet *Miombo* Woodland (Class 67), but with slightly lower overall NDVI values. It exhibits high levels from November through June, mostly greater than 0.42, decreasing through July and August to 0.25 in September. This behavior reflects the increased effect of cooler temperatures at higher altitudes in winter, followed by a very sharp increase in productivity through November (figure 13-10). The subtropical areas in South Africa and Swaziland demonstrate a similar seasonal pattern but with significantly lower NDVI values. Those above 0.25 are attained from November through April, but decline to 0.12 in August and September (figure 13-10).

In Angola, the vegetation in these areas is similar to that described as Wet *Miombo* Woodland (Class 67) and Evergreen Woodland Mosaic (Class 71), but with slightly less overall productivity. This is caused mostly by the effect of higher altitude on growth, but in some areas by varying degrees of interference. The class includes, for example, Parc Nacional da Kisama to the southeast of Luanda, parts of Parc Nacional da Biknar in Huíla Province, and high, less-accessible areas such as those on the west-facing Serra Upanda and the Serra da Chela escarpment.

In Zambia, this class is again *miombo* woodland, with vegetation types similar to those of woodlands described in Classes 67 and 71. Here, however, the class also includes parts of the "Dry Evergreen Forest" (ETC Foundation 1987), in which moisture-retaining

soils enable sufficient moisture to be reserved through the dry season to allow plant growth. It also includes some areas of intensive agriculture, such as tobacco and sugarcane near Lusaka.

The "Dry Evergreen Forests" of the Zambian Plateau now are limited in extent but include "*Parinari* Forest," which has two main canopy trees, *Parinari excelsa* and *Syzygium guineense*. Common understory trees in this type of forest are *Aidia micrantha, Chrysophyllum megalismontanum, Olea capensis, Tabernaemontana angolensis*, and *Teclea nobilis*. All of the "Dry Evergreen Forests" show signs of invasion by *miombo* woodland trees at their edges, suggesting a great ecological fragility under present climatic conditions. Disturbance by anthropogenic causes leads to formation of "Chipya Woodland." Some areas in western Zambia are protected reserves; for example, West Lunga National Park.

In Malawi, this class represents "Montane Rain Forest," which now is rare and occurs only on the wetter eastern slopes of higher mountains between 1,200 and 2,500 meters. Mean annual rainfall usually exceeds 1,250 millimeters, and any dry season effects are countered by mists. The forest is well structured, with canopy species reaching a height between 24 and 45 meters. A middle tree stratum, between 6 and 15 meters, forms a dense closed canopy with a poorly developed herb and grass layer. The rich flora includes *Aningeria adolfi–friedericii, Chrysophyllum gorungosanum, Cola greenwayi, Diospyros abyssinica, Myrianthus holstii, Ochna holstii, Olea capensis, Prunus africana*, and *Syzygium guineense*.

Apart from the general types of montane forest in Malawi, some forest areas are dominated by a single tree species. On the Nyika Plateau, pure stand forests of *Juniperus procera* and *Hagenia abyssinica* exist. The *J. procera* forest is formed on slightly drier slopes between 1,800 and 2,900 meters altitude and reaches 30 meters in height. Some evidence suggests that it is controlled by fire. The *Hagenia* forest has similar altitudinal and precipitation controls and forms a canopy 8 to 15 meters tall on the forest margin, but it is not fire tolerant.

On the Mulanje Massif, between 1,525 and 2,135 meters altitude, forests of Mulanje cedar (*Widdringtonia cupressoides*) exist, 25 to 40 meters high. The forest reserves in Malawi are mainly reservations of existing areas of "Montane Rain Forest" or closed-canopy woodland on the lowlands. Locally important exotic timber plantations exist, mostly of pine. Some of these areas are classified in this land cover class; otherwise they are classified in Classes 67 and 71.

In Mozambique, this class occurs mostly in Manica Province on the Zimbabwean border and extensively in Niassa Province (for example, near Lichinga). A few

Figure 13-10. NDVI **Profiles, Evergreen Forest (Class 82)**

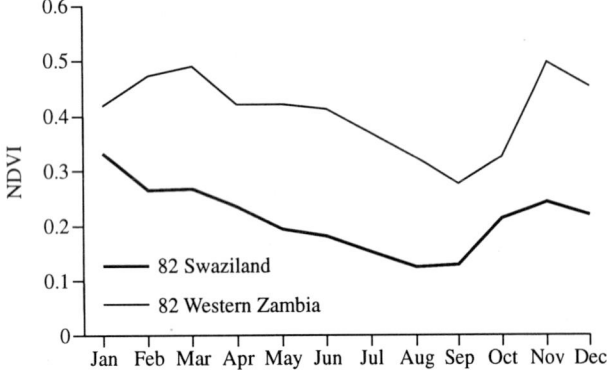

outliers exist at higher altitude near Nampula and on the Planalto do Mavia in Cabo Delgado Province. The short dry season often is offset by mists, because the areas of forest all occur above 1,200 meters.

They range from well structured to poorly structured and are floristically diverse, including conifers as well as broad-leaved tree species. In the high-altitude areas, where precipitation is greater, a well-structured forest can form with a well-developed, not dense canopy between 14 and 30 meters high. Emergents reach heights of 25 to 45 meters, with a typical height being 30 to 38 meters. A lower, dense, small-tree canopy occurs between 6 and 15 meters, and beneath the tree strata is a shrub layer of 3 to 6 meters and a sparse grass and fern herb layer. The main tree species are *Ocotea* spp., *Podocarpus* spp. (especially *P. latifolius*), *Prunus africana*, and *Xymalos monospora*.

In many areas, the montane forests are now much shorter than they were, and commonly form canopies between 8 and 15 meters. This height reduction usually is accompanied by a significant influx of secondary deciduous species, which take on a low-tree-and-shrubby habit. Large proportions of this land cover class in Mozambique are reserved, thereby restricting accessibility.

In Zimbabwe, the montane vegetation is complex and forest is rare for two reasons. First, the Zimbabwean Mountains are generally lower than elsewhere in the region, and consequently stunted ericaceous forests exist only in the Chimanimani Mountains. Second, the wet, windward slopes of the "Zimbabwean" Mountains are in Mozambique, for example near Inyanga National Park, and only the dry leeward slopes occur in Zimbabwe. Consequently, much of this class consists of other woodland forms (see Class 61) rather than true montane forest.

Above the *miombo* woodland or on wetter sites at similar altitudes, montane forests do occur. In these forests, trees rarely exceed 15 meters, but a canopy more commonly forms between 9 and 13 meters. Below the canopy, a number of other smaller trees exist, varying in height between 3.5 and 9 meters. The most common species are *Agauria salicifolia, Aphloia myrtiflora (theiformis), Apodytes dimidiata, Cussonia spicata, Diospyros whyteana, Dombeya erythroleuca, Hagenia benguelensis, Ilex mitis, Juniperus procera, Kiggelaria africana, Myrica salicifolia, Nuxia congesta, N. floribunda, Olinia usambarensis, Philippia benguelensis, Pittosporum viridiflorum, Podocarpus latifolius, Prunus africana, Pterocelastrus* spp., *Rapanea* spp., *Trichocladus ellipticus, Widdringtonia nodiflora*, and *Xymalos monospora*. Parts of this land cover class in Zimbabwe are reserved.

In Swaziland and South Africa, the majority of this class exists in plantations, although there are a few natural forests. They are exemplified by those of the western Swaziland Highveld, extending across the border into South Africa near Piet Retief. The plantations are dominated by pines, in particular Caribbean imports such as *Pinus elliottii, P. patula*, and *P. taeda*, which are mainly destined for South African pulp mills. Some of the southern Swaziland and South African plantations are of gum, especially *Eucalyptus granolis* and *E. saligna*, of black wattle (*Acacia mearnsii*), or of poplar (*Populus deltoides*).

Black wattle has been an important source of tannin, but because demand has declined, some areas in Swaziland have degenerated into "Wattle Jungle" (Millington and others 1989). *Eucalyptus* plantations have expanded, however, the wood being extensively used for poles and mining timber. In Swaziland, the coniferous plantation area has expanded steadily during the past three or four decades. Timber statistics for 1982 show that the ratio of planted to natural forest is 45.5 to 1 (Millington and others 1989). Natural forest consists mainly of *Cussonia umbellifera, Podocarpus latifolius, Rawsonia lucida*, and *Xymalos monospora*, usually only on sheltered slopes.

Woody biomass growing stock and sustainable yield are ample in most parts of this land cover class, but many places do not have any exploitable fuelwood. In parts of Angola, Zimbabwe, Mozambique, and South Africa, this class is in various reserves and thus is generally inaccessible. In most of Swaziland, and over much of South Africa, the plantations are inaccessible for fuelwood exploitation, except for the "Wattle Jungles" of Swaziland, which provide a valuable local wood resource. In Zambia, much of this class is in ecologically marginal situations (ETC Foundation 1987) and any exploitation may lead to a degraded "Chipya Woodland." Generally, many of the montane forest areas are at high altitudes and far from the centers of population.

This class has an estimated growing stock of 1,877 million tonnes, with a sustainable yield of 156 million tonnes. It is about 4 percent of the woody biomass resource of southern Africa.

Land Cover Class Tables

Tables 13-1 through 13-10, beginning on page 167, present summaries for each land cover class of the area, showing growing stock and sustainable yield for the southern African nations of Angola, Botswana, Lesotho, Malawi, Mozambique, Namibia, South Africa, Swaziland, Zambia, and Zimbabwe.

References

Every effort has been made to facilitate access to the documents listed here. Some documents, however, lack full bibliographic information because it was unavailable; also, some documents are of limited circulation.

Acocks, J. P. H. 1953. "Veld Types of South Africa." *Memoirs of the Botanical Survey of South Africa* 28: 1–192.

Acocks, J. P. H. 1975. "Veld Types of South Africa." *Memoirs of the Botanical Survey of South Africa* 40: 1–128.

Chidumayo, E. N. 1987. "Species Structure in Zambian *Miombo* Woodland." *Journal of Tropical Ecology* 3(2): 109–18.

Christopher, A. J. 1982. *South Africa.* London: Longman.

Coelho, H. V. P. 1967. "Zonagem florestal no Districto do Cuando Cubango." *Agronomicas Angolana* 26:3–27.

Edwards, I. D. 1982. "A Quantitative Description of an Area of Savannah Woodland at Nichira Mountain Conservation Area, near Blantyre." Forestry Research Institute of Malawi, Lilongwe.

ETC (Education and Training Consultants) Foundation. 1987. *Wood Energy Development: Biomass Assessment, a Study of the SADCC Region.* Leusden, Netherlands.

Fanshawe, D. B. 1969. "The Vegetation of Zambia." *Forest Research Bulletin* 7.

Millington, Andrew C., John R. G. Townshend, Pam A. Kennedy, Richard Saull, Steven D. Prince, and Robert Madams. 1989. *Biomass Assessment. Woody Biomass in the SADCC Region.* London: Earthscan.

Moll, E. J., and L. Bossi. 1984. "Assessment of the Extent of the Natural Vegetation of the *Fynbos* Biome in South Africa." *South African Journal of Science* 80 (August):355–58.

Moll, E. J., and M. L. Jarman. 1984. "Clarification of the Term *Fynbos.*" *South African Journal of Science* 80 (August):351–52.

Rutherford, M. C. 1972. "Notes on the Flora and Vegetation of the Omuverume Plateau, Mountain Waterberg, South West Africa." *Dinteria* 8:3–55.

Rutherford, M. C. 1978. "*Karoo-Fynbos* Biomass Along an Elevational Gradient in the Western Cape." *Bothalia* 12(3):555–60.

Scheepers, J. C. 1975. "The Plant Ecology of the Kroonstad and Bethlehem Areas of the High Veld Agricultural Region." Ph.D. diss., University of Pretoria, South Africa.

Stomgaard, Peter. 1985. "Biomass Estimation Equations for *Miombo* Woodland, Zambia." *Agroforestry Systems* 3(1):3–13.

Stomgaard, Peter. 1986. "Early Secondary Succession in Abandoned Shifting Cultivator's Plots in the *Miombo* of South Central Africa." *Biotropica* 18(2): 97–106.

Taylor, H. C. 1972. "Fynbos." *Veld and Flora* 2:68–75.

Taylor, H. C. 1978. "Capensis." Chapter 8 in M. J. A. Werger, ed., *Biogeography and Ecology of Southern Africa.* The Hague: Junk.

van der Merwe, J. H. 1983. *National Atlas of South West Africa.* University of Stellenbosch, South Africa: Institute for Cartographic Analysis.

Weare, P. R., and A. Yalala. 1971. "Provisional Vegetation Map of Botswana." *Botswana Notes and Records* 3.

Werger, M. J. A. 1978. *Biogeography and Ecology of Southern Africa.* The Hague: Junk.

Werger, M. J. A., and B. J. Coetzee. 1978. "The Sudano-Zambezian Region." Chapter 10 in M. J. A. Werger, ed., *Biogeography and Ecology of Southern Africa.* The Hague: Junk.

White, F. 1983. "The Vegetation of Africa." *Natural Resources Research Series* 20. Paris: UNESCO/AETFAT/UNSO (United Nations Educational, Scientific and Cultural Organization/Association pour l'Etude Taxonomique de la Flore de l'Afrique Tropicale/United Nations Sudano-Sahelian Office).

White, F., and E. J. Moll. 1978. "The Indian Ocean Coastal Belt." Chapter 13 in M. J. A. Werger, ed., *Biogeography and Ecology of Southern Africa.* The Hague: Junk.

Table 13-1. Land Cover Classes—Angola (Southern Africa Region)

Land cover class	Area		Growing stock		Sustainable yield	
	km²	*Percent*	*Thousand tonnes*	*Percent*	*Thousand tonnes per year*	*Percent*
0	20,446	1.68	0.00	0.00	0.00	0.00
11	22,922	1.89	5,203.29	0.09	229.22	0.19
14	105	0.01	23.84	0.00	1.05	0.00
1	23,027	1.90	5,227.13	0.09	230.27	0.19
24	15,967	1.32	989.95	0.02	159.67	0.13
25	5,586	0.46	1,843.38	0.03	55.86	0.05
2	21,553	1.78	2,833.33	0.05	215.53	0.18
33	33,725	2.78	33,725.00	0.60	1,686.25	1.40
35	2,002	0.16	1,121.12	0.02	100.10	0.08
3	35,727	2.94	34,846.12	0.62	1,786.35	1.48
41	1,107	0.09	1,538.73	0.03	23.25	0.02
42	5,322	0.44	9,851.02	0.17	111.76	0.09
43	71,244	5.87	120,544.85	2.13	1,496.12	1.24
4	77,673	6.40	131,934.60	2.34	1,631.13	1.35
51	25,768	2.12	57,462.64	1.02	1,623.38	1.35
5	25,768	2.12	57,462.64	1.02	1,623.38	1.35
61	127,944	10.54	376,155.36	6.66	5,629.54	4.67
66	450,070	37.09	2,503,739.41	44.32	40,056.23	33.22
67	150,234	12.38	1,787,784.60	31.65	19,981.12	16.57
6	728,248	60.01	4,667,679.37	82.63	65,666.89	54.46
71	96,222	7.93	49,361.89	0.87	1,828.22	1.52
72	99,172	8.17	167,005.65	2.96	1,884.27	1.56
7	195,394	16.10	216,367.54	3.83	3,712.49	3.08
82	82,573	6.80	495,438.00	8.77	41,121.35	34.10
85	2,002	0.16	25,581.56	0.45	2,962.96	2.46
87	1,107	0.09	11,611.32	0.21	1,638.36	1.36
8	85,682	7.06	532,630.88	9.43	45,722.67	37.92
Total	1,213,518	100.00	5,648,981.61	100.00	120,588.71	100.00
(Percentage of region)	(20.45)		(30.24)		(26.90)	

Note: In the following tables, details may not add to totals because of rounding.
Source: Authors' calculations from data bases derived from land cover classification and table 4-1.

Table 13-2. Land Cover Classes—Botswana (Southern Africa Region)

Land cover class		Area km²	Percent	Growing stock Thousand tonnes	Percent	Sustainable yield Thousand tonnes per year	Percent
0		29,457	5.01	0.00	0.00	0.00	0.00
	11	53,380	9.07	12,117.26	2.61	533.80	3.13
	14	105	0.02	23.84	0.01	1.05	0.01
1		53,485	9.09	12,141.10	2.62	534.85	3.14
	24	240,607	40.90	14,917.63	3.22	2,406.07	14.10
2		240,607	40.90	14,917.63	3.22	2,406.07	14.10
	33	316	0.05	316.00	0.07	15.80	0.09
	34	52,643	8.95	52,643.00	11.35	2,632.15	15.43
	35	60,547	10.29	33,906.32	7.31	3,027.35	17.75
3		113,506	19.30	86,865.32	18.74	5,675.30	33.27
	42	158	0.03	292.46	0.06	3.32	0.02
	43	10,117	1.72	17,117.96	3.69	212.46	1.25
4		10,275	1.75	17,410.42	3.76	215.78	1.26
	51	97,012	16.49	216,336.76	46.66	6,111.76	35.83
5		97,012	16.49	216,336.76	46.66	6,111.76	35.83
	61	24,767	4.21	72,814.98	15.71	1,089.75	6.39
	66	1,528	0.26	8,500.26	1.83	135.99	0.80
6		26,295	4.47	81,315.24	17.54	1,225.74	7.19
	72	16,441	2.79	27,686.64	5.97	312.38	1.83
7		16,441	2.79	27,686.64	5.97	312.38	1.83
	82	1,159	0.20	6,954.00	1.50	577.18	3.38
8		1,159	0.20	6,954.00	1.50	577.18	3.38
Total		588,237	100.00	463,627.12	100.00	17,059.05	100.00
(Percentage of region)		(9.91)		(2.48)		(3.80)	

Source: Authors' calculations from data bases derived from land cover classification and table 4-1.

Table 13-3. Land Cover Classes—Lesotho (Southern Africa Region)

Land cover class		Area km²	Percent	Growing stock Thousand tonnes	Percent	Sustainable yield Thousand tonnes per year	Percent
	11	25,136	73.27	5,705.87	56.83	251.36	68.78
	14	7,535	21.96	1,710.45	17.04	75.35	20.62
1		32,671	95.23	7,416.32	73.87	326.71	89.40
	24	369	1.08	22.88	0.23	3.69	1.01
2		369	1.08	22.88	0.23	3.69	1.01
	43	896	2.61	1,516.03	15.10	18.82	5.15
4		896	2.61	1,516.03	15.10	18.82	5.15
	61	369	1.08	1,084.86	10.81	16.24	4.44
6		369	1.08	1,084.86	10.81	16.24	4.44
Total		34,305	100.00	10,040.09	100.00	365.45	100.00
(Percentage of region)		(0.58)		(0.05)		(0.08)	

Source: Authors' calculations from data bases derived from land cover classification and table 4-1.

Table 13-4. Land Cover Classes—Malawi (Southern Africa Region)

Land cover class		Area		Growing stock		Sustainable yield	
		km^2	Percent	Thousand tonnes	Percent	Thousand tonnes per year	Percent
	11	316	0.24	71.73	0.01	3.16	0.03
1		316	0.24	71.73	0.01	3.16	0.03
	24	53	0.04	3.29	0.00	0.53	0.00
	25	158	0.12	52.14	0.01	1.58	0.01
2		211	0.16	55.43	0.01	2.11	0.02
	33	2,477	1.92	2,477.00	0.50	123.85	1.02
3		2,477	1.92	2,477.00	0.50	123.85	1.02
	41	158	0.12	219.62	0.04	3.32	0.03
	43	32,408	25.07	54,834.34	11.09	680.57	5.58
4		32,566	25.19	55,053.96	11.13	683.89	5.61
	51	474	0.37	1,057.02	0.21	29.86	0.24
5		474	0.37	1,057.02	0.21	29.86	0.24
	61	4,005	3.10	11,774.70	2.38	176.22	1.45
	65	53	0.04	141.51	0.03	2.54	0.02
	66	46,425	35.92	258,262.27	52.21	4,131.82	33.90
	67	7,694	5.95	91,558.60	18.51	1,023.30	8.40
6		58,177	45.01	361,737.08	73.13	5,333.89	43.76
	71	2,635	2.04	1,351.76	0.27	50.06	0.41
	72	263	0.20	442.89	0.09	5.00	0.04
7		2,898	2.24	1,794.65	0.36	55.06	0.45
	82	11,751	9.09	70,506.00	14.25	5,852.00	48.01
	83	211	0.16	1,873.26	0.38	105.08	0.86
8		11,962	9.25	72,379.26	14.63	5,957.08	48.87
Lakes		20,182	15.61	0.00	0.00	0.00	0.00
Total		129,263	100.00	494,626.12	100.00	12,188.90	100.00
(Percentage of region)		(2.18)		(2.65)		(2.72)	

Source: Authors' calculations from data bases derived from land cover classification and table 4-1.

Table 13-5. Land Cover Classes—Mozambique (Southern Africa Region)

Land cover class	Area km²	Area Percent	Growing stock Thousand tonnes	Growing stock Percent	Sustainable yield Thousand tonnes per year	Sustainable yield Percent
11	6,271	0.82	1,423.52	0.04	62.71	0.07
14	474	0.06	107.60	0.00	4.74	0.01
1	6,745	0.88	1,531.12	0.04	67.45	0.07
24	16,388	2.14	1,016.06	0.03	163.88	0.18
2	16,388	2.14	1,016.06	0.03	163.88	0.18
33	2,529	0.33	2,529.00	0.06	126.45	0.14
35	4,743	0.62	2,656.08	0.07	237.15	0.26
3	7,272	0.95	5,185.08	0.13	363.60	0.40
41	53	0.01	73.67	0.00	1.11	0.00
42	316	0.04	584.92	0.01	6.64	0.01
43	75,038	9.78	126,964.30	3.17	1,575.80	1.75
4	75,047	9.83	127,622.89	3.18	1,583.55	1.76
51	31,143	4.06	69,448.89	1.74	1,962.01	2.18
5	31,143	4.06	69,448.89	1.74	1,962.01	2.18
61	26,295	3.43	77,307.30	1.93	1,156.98	1.28
66	360,646	47.02	2,006,273.70	50.15	32,097.49	35.61
67	102,861	13.41	1,224,045.90	30.60	13,680.51	15.18
6	489,802	63.86	3,307,626.90	82.68	46,934.99	52.08
71	59,071	7.70	30,303.42	0.76	1,122.35	1.25
72	475	0.06	798.22	0.02	9.01	0.01
7	59,545	7.76	31,101.64	0.78	1,131.36	1.26
82	76,039	9.91	456,234.00	11.40	37,867.42	42.02
83	105	0.01	932.19	0.02	52.29	0.06
8	76,144	9.93	457,166.19	11.43	37,919.71	42.07
Lakes	4,584	0.60	0.00	0.00	0.00	0.00
Total	767,030	100.00	4,000,698.75	100.00	90,126.54	100.00
(Percentage of region)	(12.93)		(21.42)		(20.10)	

Source: Authors' calculations from data bases derived from land cover classification and table 4-1.

Table 13-6. Land Cover Classes—Namibia (Southern Africa Region)

Land cover class	Area km²	Percent	Growing stock Thousand tonnes	Percent	Sustainable yield Thousand tonnes per year	Percent
0	221,478	27.13	0.00	0.00	0.00	0.00
11	15,282	1.87	3,469.01	0.41	152.82	0.64
14	211	0.03	47.90	0.01	2.11	0.01
1	15,493	1.90	3,516.91	0.42	154.93	0.65
24	161,458	19.78	10,010.40	1.19	1,614.58	6.74
25	211	0.03	69.63	0.01	2.11	0.01
2	161,669	19.81	10,080.03	1.20	1,616.69	6.75
33	7,493	0.92	7,483.00	0.89	374.15	1.56
34	63,603	7.79	63,603.00	7.59	3,180.15	13.28
35	50,061	6.13	28,034.16	3.34	2,503.05	10.45
3	121,147	14.84	99,120.16	11.82	6,057.35	25.30
41	2,213	0.27	3,076.07	0.37	46.47	0.19
42	2,108	0.26	3,901.91	0.47	44.27	0.18
43	13,964	1.71	23,627.09	2.82	293.24	1.22
4	18,285	2.24	30,605.07	3.66	383.99	1.60
51	166,464	20.39	371,214.72	44.28	10,487.23	43.79
5	166,464	20.39	371,214.72	44.28	10,487.23	43.79
61	99,278	12.16	291,877.32	34.82	4,368.23	18.24
66	1,581	0.19	8,795.10	1.05	140.71	0.59
6	100,859	12.36	300,672.42	35.87	4,508.94	18.83
72	9,801	1.20	16,504.88	1.97	186.22	0.78
7	9,801	1.20	16,504.88	1.97	186.22	0.78
82	1,107	0.14	6,642.00	0.79	551.29	2.30
8	1,107	0.14	6,642.00	0.79	551.29	2.30
Total	816,303	100.00	838,356.19	100.00	23,946.63	100.00
(Percentage of region)	(13.76)		(4.49)		(5.34)	

Source: Authors' calculations from data bases derived from land cover classification and table 4-1.

Table 13-7. Land Cover Classes—South Africa (Southern Africa Region)

Land cover class		Area km²	Percent	Growing stock Thousand tonnes	Percent	Sustainable yield Thousand tonnes per year	Percent
0		79,148	6.48	0.00	0.00	0.00	0.00
	11	214,364	17.56	48,660.63	2.92	2,143.64	3.98
	14	6,587	0.45	1,495.25	0.09	65.87	0.12
1		220,951	18.10	50,155.88	3.01	2,209.51	4.10
	24	171,365	14.04	10,624.63	0.64	1,713.65	3.18
	25	1,370	0.11	452.10	0.03	13.70	0.03
2		172,735	14.15	11,076.73	0.67	1,727.35	3.21
	31	4,005	0.33	3,424.28	0.21	200.25	0.37
	32	24,978	2.05	11,240.10	0.68	1,248.90	2.32
	33	70,032	5.74	70,032.00	4.21	3,501.60	6.50
	34	129,367	10.60	129,367.00	7.77	6,468.35	12.01
	35	123,781	10.14	69,317.36	4.16	6,189.05	11.49
3		352,163	28.86	283,380.74	17.03	17,608.15	32.70
	42	17,021	1.39	31,505.87	1.89	357.44	0.66
	43	111,767	9.15	189,109.76	11.36	2,347.11	4.36
4		128,788	10.54	220,615.63	13.25	2,704.55	5.02
	51	142,277	11.65	317,277.71	19.06	8,963.45	16.64
5		142,277	11.65	317,277.71	19.06	8,963.45	16.64
	61	19,023	1.56	55,927.62	3.36	837.01	1.55
	66	54,065	4.43	300,763.59	18.07	4,811.78	8.94
	67	23,344	1.91	277,793.60	16.69	3,104.75	5.77
6		96,432	7.90	634,484.81	38.12	8,753.54	16.26
	71	474	0.04	243.16	0.01	9.01	0.02
	72	5,270	0.43	8,874.68	0.53	100.13	0.19
7		5,744	0.47	9,117.84	0.55	109.14	0.21
	82	22,237	1.82	133,422.00	8.02	11,074.03	20.56
	87	474	0.04	4,971.79	0.30	701.52	1.30
8		22,711	1.86	138,393.13	8.31	11,775.55	21.86
Total		1,220,949	100.00	1,664,503.13	100.00	53,851.24	100.00
(Percentage of region)		(20.58)		(8.91)		(12.01)	

Source: Authors' calculations from data bases derived from land cover classification and table 4-1.

Table 13-8. Land Cover Classes—Swaziland (Southern Africa Region)

Land cover class		Area		Growing stock		Sustainable yield	
		km^2	Percent	Thousand tonnes	Percent	Thousand tonnes per year	Percent
2	24	474	2.64	29.39	0.05	4.74	0.23
		474	2.64	29.39	0.05	4.74	0.23
3	35	105	0.58	58.80	0.09	5.25	0.26
		105	0.58	58.80	0.09	5.25	0.26
4	43	7,588	42.23	12,838.90	20.25	159.35	7.82
		7,588	42.23	12,838.90	20.25	159.35	7.82
	61	2,319	12.91	6,817.86	10.75	102.04	5.01
	66	4,479	24.93	24,916.68	39.29	398.63	19.56
	67	211	1.17	2,510.90	3.96	28.06	1.38
6		7,009	39.01	34,245.44	54.01	528.73	25.95
	71	53	0.29	27.19	0.04	1.01	0.05
	72	53	0.29	89.25	0.14	1.01	0.05
7		106	0.59	116.44	0.18	2.01	0.01
8	82	2,687	14.95	16,122.00	25.42	1,338.13	65.65
		2,687	14.95	16,122.00	25.42	1,338.13	65.65
Total		17,969	100.00	63,410.96	100.00	2,038.21	100.00
(Percentage of region)		(0.30)		(0.34)		(0.45)	

Source: Authors' calculations from data bases derived from land cover classification and table 4-1.

Table 13-9. Land Cover Classes—Zambia (Southern Africa Region)

Land cover class		Area		Growing stock		Sustainable yield	
		km^2	Percent	Thousand tonnes	Percent	Thousand tonnes per year	Percent
1	11	1,686	0.23	382.72	0.01	16.86	0.02
		1,686	0.23	382.72	0.01	16.86	0.02
	24	25,294	3.38	1,568.23	0.04	252.94	0.27
	25	2,108	0.28	695.64	0.02	21.08	0.02
2		27,402	3.67	2,263.87	0.06	274.02	0.29
3	33	843	0.11	843.00	0.02	42.15	0.05
		843	0.11	843.00	0.02	42.15	0.05
	42	738	0.10	1,366.04	0.03	15.50	0.02
	43	55,067	7.37	93,173.36	2.30	1,156.41	1.24
4		55,805	7.47	94,539.40	2.34	1,171.91	1.26
5	51	9,116	1.22	20,328.68	0.50	574.31	0.62
		9,116	1.22	20,328.68	0.50	574.31	0.62
	61	72,561	9.71	213,329.34	5.27	3,192.68	3.42
	66	344,891	46.15	1,918,628.63	47.42	30,695.30	32.88
	67	102,967	13.78	1,225,307.30	30.28	13,694.61	14.67
6		520,419	69.64	3,357,265.27	82.97	47,582.59	50.97
	71	6,060	0.81	3,108.78	0.08	115.14	0.12
	72	29,193	3.91	49,161.01	1.21	554.67	0.59
7		35,253	4.72	52,269.79	1.29	669.81	0.71
8	82	86,420	11.56	518,520.00	12.81	43,037.16	46.09
		86,420	11.56	518,520.00	12.81	43,037.16	46.09
Lakes		10,381	1.39	0.00	0.00	0.00	0.00
Total		747,325	100.00	4,046,412.74	100.00	93,368.80	100.00
(Percentage of region)		(12.59)		(21.66)		(20.82)	

Source: Authors' calculations from data bases derived from land cover classification and table 4-1.

Table 13-10.   Land Cover Classes—Zimbabwe (Southern Africa Region)

Land cover class	Area km²	Area Percent	Growing stock Thousand tonnes	Growing stock Percent	Sustainable yield Thousand tonnes per year	Sustainable yield Percent
11	1,212	0.30	275.12	0.02	12.12	0.03
14	422	0.11	95.79	0.01	4.22	0.01
1	1,634	0.41	370.91	0.03	16.34	0.04
24	25,610	6.42	1,587.82	0.11	256.10	0.74
2	25,610	6.42	1,587.82	0.11	256.10	0.74
35	3,425	0.86	1,918.00	0.13	171.25	0.49
3	3,425	0.86	1,918.00	0.13	171.25	0.49
42	474	0.12	877.37	0.06	9.95	0.03
43	63,498	15.92	107,438.62	7.43	1,333.46	3.83
4	63,972	16.04	108,315.99	7.49	1,343.41	3.86
51	47,637	11.94	106,230.51	7.34	3,001.13	8.62
5	47,637	11.94	106,230.51	7.34	3,001.13	8.62
61	115,455	28.94	339,437.70	23.46	5,080.02	14.59
66	78,252	19.62	435,315.88	30.09	6,964.43	20.00
67	22,606	5.67	269,011.40	18.59	3,006.60	8.64
6	216,313	54.23	1,043,764.98	72.14	15,051.05	43.23
71	1,265	0.32	648.95	0.04	24.04	0.07
72	2,582	0.65	4,348.09	0.30	49.06	0.14
7	3,847	0.96	4,997.04	0.34	73.10	0.21
82	29,931	7.50	179,586.00	12.41	14,905.64	42.81
8	29,931	7.50	179,586.00	12.41	14,905.64	42.81
Lakes	6,534	1.64	0.00	0.00	0.00	0.00
Total	398,903	100.00	1,446,771.25	100.00	34,818.01	100.00
(Percentage of region)	(6.72)		(7.75)		(7.77)	

Source: Authors' calculations from data bases derived from land cover classification and table 4-1.

Glossary

acheb	Local West African term for herbaceous vegetation growth following rainfall in arid areas.
alfisol(s)	A relatively productive, and therefore agriculturally important, type of soil commonly occurring in West, East, and southern Africa; characterized by downward movement of water, clay minerals, and cations.
anthropogenic	Vegetation that has formed or is controlled mainly by human activity (for example, anthropogenic savannas and anthropogenic grasslands). See also DERIVED SAVANNA.
arborescent	Having a treelike form.
Archaean complex	Precambrian basement rocks; some occur in various parts of Africa. See also PRECAMBRIAN.
basement complex	See ARCHAEAN COMPLEX.
base-rich	Soils that have high cation content (for example, calcium, magnesium, potassium) and are, therefore, usually highly productive.
bolilands	The region in central Sierra Leone where the landscape is dominated by *bolis*.
bolis	Seasonally flooded, low-lying areas in central Sierra Leone. The edaphic climax vegetation of such areas is grassland and swamp forest, but they often are cleared for rice cultivation. They are somewhat similar to the *dambos* of East and central Africa.
bush fallowing	A farming system common in humid tropical Africa in which forest or woodland is cut, dried, and burned, then cultivated for one to four years and allowed to revert to a fallow of secondary woody vegetation. See also CHITEMENE.
calcrete	A sedimentary pedogenic deposit composed of rock fragments cemented together by calcium carbonate.
Cape Domain	A regional center of endemism covering the southwestern and southern part of Cape Province, South Africa (after White 1983).
Capensis Realm	The geographic area recognized as the Fynbos Biome, equivalent to Cape Floristic Region or Cape Floral Kingdom (after Werger 1978).
chipya	A local Zambian name for a variety of woody vegetation regrowth.
chitemene	A type of bush-fallowing farming system practiced by the Bemba in Zambia. See also BUSH FALLOWING.
cirrus cloud	A high-level, white, wispy cloud occurring at an altitude between 5,000 and 13,700 meters.
commercial wood	Wood useful in construction or manufacturing, which gives it a value such that it is not used as firewood.
Cretaceous	Geologic period extending from 65 million to 135 million years ago.

175

dambo	A southern African term for a seasonally waterlogged streamless hollow with vegetation differing from the surrounding woodland or savanna, commonly occurring in East, central, and southern Africa.
dbh	Diameter at breast height: a measurement of tree girth at 1.3 meters above the ground.
derived savanna	Savanna vegetation formed and sustained by human activity, resulting from disturbance of a different type of vegetation, usually forest. See also ANTHROPOGENIC.
edaphic	Vegetation that is adapted to, and controlled by, the soil and groundwater conditions of a site; for example, edaphic grasslands.
elfin thicket	Thickets of stunted and gnarled woody vegetation found at high altitudes.
ephemeral	Vegetation which grows only occasionally in response to favorable conditions, usually wet. Common in arid and semiarid areas.
ephemerals	Ephemeral plant species.
erg	An arid (desert) landscape consisting of extensive sand cover.
ericoid	A descriptive term for plants having a shrubby growth habit, like heathers.
fire resistant	Plants with thick bark or other adaptations which enable them often to survive fire.
fire tolerant	Plants with a limited tolerance to fire, better able to survive fire than nontolerant plants, but less able to survive fire than fire-resistant species.
fynbos	A Mediterranean evergreen sclerophyllous shrubland occurring in South Africa.
gallery forest	A dense growth of tropical forest that follows the course of a river.
gizzu	A local term for the flush of grass-dominated vegetation that grows in the Libyan Desert following rainfall events.
ground resolution element	The smallest area of the Earth's surface that a sensor can resolve. For each ground resolution element, one measurement of reflection per channel occurs on a sensor.
growing stock	Air-dried, above-ground woody biomass, expressed in tonnes per hectare.
Guineo-Congolian Domain	A regional center of endemism found as a broad band north and south of the equator from the Atlantic, through the Zaire Basin, to the western slopes of the *dorsale du Kivu* (after White 1983).
halophyte	A plant that is adapted to and favors growth in soils with a high content of soluble salt.
Hammer-Aitoff Conic Equal Area Projection	The map projection used in the construction of AVHRR NDVI data.
hardpan(s)	A subsurface soil horizon that is often hard and rocklike. In the humid tropics, hardpan often forms by precipitation of iron and aluminum, which then harden to form a ferricrete (laterite). Hardpans usually form impermeable horizons in the soil.
harmattan	A dust-laden wind blowing off the Sahara from December to February.
high forest	Closed-canopy climax forest.
illite	A clay mineral derived from the weathering of micas and feldspars under alkaline conditions.
isenga	Sand patches on the Zambian Plateau. They lead to the growth of a distinct form of *miombo* woodland.
itigi thicket	An East African vegetation type containing many deciduous woody species which form a dense, intertangled thicket.
Kalahari Sands	Areas of Quaternary aeolian sands from southern Africa northward into southern Zaire.
kaolinite (kaolinitic)	A clay mineral produced by the weathering of feldspar-rich rocks under acid conditions, particularly common in humid tropical soils.

Karroo Domain	The area of summer rainfall east of the Western Cape Domain and south of the Namaland Domain (after Werger 1978).
kaukau veld	A *veld* grassland type occurring on the Namibia-Botswana border (Kaokoland).
little dry season	A period of approximately 4 to 6 weeks, usually in August, during the wet season in West Africa when rainfall amounts decline markedly. The phenomenon is related to the northward movement of the Intertropical Convergence Zone (ITCZ) and is restricted to the southernmost parts of the West African coast.
macchia	A term used in South Africa to describe heathlands and evergreen sclerophyllous shrubland.
mguba	A local East African name for a VERTISOL.
miombo	An open woodland land cover occurring in southern Africa.
mist belt	An area characterized by OCCULT PRECIPITATION.
montmorillonite	A clay mineral derived from the weathering of basic rocks. Montmorillonite-type clay minerals absorb moisture and swell during wet periods and contract during dry periods.
munga	Open, parklike savanna woodland occurring in Zambia.
Namaland Domain	A region of southern Africa covering the narrow escarpment belt inland of the Namib, broadening southward to include the southern Kalahari Plateau (after Werger 1978).
Namib Domain	A region of southern Africa comprising the arid coastal strip of Namibia between the Atlantic and the escarpment (after Werger 1978).
occult precipitation	Precipitation that occurs directly on vegetation in the form of condensation. It commonly occurs when humid onshore winds are forced to rise up high escarpments along the coast.
piosphere	An area of degraded soil and vegetation around a waterhole or wells, occurring in arid and semiarid areas.
Precambrian	The entire span of geological time before the start of the Cambrian Period at approximately 600 million years before the present.
proteoid	A term applied to plants that have leaves which are thickened, as in the genus *Protea*.
qoz	Stabilized wind-blown sands of Quaternary age in Sudan.
Quaternary	The current geological period, which began approximately 1.8 million to 2 million years before the present.
raster	A means of representing spatial data in the form of a grid.
reg	An arid (desert) landscape consisting of extensive tracts of gravel.
restoid	Having a low, cushionlike form similar to that of the southern African genus *Restio*.
sclerophyllous	A term describing evergreen shrubs and trees that have adapted to lengthy seasonal droughts by producing tough, leathery leaves.
senescence	Ripening and/or dieback of vegetation.
seral succession	Sequential development of a plant community or communities over time.
shallahalla	A term used in Lesotho to describe shrub woodland growing on overgrazed, degraded pasture.
shifting cultivation	See BUSH FALLOWING.
stem volume	Volume of main woody stem derived from height and diameter measurements, expressed in cubic meters.
stère	A stacked volume of wood, expressed in cubic meters.
subcanopy	The layer or layers of vegetation beneath the canopy.

succulent	A plant with thick, fleshy leaves or stems that retain large amounts of moisture; common in arid areas.
suffrutex	An understory shrub.
sustainable yield	Mean annual increment of air-dried, above-ground woody growth, expressed in tonnes per hectare per year.
swartveld	A gently undulating landscape in Cape Province, South Africa, extensively cultivated for wheat.
Tertiary	Geologic period extending between 2 million and 65 million years before the present.
thionic fluvisols	Alluvial soils, occurring mainly in estuaries, in which subsoil sulfuric acid builds up when the soils are dried.
tonne	Metric ton; 1,000 kilograms or 2,204.62 avoirdupois pounds.
veld	(1) A southern African term describing almost treeless grassland. (2) A southern African term used to define a region regardless of land cover (for example, the Highveld of Swaziland).
vertisol	A black soil commonly found in semiarid areas which, because of its high montmorillonite content, swells during the wet season and shrinks and cracks open in the dry season. Often described as having "cracking clays."
vloere	A local Afrikaans name for brackish, low-lying areas in the Karroo.
wadi	A valley in a desert that is wet only after heavy rain and will briefly support vegetation.
wattle jungle	A dense growth of previously well-managed, but now abandoned, plantations of *Acacia mearnsii* in Swaziland.
Western Cape Domain	A region of southern Africa including the coastal strip and escarpment mountains in South Africa south of the Namib (after Werger 1978).
Wild Coast	A local name for a remote stretch of the Transkei coast between East London and Port Alfred.
Woing Dega	Local name given to the Ethiopian highlands that experience a temperate climate.
woodfuels	Fuelwood and charcoal.
woody biomass	Ligneous plant material.
Zambezi teak	Name used by foresters for the important timber tree *Baikiaea plurijuga* in southern Africa.
Zambezian Domain	A region extending from the Atlantic to the Indian Ocean including Zambia, Malawi, and Zimbabwe, as well as large parts of Angola, Tanzania, and Mozambique.
Zone de l'igname	An agricultural zone extending across the central Côte d'Ivoire.

Index of Botanical Names

Index of Place Names

Legend for Land Cover Class Maps

DESERT

Desert

GRASSLAND

Veld grassland

Hydromorphic grassland

Ethiopian montane steppe

Montane grassland and heathland

WOODED GRASSLAND

Semi-desert wooded grassland

Acacia wooded grassland

Plateau wooded grassland

Transitional wooded grassland

Edaphic wooded grassland

SHRUBLAND

Veld shrubland

Hill shrubland

Bushy shrubland

Kalahari shrubland

Wooded shrubland

BUSHLAND & THICKET

Dry *Acacia-Commiphora* bushland and thicket

Fynboc thicket

Moist *Acacia-Commiphora* bushland and thicket

Sahel-Sudanian *Acacia* wooded bushland

Escarpment wooded thicket

LOW WOODY BIOMASS MOSAICS

Acacia woodland mosaic

East African low woody biomass mosaic

WOODLAND

Open woodland

Dry Sudanian woodland

Sudan/Ethiopian woodland and thicket

Sudanian woodland

Moist Sudanian woodland

Seasonal miombo

Wet miombo

HIGH WOODY BIOMASS MOSAICS

Evergreen woodland mosaic

Cultivation and forest/woodland mosaic

Cultivation and forest regrowth mosaic

Guinean woodland

High productivity West African cultivation and forest

Medium productivity West African cultivation and forest

Highland cultivation mosaic

FOREST

Mangrove

Evergreen forest

Coastal and gallery forest

Montane forest

Mesophilous humid tropical forest

Humid tropical swamp forest

Ombrophilous humid tropical forest

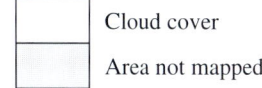

Cloud cover

Area not mapped

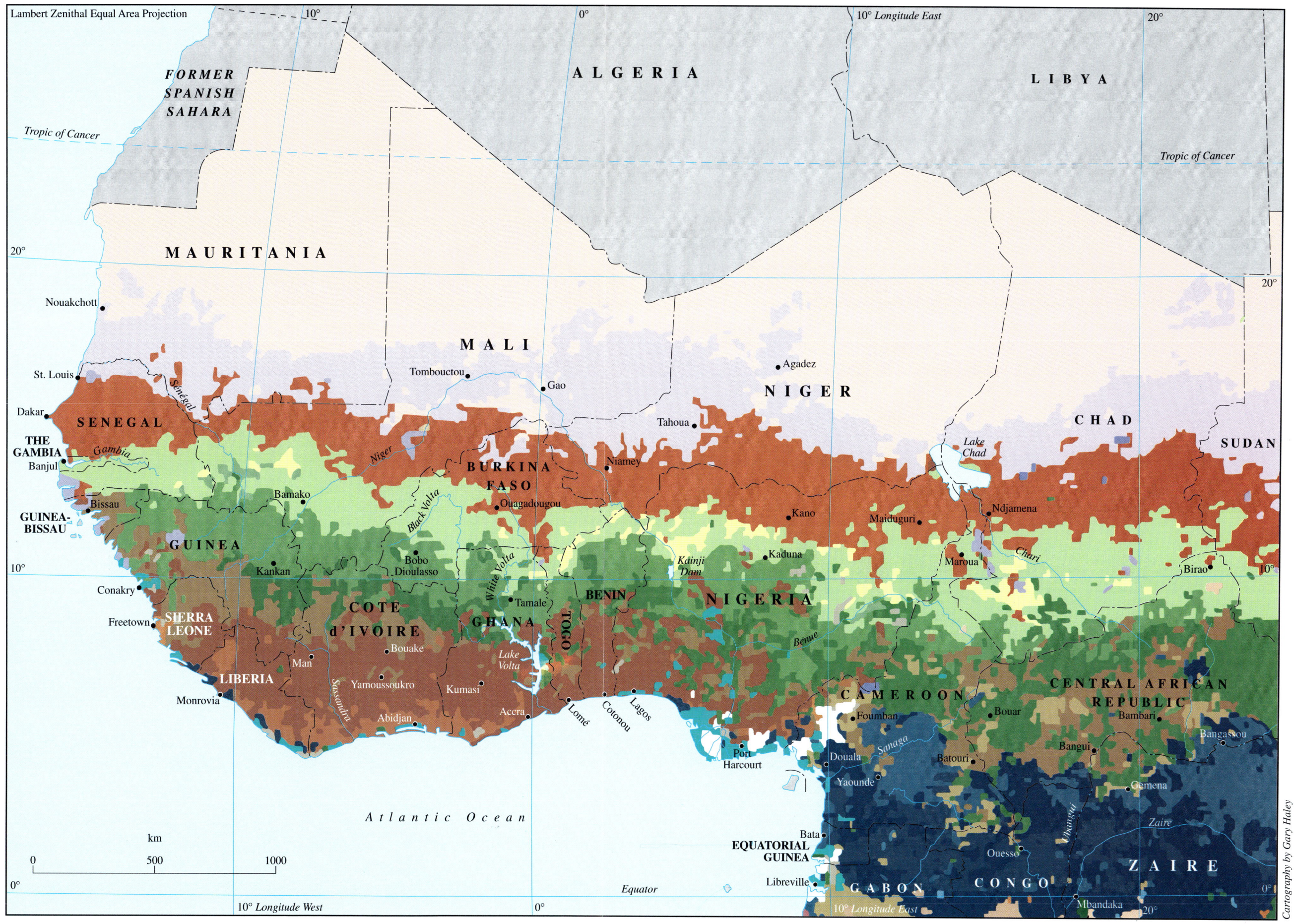

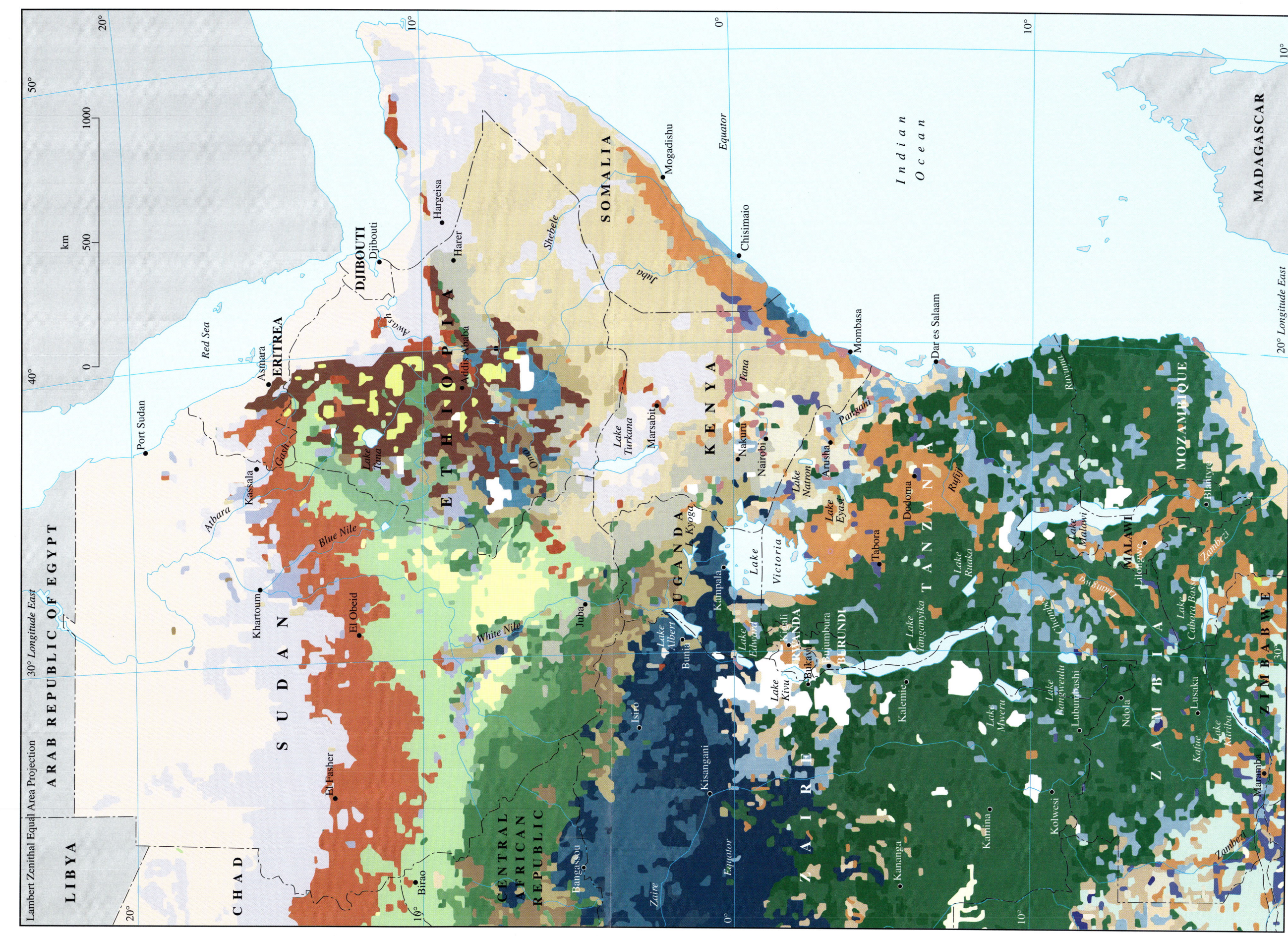

Lambert Zenithal Equal Area Projection

ARAB REPUBLIC OF EGYPT

LIBYA

CHAD

SUDAN

ERITREA

DJIBOUTI

ETHIOPIA

SOMALIA

CENTRAL AFRICAN REPUBLIC

UGANDA

KENYA

ZAIRE

RWANDA

BURUNDI

TANZANIA

ZAMBIA

MALAWI

MOZAMBIQUE

ZIMBABWE

MADAGASCAR

Red Sea

Indian Ocean

Equator

Port Sudan

Asmara

Khartoum

Kassala

El Obeid

El Fasher

Birao

Bangassou

Isiro

Kisangani

Juba

Kananga

Kamina

Kolwezi

Lubumbashi

Ndola

Lusaka

Kalemie

Djibouti

Hargeisa

Harer

Addis Ababa

Mogadishu

Chisimaio

Marsabil

Nakuru

Nairobi

Mombasa

Arusha

Dar es Salaam

Dodoma

Tabora

Lilongwe

Blantyre

Maramba

Kampala

Buna

Bukavu

Bujumbura

Kigali

Atbara

Blue Nile

White Nile

Gash

Awash

Takkaze

Abbai

Omo

Shebele

Juba

Tana

Pangani

Ruvuma

Rufiji

Zambezi

Zaire

Lake Tana

Lake Turkana

Lake Victoria

Lake Albert

Lake Edward

Lake Kivu

Lake Tanganyika

Lake Rukwa

Lake Mweru

Lake Bangweulu

Lake Malawi

Lake Cabora Bassa

Lake Kariba

Lake Kafue

Lake Natron

Lake Eyasi

Lake Kyoga

km

0   500   1000

20°   30° Longitude East   40°   50°

10°   0°   10°   20° Longitude East

Equator

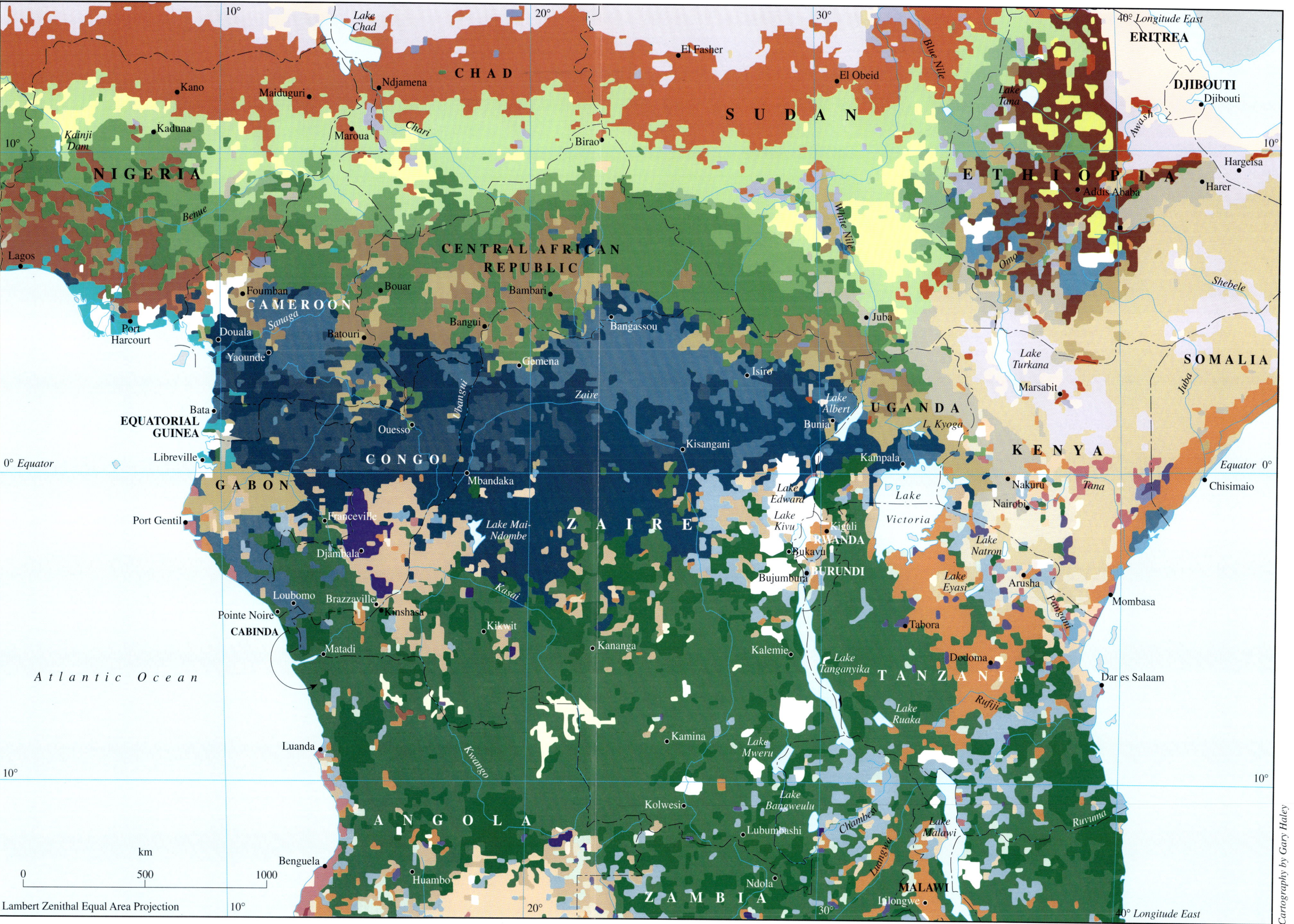

Lambert Zenithal Equal Area Projection

km
0   500   1000

ERITREA

DJIBOUTI
Djibouti

Hargeisa

Harer

ETHIOPIA
Addis Ababa

SOMALIA

KENYA
Nakuru
Nairobi
Marsabit

Lake
Turkana

Chisimaio

Mombasa

Dar es Salaam

SUDAN
El Fasher
El Obeid
Birao
Juba

CHAD
Kano
Maiduguri
Ndjamena
Maroua
Lake
Chad

NIGERIA
Kaduna
Lagos
Port
Harcourt

Kainji
Dam

Benue

**CENTRAL AFRICAN
REPUBLIC**
Bouar
Bambari
Bangui
Bangassou
Gomena

CAMEROON
Foumban
Douala
Yaounde
Batouri

**EQUATORIAL
GUINEA**
Bata

GABON
Libreville
Port Gentil
Franceville
Djambala

CONGO
Ouesso
Brazzaville
Loubomo
Pointe Noire
CABINDA
Matadi

ZAIRE
Mbandaka
Kisangani
Isiro
Bunia
Kinshasa
Kikwit
Kananga
Kamina
Kolwesi
Lubumbashi
Kalemie

Lake Mai-
Ndombe

UGANDA
Kampala
Lake
Albert
Lake
Edward
Lake
Kivu

RWANDA
Kigali
Bukavu
BURUNDI
Bujumbura

L. Kyoga

Lake
Victoria

TANZANIA
Tabora
Dodoma
Arusha
Kalemie
Lake
Tanganyika
Lake
Natron
Lake
Eyasi

Lake
Mweru
Lake
Bangweulu
Lake
Ruaka

ANGOLA
Luanda
Benguela
Huambo

ZAMBIA
Ndola

MALAWI
Lilongwe
Lake
Malawi

Atlantic Ocean

0° Equator
10°
20°
30°
40° Longitude East
10°
0° Equator
10°

Blue Nile
White Nile
Awash
Shebele
Juba
Tana
Omo
Ubangui
Zaire
Sanaga
Chari
Kasai
Kwango
Kwilu
Chambeshi
Rufiji
Ruvuma
Luangwa
Lake
Tana

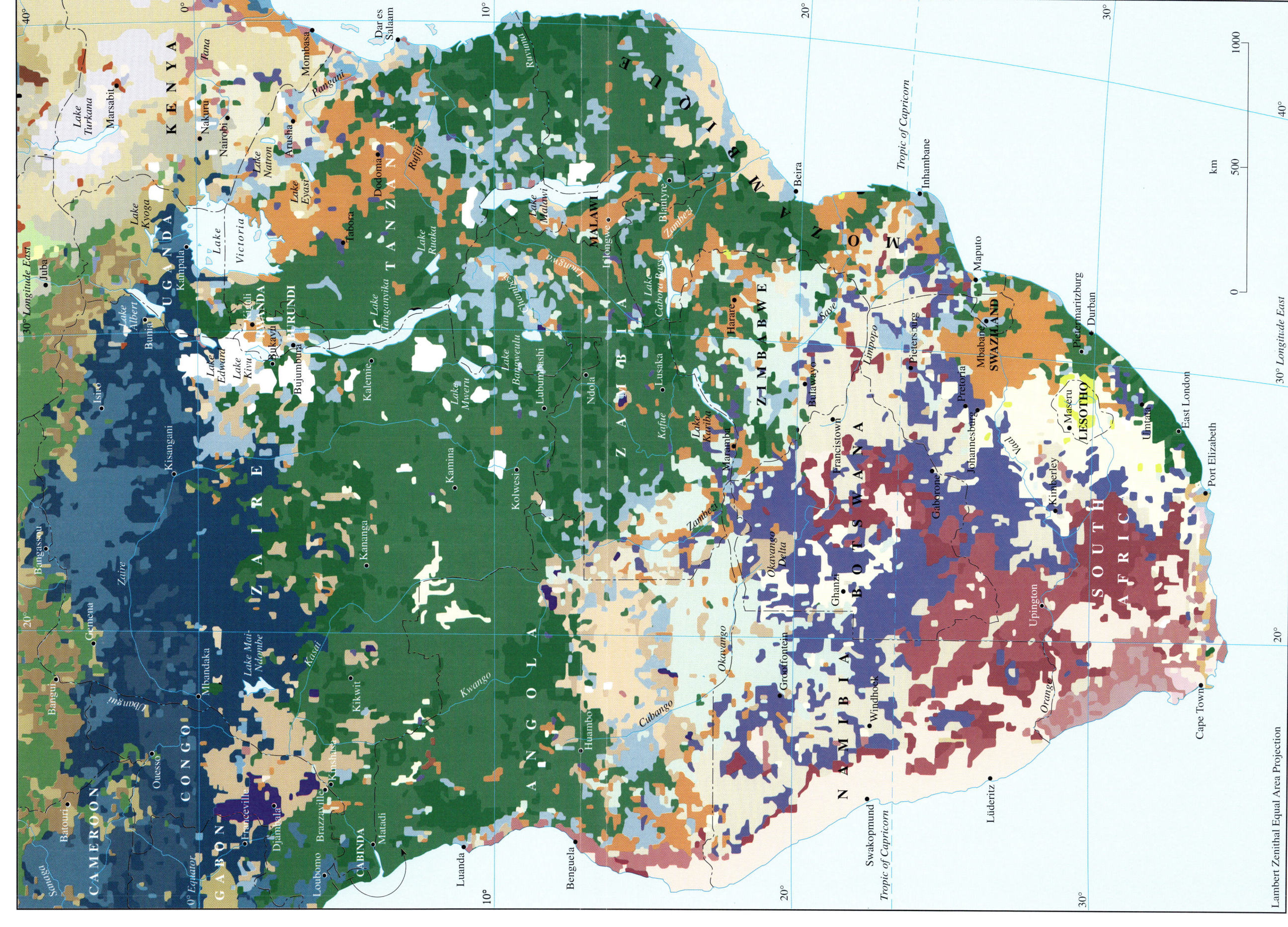

KENYA

Lake
Turkana

Marsabit

Nakuru

Nairobi

Mombasa

Dar es
Salaam

Tana

Pangani

UGANDA

Lake
Kyoga

Lake
Albert

Lake
Edward

Lake
Kivu

Kampala

Jinja

Juba

Bukavu

Bujumbura

RWANDA

BURUNDI

Kigali

Lake
Victoria

Lake
Natron

Lake
Eyasi

Arusha

Dodoma

Tabora

T A N Z A N I A

Lake
Rukwa

Rufiji

Lake
Tanganyika

Lake
Malawi

MALAWI

Lilongwe

Blantyre

Zambezi

Rovuma

M O Z A M B I Q U E

Beira

Inhambane

Ruvuma

Kalemie

Lake
Mweru

Lake
Bangweulu

Lubumbashi

Ndola

Lusaka

Kabwe

Harare

*Lake
Kariba*

Cahora Bassa

ZIMBABWE

Bulawayo

Save

Limpopo

Masvingo

Pieters'rg

Pretoria

Johannesburg

Maputo

Mbabane

SWAZILAND

Pietermaritzburg

Durban

Z A I R E

Kisangani

Kananga

Kamina

Kolwezi

Mbuji-Mayi

Z A M B I A

Kafue

Zambezi

Francistown

Ghanzi

B O T S W A N A

Gaborone

Kimberley

Upington

Orange

Vaal

SOUTH
AFRICA

East London

Umtata

Port Elizabeth

Maseru

LESOTHO

Bangassou

CAMEROON

Sanaga

Batouri

Bertoua

Ouesso

CONGO

GABON

Ogooué

Franceville

Djambala

Brazzaville

Loubomo

Ubangi

Mbandaka

Lake Mai-
Ndombe

Kasai

Kikwit

Kinshasa

Matadi

CABINDA

Luanda

Benguela

A N G O L A

Huambo

Kwango

Cubango

Cuanza

Okavango

*Okavango
Delta*

Cubango

N A M I B I A

Windhoek

Swakopmund

Lüderitz

Keetmanshoop

Grootfontein

Lake
Turkana

Isiro

Goma

Mbala

Zumbo

0° Equator

Tropic of Capricorn

Tropic of Capricorn

0°

10°

20°

30°

40°

10°

20°

30°

40°

10°

20°

30°

30° Longitude East

30° Longitude East

1000

500

km

0

Lambert Zenithal Equal Area Projection

Cartography by Gary Haley